The Insect World
of J. Henri Fabre

The Insect World of J. Henri Fabre

Translated by Alexander Teixeira de Mattos

With Introduction and Interpretive Comments by
Edwin Way Teale

Foreword by
Gerald Durrell

Beacon Press
Boston

Beacon Press
25 Beacon Street
Boston, Massachusetts 02108

Beacon Press books
are published under the auspices of
the Unitarian Universalist Association of Congregations.

Library of Congress Cataloging-in-Publication Data
Fabre, Jean-Henri, 1823–1915.
   The insect world of J. Henri Fabre / translated by Alexander
Teixeira de Mattos : with introduction and interpretive comments by
Edwin Way Teale : foreword by Gerald Durrell.
      p. cm.
   Reprint. Originally published: New York : Dodd, Mead, 1949.
   Includes index.
   ISBN 0–8070–8512–X (cloth)
   ISBN 0–8070–8513–8 (paper)
   1. Insects—Behavior. 2. Insect societies. 3. Social behavior in
animals. 4. Insects. I. Teale, Edwin Way, 1899–   . II. Title.
QL496.F12   1991
595.7′051—dc20   90-20516
                                                         CIP

CONTENTS

FOREWORD TO THE 1991 EDITION

Jean Henri Fabre has played an important part in my life since I was eight years old. At that time I was enjoying a wonderful childhood on the island of Corfu, a magical place for me because there were so many animals living there that I had never seen before. I was particularly fascinated by the insects which—in the days before DDT and other noxious sprays—were to be found in great abundance. However, my childhood was a frustrating time for me as well. Deeply interested in the insect life around me, I had no books to tell me what the insects were or what they did. This was because my family was not zoologically oriented and had no idea of the sort of books I needed. So I watched the insects with irritated ignorance. Why did a large and obviously predatory wasp sting spiders and then carry them off—why not eat them on the spot? And where did it carry them? How did cicadas make that thrumming noise that trepanned your skull on hot afternoons? Why did scarab beetles collect dung, roll it into balls, and then take it into tunnels? These and a myriad other questions surrounded me and I had no means of answering them. I was surrounded on every side by the word why.

Then one day my elder brother—who had obviously been giving the problem some thought—presented me with the collected works of J. H. Fabre. If someone had presented me with the touchstone that turns everything to gold I could not have been more delighted. From that moment Fabre became my personal friend. He wrote for me in simple but poetic language, and in his gentle, enthusiastic voice (which I could hear whispering in my brain) he unraveled the many mysteries that surrounded me and showed me miracles and how they were performed. Through his entrancing prose I became the hunting wasp, the paralyzed spider, the cicada, the burly, burnished scarab beetle, and a host of other creatures as well.

Fabre was an extraordinary man, every bit as exciting as people who explored the mighty Amazon, or plunged into what was then called Darkest Africa. He was an explorer in Lilliput, the vast world that lies under our shoes and which we so seldom notice. Anyone interested in animals has at one time or another visited the strange backwaters of a natural history museum, catacombs smelling of camphor, ether, formaldehyde, of fur and feather and chiton. Here the mammals and birds lie in serried ranks, each with a label attached to its leg, like humans in a mortuary; snakes and frogs floating immobile in bottles of spirit, insects carefully crucified on cork. But Fabre did more than explore the museum; he stepped outside it.

There are three vital questions to be asked about the living world around us: what? where? and why? Museums answer the first two queries but not the third. It was Fabre who took us from the dim recesses of the museums, redolent of death, out into the brilliant countryside, there to point out to us his beloved insects and pose the question *why*. Why did Pine Processionary caterpillars walk, head to tail, in long caravans? Why, if you attached the leader to the hindmost caterpillar, would they walk in a circle until they died of starvation? Were cicadas, noisiest of the insects in his area of France, deaf? To try and find out, he dragged the village cannon into his garden and fired off a salvo which did not prevent the cicadas from continuing their monotonous song. It proved, at least, that cicadas were not affected by cannon fire.

Fabre's pursuit of the truth was insatiable. He would lie in a ditch in the stunning sunshine of Provence, watching and recording the lives of ants, beetles, wasps, and countless other tiny inhabitants of the region and then write about their fascinating lives so enthusiastically, so poetically, that even someone not interested in insects became entranced. If these people were entranced, you can imagine the magical effect the discovery of Fabre had on a budding eight-year-old naturalist.

When I acquired a house in Provence, it was within easy distance of Fabre's old home, now a museum. On my first visit I was overwhelmed. The main rooms of the house were as he had left them, his workroom looking exactly as I imagined and as though he had only that moment slipped out to interview a beetle. The garden was peaceful and delightful, planted with aromatic herbs such as thyme, lavender, and rosemary, to attract the insects, and boasting a round, deep pond full of water lilies and frogs, where the insects could come and drink and gather moisture for their multifarious building operations.

I was reluctant to make my way to the tiny village cemetery where Fabre was buried, for I feared I would find that his small, frail body had been interred beneath a large unseemly tombstone. But to my relief the gravestone was a simple slab, decorated only with a granite urn. It was a tasteful grave worthy of the man. Then I saw, to my delight, that a Mason wasp had built its little grey, multicelled mud nest under one of the handles of the urn. What better tribute to the great man could there be? No plastic flowers under plastic domes, but the nursery of one of the insects he loved. He would have been enchanted.

Gerald Durrell

FOREWORD TO THE 1949 EDITION

In the field of insect-study, the works of J. Henri Fabre are classics; in the field of literature, they hold a special place of their own. The present volume brings into the compass of a single book the most famous of Fabre's studies, many of them now out of print. In some instances, material has been shortened but nothing has been added. My own comments are confined to the italicized sections at the beginning of each chapter.

For assistance of various kinds, I am deeply indebted to Dr. Mont A. Cazier, William P. Comstock, Dr. C. H. Curran, Dr. J. McDunnough, John Pallister, Herbert F. Schwarz, Dr. W. J. Gertsch, Dr. T. C. Schnierla and Miss Hazel Gay, Librarian, all of the American Museum of Natural History, New York City.

In the matter of style and spelling in the text, I have followed the original translation. This, in all cases, was the work of Alexander Teixeira de Mattos, whose rendering into English of Fabre's prose cannot be too highly commended. A single instance will illustrate the feeling and skill he contributed to the work. Where another translator expressed Fabre's words as: ". . . to see whether he was rolling his manure ball, the image of the world for ancient Egypt," De Mattos phrases it: ". . . to see whether he was rolling that pellet of dung in which ancient Egypt beheld an image of the world." Without changing the meaning by a hair, his phrasing gave the line the nobility of literature.

Edwin Way Teale

June 1, 1949.

INTRODUCTION

It is the fate of a few men to become symbols of a way of life or an attitude of mind. Diogenes with his lantern, Ponce de León and his search for eternal youth, Thoreau living the simple life in his hut by Walden Pond, such men epitomize philosophies that cut across time and are found recurring in all generations. To this small group belongs that humble chronicler of the commonplace, J. Henri Fabre, of France.

The very antithesis of Ulysses of Tennyson's poem, "always roaming with a hungry heart," Fabre found the delights of a lifetime—adventure and fame, as well—in observing the near-at-home. He was an explorer whose jungles were weed-lots and whose deserts were sandbanks. There, with tireless enthusiasm, he studied "all those vague, unconscious, rudimentary and almost nameless little lives," as Maeterlinck called the insects. Throughout most of his long life—a life that stretched to within eight years of the century mark, from 1823 to 1915, from two years after the death of Napoleon Bonaparte on St. Helena to the days of the first World War—Fabre never lost his zest for the insect world with its multitudinous mysteries. And no man ever lived who transmitted that enthusiasm to more of his fellow-men.

In one of his essays, Clarence Day observes that "the first thing the world does to a genius is to make him lose all his youth. Fabre, after losing all his youth, and his middle age, too, finally won out at sixty." That is almost literally true. All of Fabre's years were shadowed by penury and his life was an epic of struggle against the worries and irritations that beset the poor. He knew nothing of his peasant ancestors beyond the second generation. His mother could neither read nor write. Reminiscences of his youth, little biographical glimpses, sidelights on his prolonged and lonely labors, these illuminate the pages of his books. The man, Fabre, becomes as interesting to readers as the small protagonists of his insect dramas.

The author's father, Antoine Fabre, was the first of his line to leave the land for the city. An unsuccessful keeper of small cafés, he moved from failure to failure. The family was living at Saint-Léons, market town and administrative center of the canton of Vezins, on December 22, 1823, when Jean Henri Casimer Fabre was born. To make one mouth less to feed, the child was soon sent to live with his maternal grandparents at their farm at Malaval, on "the cold granite ridge of the Rouergue tableland." He was six when he returned to Saint-Léons to begin his schooling; ten when his father moved to Rodez. He never saw his birthplace again. Throughout the rest of his life, Fabre always looked back on Malaval and Saint Léons as the scenes of his happiest years.

At Rodez, the ten-year-old earned free instruction at the Lycée by serving mass in the chapel. A few years passed and he was uprooted by a fresh disaster. For a time, working as a laborer, doing odd jobs, selling lemonade at fairs, he supported himself and struggled toward a diploma. In 1842, when he was eighteen years old, Fabre left the Normal College of Avignon with this goal achieved and began his teaching career as a primary schoolmaster at Carpentras. Here his meager salary was often in arrears. At Ajaccio, Corsica, where he taught science for a few years, malaria forced him to return to the mainland. Finally, in 1852, he became a professor at the Lycée of Avignon. Here he labored for nearly twenty years at a salary that never exceeded $320 a year. When he left, his rank, his title, his salary were the same as when he began.

During those years, his precious Thursday afternoons—the traditional half-holiday of the French school system—and those "thrice-blest summer holidays" when the schoolmaster became the schoolboy again, were devoted to the study of insects. Hours spent along the sunken road near Carpentras, at the Bois des Issarts, along the banks of the Rhone, filled his notebooks with entries.

These activities—which did little to advance him with his school superiors—gave him a local reputation in this strange field of endeavor. Louis Pasteur was sent to see Fabre when he began his study of the silkworm disease. Victor Duruy, energetic minister of public instruction during the reign of Napoleon III, was so impressed by the obscure Provençal teacher, whom he met on a tour of southern departments, that he invited him to Paris. There is more than a hint that Duruy hoped Fabre might become the tutor of the imperial family. He conferred upon him the ribbon of a chevalier of the Legion of Honor and introduced him to the Emperor. But nothing came of his plans. The simple son of the Rouergue peasantry was ill-fitted for life in the royal court. He fled from the great city where, he said, he "never felt such loneliness before."

Back at Avignon, clouds were gathering. Some of Fabre's advanced ideas in teaching, which had attracted Duruy, outraged his superiors. When he admitted girls to his science classes, the storm broke. The clergy denounced him from the pulpit. In 1870, when the German armies were overrunning France, Fabre was dismissed from the Lycée. His slender financial support was severed. He was ejected from his house with his wife and five small children. In that bitterest hour, without money, without home, with the community turned against him, Fabre was saved by a great English friend, then living in Avignon, the economist and philosopher, John Stuart Mill. Without security and without interest, Mill advanced Fabre $600 to see him through the crisis. The two had made a number of excursions together into the fields around Avignon, drawn together by a mutual interest in botany.

Feeling abandoned and alone, Fabre found sanctuary in a house that faced wide fields on the edge of Orange. There, depending upon his ink-pot for a livelihood, he supported his family for nearly a decade and paid back Mill's never-to-be-forgotten loan. Books of popularized science, books on the stars, the earth, the trees, on a host of subjects, books that were simple, interesting, instructional, were turned out during those years. Many were pot-boilers; Fabre was often miserably paid; but the time was not wasted. Writing for the young, making expository writing interesting—this disciplined his style and prepared him for his great work later on.

During the nine years he remained at Orange, he spent his hours of relaxation wandering in the fields or along the banks of the Rhone, observing the life of the insects. His dream, "a dream ever receding into the mists of tomorrow," was the possession of a piece of ground where, without being bothered by the passerby or exhausted by long walks, he could observe the actions of living insects under natural conditions, where he might maintain a laboratory of the open fields. "Long ago," he wrote, "I should have thrown the helve after the axhead had I not had for my encouragement the consciousness of being engaged in the continual search for truth in the little world of which I had made myself the historian."

In 1879, when Fabre was already fifty-five years old, his long dream came true. From his labors with paper and pen, he was able to buy a small foothold of earth, sun-scorched and thistle-ridden, unfit for grazing or agriculture, an area known locally as a *harmas,* at the edge of the village of Sérignan. It was the first bit of land Fabre had owned in his life. To him, the stony soil, arid and rusty-red, formed an Eden.

It was inhabited by wasps and wild bees and all those small creatures to which he wished to devote uninterrupted study during the remaining years of his life.

For a quarter of a century, he had been assembling material, shaping and planning, preparing in fragments a great work on the lives of the insects. It was to be called *Souvenirs Entomologiques.* The initial volume appeared the very year that Fabre reached his *harmas* Eden. Nearly thirty years were to elapse before the tenth and final volume would come from the press in 1907. Oftentimes, Fabre felt that he had reached the end of his strength, that his grand scheme would fail of fulfillment. "Dear insects," he wrote in the final paragraph of Volume III, "my study of you has sustained me in my heaviest trials. I must take leave of you for today. The ranks are thinning around me and the long hopes have fled. Shall I be able to speak to you again?" Often despairing but working on in despair, Fabre continued through seven more volumes, containing more than 2,500 pages and nearly 850,000 words.

He reached his *harmas* almost overwhelmed with grief and care. His best-loved son, Jules, a promising fifteen-year-old boy who was his helper and companion, had just died. He, himself, had barely escaped death in a desperate battle with pneumonia. His aged father, the unsuccessful keeper of cafés, had come to spend his last years with him. Then, soon after he reached Sérignan, death came to his wife. Troubles and anxieties multiplied around him. When he was past sixty, Fabre married a second time, choosing a capable and industrious young woman of Sérignan. She managed the house, relieved him of responsibilities, enabled him to devote himself to his insect researches. She also bore him three children. Surrounded by this second family of his, the aging historian of the insects continued his labors.

His routine became fixed. He arose at six, walked about his *harmas,* ate a sparing breakfast and retired to his laboratory—a long room with whitewashed walls, a tile floor and a great walnut table supporting home-made rearing cages and the simple equipment of the experimenter. He even lacked a microscope for many years, relying on a pocket lens in his work. Late in his life, long after he was too old to use it, the state bestowed upon him the elaborate equipment of a modern laboratory. His best instruments, Fabre used to say, were Time and Patience.

Evidence of ample use of both remained when Fabre died. In the floor of his laboratory, a path or groove had been worn in the tiles by Fabre's years of circling his table in his heavy peasant shoes. Once in his laboratory, he shut himself up "like a snail." He rarely saw callers. He rarely answered letters. To a friend who complained of a long-neglected reply to a letter, Fabre wrote: "Once I have mounted my hobby-horse, goodbye to letters, goodbye to replies, goodbye to everything."

People in the village hardly ever saw him. He became something of a recluse and a mystery, a kind of scientific hermit lost in another world, the world of the insects. Clad in a linen jacket, wearing indoors and out a broad-brimmed black felt hat, smoking a briar pipe that was eternally going out, he continued his lonely labors. His life was almost ascetic in its simplicity. At mealtime, he ate sparingly of fruits and vegetables, avoiding all meats. He slept in a low bed by a window that overlooked the garden in a little chamber with bare walls and a bare tile floor. A small black desk, scarred and stained by more than half a century of use, a desk "no larger than a pocket handkerchief," provided him with the place on which he wrote—using a penny bottle of ink and setting down his words in a clear, legible hand with few erasures—all those pages which were later collected into the books on which his fame rests. All of the thirty-seven years that remained to Fabre were spent at his hermitage at Sérignan. "Here I am," he wrote a friend, "run wild, and I shall be so till the end."

As the successive volumes of *Souvenirs Entomologiques* appeared, they attracted only mild attention. Fabre worked alone. He had no connection with any large institution. He lived, in truth, more in the world of insects than in the world of men. His publisher, moreover, issued the original editions in a heavy, unattractive format. Thus, overlooked and ignored, Fabre struggled on "without masters, without guides, often without books . . . with one aim: to add a few pages to the history of the insects."

His long years of neglect, his decades of painstaking labor, ended in five exhausting years of fame. Fabre was eighty-four when the last of the ten volumes of his *magnum-opus* appeared. Soon afterwards, he was suddenly discovered by such eminent literary figures as Maurice Maeterlinck, Edmund Rostand and Romain Rolland. A jubilee in his honor was held at Sérignan in 1910. Government officials and representatives of scientific institutions and societies did him homage. A statue was erected in the village. "I shall see myself," Fabre observed to an old friend, "but shall I recognize myself?" What inscription did he want on the base of the statue? One word, he said, *Laboremus*.

People who had never heard of the "Insect's Homer" before, began reading his books at home and abroad. Scientific societies in London, Brussels, Stockholm, Geneva and St. Petersburg elected him to membership. The government bestowed upon him an annual pension of $400. The President of France journeyed to Sérignan to meet its first citizen. After the long years of poverty, of labor, of niggardly recognition, Fabre, nearing his ninetieth year, saw with his failing eyesight the sunshine of brilliant acclaim.

But, even then, tension and sadness darkened the sunshine. Fabre's second wife died in 1912. His country began fighting for its life in 1914. As the Germans had overrun his homeland in the year of his despair at Avignon, so in the year of his death, the armies from across the Rhine were sweeping toward Paris. On the 11th of October, 1915, fighting was heavy in the Argonne and Belgrade, in the Balkans, had just fallen to the Germans. On that day, in the village quiet of Sérignan, the long life of J. Henri Fabre came to an end.

Looking back on Fabre's work today, it is easy to see the mistakes he made, blind spots in his approach to the larger aspects of biological research. He never accepted, for example, the theory of evolution. Although Charles Darwin called Fabre an "incomparable observer" and suggested experiments for him to do and although Fabre had the highest personal regard for the English scientist, he was so little interested in the *Origin of Species* that he never read more than a few pages. He was a realist opposed to hypotheses. His harvest was a harvest of facts rather than a harvest of conclusions. "I observe, I experiment and I let the facts speak for themselves," he used to say. But this attitude has its limitations. Sometimes, Fabre's facts reveal only a segment of the whole picture. Again, Fabre was dominated by the general rule and failed to give as much importance to exceptions to the rule as modern research workers have found they have to do. He was far surer than present-day scientists of the machine-like character of instinct. Fabre tended to over-emphasize the belief that instinct sticks to its course like a train on its rails; that the creature has almost no choice in the matter. The Peckhams, experimenting with solitary wasps in America, demonstrated that instinct provides considerable latitude in the actions of many insects, a finding that has been substantiated by numerous more recent investigations.

Because he was isolated, remote from centers of research, narrow in his knowledge of entomological literature, Fabre frequently repeated experiments that had already been reported by others without apparently knowing they had been made. Fabre was not a trained entomologist; he depended upon others for the identification of his insects. In some cases, in his writings, his description of an insect is not sufficiently detailed for its identification. Nobody knows which of several species he was experimenting with at the time. He was primarily interested in the living animal, not in the collected specimen for identification; nevertheless, the doubt which remains as to the species he is referring to mars several of his studies.

But, when all these weaknesses are acknowledged, the greatness of Fabre towers above them. He was a pioneer, working alone, undervalued, poverty-cramped, using crude equipment that would cause a modern research worker to throw up his hands in despair. Yet, by repeating an experiment over and over again, by the endless inventiveness of his mind, he made up for the lack of adequate equipment. He produced some of the basic studies of the nature of instinct. All students of insect behavior, of comparative psychology, of experimental biology are indebted to Fabre. His harvest of facts is invaluable still.

Fabre went direct to Nature. He interrogated the insects themselves —"through the language of experiment." He led in the study of *living entomology* at a time when the science seemed preempted by those whose horizon of interest was limited to the dead insect and the pinned specimen. And this indefatigable self-teacher possessed to a remarkable degree that divine gift of all great teachers, the ability of transmitting his enthusiasm to others. Each experiment becomes an adventure. The reader advances, "by the light of facts," into unknown territory. His companion—this modest man of simple tastes and humane viewpoint— is a philosopher who never loses sight of humanity in his writings. There is feeling as well as seeing, poetry as well as science, in his pages.

And all is set down in simple, durable prose. Alexander Teixeira de Mattos, translator of Fabre's works into English, once wrote: "I have always thought that this quality, the quality of being readily rendered into a foreign tongue, is a test of good writing. It is the tortured, laboured, fantastic, would-be 'original' style that hampers the translator. Fabre's style is invariably straightforward." It also possesses a charm that defies definition. It is the charm of the man himself, compounded of his humanity, his sincerity, his enthusiasm, qualities that endured to the end.

In the autumn of 1911, Fabre penned a short introduction for Dr. C. V. Legros' biography, *Fabre, Poet of Science*. "It seems to me," he wrote, "that in the depths of my being I can still feel rising in me all the fever of my early years, all the enthusiasm of long ago, and that I should still be no less ardent a worker were not the weakness of my eyes and the failure of my strength today an insurmountable obstacle." At the time, J. Henri Fabre was eighty-eight years old. His sense of wonder outlived his sense of sight; his interest and enthusiasm outlasted his strength.

"To scrutinize and to follow in their enchantment the mysterious habits of the insects," Dr. E. L. Bouvier, of the Museum of Natural History, in Paris, wrote after Fabre's death, "it does not suffice to be an ingenious experimenter; there must be a keen observation, a patience that cannot be discouraged and an extraordinary intuitive power. Fabre had these qualities up to the point of genius."

There is an epic character, something universal and symbolic, in the story of this simple man's life, in his years of unremitting struggle to solve mysteries of the commonplace. With so little, Fabre did so much.

A LABORATORY OF
THE OPEN FIELDS

« 1 »

The village of Sérignan, in the Department of Vaucluse, lies in the Rhone valley about fifty miles north of the Mediterranean. The nearest town is Orange. At the time of Fabre's death, the population of Sérignan was about 1,000. The soil of the region is rust-red, stony and arid, supporting, here and there, a vineyard or olive-grove. The Lygues River, flowing by the village on its way to the Rhone, is a muddy torrent in spring, a dry river of stones and pebbles in summer. The horizon to the east is formed by the foothills of the Alps.

Fabre's harmas had an area of 2.47 acres, almost exactly a Roman hectare. It was roughly rectangular in shape, with the flat-roofed, two-story house set at one end. A wall provided seclusion and paths wound among the tangles of his "Eden." Beyond the wall, a dusty road, lined with plane trees, led to Orange where the family did its weekly marketing.

Among the insects named by Fabre in this chapter, the Cicadella is a froghopper; the Scarab, Lamellicorns, Orycetes and Cetoniae are beetles; the Sphex, Ammophila, Cerceris, Stizus, Pompilus, Pelopaeus, Polistes, Eumenes, Scoliae and Bembeces are wasps; the Ephippiger is a green grasshopper and the Halicti, Megachiles, Chalcicodomae, Osmiae, Anthophorae, Anthidium, Eucerae, Macrocerae, Dasypodae and Andrenae are all wild bees. The first two paragraphs of this chapter are taken from THE GLOW-WORM; *the rest forms Chapter I of* THE LIFE OF THE FLY.

To travel the world, by land and sea, from pole to pole; to cross-question life, under every clime, in the infinite variety of its manifestations: that surely would be glorious luck for him that has eyes to see; and it formed the radiant dreams of my young years, at the time when ROBINSON CRUSOE was my delight. These rosy illusions, rich in voyages, were soon suc-ceeded by dull, stay-at-home reality. The jungles of India, the virgin forests of Brazil, the towering crests of the Andes, beloved by the Con-dor, were reduced, as a field for exploration, to a patch of pebbles en-closed within four walls.

I go the circuit of my enclosure over and over again, a hundred times, by short stages; I stop here and I stop there; patiently, I put questions and, at long intervals, I receive some scrap of a reply. The smallest insect village has become familiar to me: I know each fruit-branch where the Praying Mantis perches; each bush where the pale Italian Cricket strums amid the calmness of the summer nights; each downy plant scraped by the Anthidium, that maker of cotton bags; each cluster of lilac worked by the Megachile, the leaf-cutter.

This is what I wished for, *hoc erat in votis:* a bit of land, oh, not so very large, but fenced in, to avoid the drawbacks of a public way; an aban-doned, barren, sun-scorched bit of land, favoured by thistles and by Wasps and Bees. Here, without fear of being troubled by the passers-by, I could consult the Ammophila and the Sphex and engage in that diffi-cult conversation whose questions and answers have experiment for their language; here, without distant expeditions that take up my time, without tiring rambles that strain my nerves, I could contrive my plans of attack, lay my ambushes and watch their effects at every hour of the day. *Hoc erat in votis.* Yes, this was my wish, my dream, always cher-ished, always vanishing into the mists of the future.

And it is no easy matter to acquire a laboratory in the open fields, when harassed by a terrible anxiety about one's daily bread. For forty years have I fought, with steadfast courage, against the paltry plagues of life; and the long-wished-for laboratory has come at last. What it has cost me in perseverance and relentless work I will not try to say. It has come; and, with it—a more serious condition—perhaps a little leisure. I say perhaps, for my leg is still hampered with a few links of the convict's chain.

The wish is realized. It is a little late, O my pretty insects! I greatly fear that the peach is offered to me when I am beginning to have no teeth wherewith to eat it. Yes, it is a little late: the wide horizons of the outset have shrunk into a low and stifling canopy, more and more straitened day by day. Regretting nothing in the past, save those whom I have lost; regretting nothing, not even my first youth; hoping nothing either, I have reached the point at which, worn out by the experience of things, we ask ourselves if life be worth the living.

Amid the ruins that surround me, one strip of wall remains standing, immovable upon its solid base: my passion for scientific truth. Is that enough, O my busy insects, to enable me to add yet a few seemly pages to your history? Will my strength not cheat my good intentions? Why, indeed, did I forsake you so long? Friends have reproached me for it. Ah, tell them, tell those friends, who are yours as well as mine, tell them that it was not forgetfulness on my part, not weariness, nor neglect: I thought of you; I was convinced that the Cerceris cave had more fair secrets to reveal to us, that the chase of the Sphex held fresh surprises in store. But time failed me; I was alone, deserted, struggling against misfortune. Before philosophizing, one had to live. Tell them that; and they will pardon me.

Others again have reproached me with my style, which has not the solemnity, nay, better, the dryness of the schools. They fear lest a page that is read without fatigue should not always be the expression of the truth. Were I to take their word for it, we are profound only on condition of being obscure. Come here, one and all of you—you, the sting-bearers, and you, the wing-cased armour-clads—take up my defence and bear witness in my favour. Tell of the intimate terms on which I live with you, of the patience with which I observe you, of the care with which I record your actions. Your evidence is unanimous: yes, my pages, though they bristle not with hollow formulas nor learned smatterings, are the exact narrative of facts observed, neither more nor less; and whoso cares to question you in his turn will obtain the same replies.

And then, my dear insects, if you cannot convince those good people, because you do not carry the weight of tedium, I, in my turn, will say to them:

"You rip up the animal and I study it alive; you turn it into an object of horror and pity, whereas I cause it to be loved; you labour in a torture-chamber and dissecting-room, I make my observations under the blue sky to the song of the Cicadas, you subject cell and protoplasm to chemical tests, I study instinct in its loftiest manifestations; you pry into death, I pry into life. And why should I not complete my thought: the boars have muddied the clear stream; natural history, youth's glorious study, has, by dint of cellular improvements, become a hateful and repulsive thing. Well, if I write for men of learning, for philosophers, who, one day, will try to some extent to unravel the tough problem of instinct, I write also, I write above all things for the young. I want to make them love the natural history which you make them hate; and that is why, while keeping strictly to the domain of truth, I avoid your scientific prose, which too often, alas seems borrowed from some Iroquois idiom!"

But this is not my business for the moment: I want to speak of the bit of land long cherished in my plans to form a laboratory of living entomology, the bit of land which I have at last obtained in the solitude of a little village. It is a *harmas,* the name given, in this district, to an untilled, pebbly expanse abandoned to the vegetation of the thyme. It is too poor to repay the work of the plough; but the sheep passes there in spring, when it has chanced to rain and a little grass shoots up.

My harmas, however, because of its modicum of red earth swamped by a huge mass of stones, has received a rough first attempt at cultivation: I am told that vines once grew here. And, in fact, when we dig the ground before planting a few trees, we turn up, here and there, remains of the precious stock, half-carbonized by time. The three-pronged fork, therefore, the only implement of husbandry that can penetrate such a soil as this, has entered here; and I am sorry, for the primitive vegetation has disappeared. No more thyme, no more lavender, no more clumps of kermes-oak, the dwarf oak that forms forests across which we step by lengthening our stride a little. As these plants, especially the first two, might be of use to me by offering the Bees and Wasps a spoil to forage, I am compelled to reinstate them in the ground whence they were driven by the fork.

What abounds without my mediation is the invaders of any soil that is first dug up and then left for a long time to its own resources. We have, in the first rank, the couch-grass, that execrable weed which three years of stubborn warfare have not succeeded in exterminating. Next, in respect of number, come the centauries, grim-looking one and all, bristling with prickles or starry halberds. They are the yellow-flowered centaury, the mountain centaury, the star-thistle and the rough centaury: the first predominates. Here and there, amid their inextricable confusion, stands, like a chandelier with spreading, orange flowers for lights, the fierce Spanish oyster-plant, whose spikes are strong as nails. Above it, towers the Illyrian cotton-thistle, whose straight and solitary stalk soars to a height of three to six feet and ends in large pink tufts. Its armour hardly yields before that of the oyster-plant. Nor must we forget the lesser thistle-tribe, with first of all, the prickly or "cruel" thistle, which is so well armed that the plant-collector knows not where to grasp it; next, the spear-thistle, with its ample foliage, ending each of its veins with a spear-head; lastly, the black knap-weed, which gathers itself into a spiky knot. In among these, in long lines armed with hooks, the shoots of the blue dewberry creep along the ground. To visit the prickly thicket when the Wasp goes foraging, you must wear boots that come to mid-leg or else resign yourself to a smarting in the calves. As long as the ground retains a few remnants of the vernal rains, this rude vegetation does not lack a certain charm, when the pyramids of the oyster-plant and the slender branches of the cotton-thistle rise above the wide carpet formed by the yellow-flowered centaury saffron heads; but let the droughts of summer come and we see but a desolate waste, which the flame of a match would set ablaze from one end to the other. Such is, or rather was, when I took possession of it, the Eden of bliss where I mean to live henceforth alone with the insect. Forty years of desperate struggle have won it for me.

Eden, I said; and, from the point of view that interests me, the expression is not out of place. This cursed ground, which no one would have had at a gift to sow with a pinch of turnip-seed, is an earthly paradise for the Bees and Wasps. Its mighty growth of thistles and centauries draws them all to me from everywhere around. Never, in my insect-hunting memories, have I seen so large a population at a single spot; all the trades have made it their rallying-point. Here come hunters of every kind of game, builders in clay, weavers of cotton goods, collectors of pieces cut from a leaf or the petals of a flower, architects in pasteboard, plasterers mixing mortar, carpenters boring wood, miners digging underground galleries, workers handling goldbeater's skin and many more.

Who is this one? An Anthidium. She scrapes the cobwebby stalk of the yellow-flowered centaury and gathers a ball of wadding which she carries off proudly in the tips of her mandibles. She will turn it, under ground, into cotton-felt satchels to hold the store of honey and the egg. And these others, so eager for plunder? They are Megachiles, carrying under their bellies their black, white or blood-red reaping-brushes. They will leave the thistles to visit the neighbouring shrubs and there cut from the leaves oval pieces which will be made into a fit receptacle to contain the harvest. And these, clad in black velvet? They are Chalcicodomæ, who work with cement and gravel. We could easily find their masonry on the stones in the harmas. And these, noisily buzzing with a sudden flight? They are the Anthophoræ, who live in the old walls and the sunny banks of the neighbourhood.

Now come the Osmiæ. One stacks her cells in the spiral staircase of an empty snail-shell; another, attacking the pith of a dry bit of bramble, obtains for her grubs a cylindrical lodging and divides it into floors by means of partition-walls; a third employs the natural channel of a cut reed; a fourth is a rent-free tenant of the vacant galleries of some Mason-bee. Here are the Macroceræ and the Euceræ, whose males are proudly horned; the Dasypodæ, who carry an ample brush of bristles on their hind-legs for a reaping implement; the Andrenæ, so manifold in species; the slender-bellied Halicti. I omit a host of others. If I tried to continue this record of the guests of my thistles, it would muster almost the whole of the honey-yielding tribe. A learned entomologist of Bordeaux, Professor Pérez, to whom I submit the naming of my prizes, once asked me if I had any special means of hunting, to send him so many rarities and even novelties. I am not at all an experienced and, still less, a zealous hunter, for the insect interests me much more when engaged in its work than when struck on a pin in a cabinet. The whole secret of my hunting is reduced to my dense nursery of thistles and centauries.

By a most fortunate chance, with this populous family of honey-gatherers was allied the whole hunting tribe. The builders' men had distributed here and there in the harmas great mounds of sand and heaps of stones, with a view to running up some surrounding walls. The work dragged on slowly; and the materials found occupants from the first year. The Mason-bees had chosen the interstices between the stones as a dormitory where to pass the night, in serried groups. The powerful Eyed Lizard, who, when close-pressed, attacks both man and dog, wide-mouthed, had selected a cave wherein to lie in wait for the passing Scarab; the Black-eared Chat, garbed like a Dominican, white-frocked with black wings, sat on the top stone, singing his short rustic lay: his nest, with its sky-blue eggs, must be somewhere in the heap. The little Dominican disappeared with the loads of stones. I regret him: he would have been a charming neighbour. The Eyed Lizard I do not regret at all.

The sand sheltered a different colony. Here, the Bembeces were sweeping the threshold of their burrows, flinging a curve of dust behind them; the Languedocian Sphex was dragging her Ephippigera by the antennæ; a Stizus was storing her preserves of Cicadellæ. To my sorrow, the masons ended by evicting the sporting tribe; but, should I ever wish to recall it, I have but to renew the mounds of sand: they will soon all be there.

Hunters that have not disappeared, their homes being different, are the Ammophilæ, whom I see fluttering, one in spring, the others in autumn, along the garden-walks and over the lawns, in search of a Caterpillar; the Pompili, who travel alertly, beating their wings and rummaging in every corner in quest of a Spider. The largest of them waylays the Narbonne Lycosa, whose burrow is not infrequent in the harmas. This burrow is a vertical well, with a curb of fescue-grass inter‹ twined with silk. You can see the eyes of the mighty Spider gleam at the bottom of the den like little diamonds, an object of terror to most. What a prey and what dangerous hunting for the Pompilus! And here, on a hot summer afternoon, is the Amazon-ant, who leaves her barrack-rooms in long battalions and marches far afield to hunt for slaves. We will follow her in her raids when we find time. Here again, around a heap of grasses turned to mould, are Scoliæ an inch and a half long, who fly gracefully and dive into the heap, attracted by a rich prey, the grubs of Lamellicorns, Oryctes and Cetoniæ.

What subjects for study! And there are more to come. The house was as utterly deserted as the ground. When man was gone and peace assured, the animal hastily seized on everything. The Warbler took up his abode in the lilac-shrubs; the Greenfinch settled in the thick shelter of the cypresses; the Sparrow carted rags and straw under every slate; the Serin-finch, whose downy nest is no bigger than half an apricot, came and chirped in the plane-tree-tops; the Scops made a habit of uttering his monotonous, piping note here, of an evening; the bird of Pallas Athene, the Owl, came hurrying along to hoot and hiss.

In front of the house is a large pond, fed by the aqueduct that supplies the village-pumps with water. Here, from half a mile and more around, come the Frogs and Toads in the lovers' season. The Natterjack, sometimes as large as a plate, with a narrow stripe of yellow down his back, makes his appointments here to take his bath; when the evening twilight falls, we see hopping along the edge the Midwife Toad, the male, who carries a cluster of eggs, the size of peppercorns, wrapped round his hindlegs: the genial paterfamilias has brought his precious packet from afar, to leave it in the water and afterwards retire under some flat stone, whence he will emit a sound like a tinkling bell. Lastly, when not croaking amid the foliage, the Tree-frogs indulge in the most graceful dives. And so, in May, as soon as it is dark, the pond becomes a deafening orchestra: it is impossible to talk at table, impossible to sleep. We had to remedy this by means perhaps a little too rigorous. What could we do? He who tries to sleep and cannot needs becomes ruthless.

Bolder still, the Wasp has taken possession of the dwelling-house. On my door-sill, in a soil of rubbish, nestles the White-banded Sphex: when I go indoors, I must be careful not to damage her burrows, not to tread upon the miner absorbed in her work. It is quite a quarter of a century since I last saw the saucy Cricket-hunter. When I made her acquaintance, I used to visit her at a few miles' distance: each time, it meant an expedition under the blazing August sun. To-day, I find her at my door; we are intimate neighbours. The embrasure of the closed window provides an apartment of a mild temperature for the Pelopæus. The earth-built nest is fixed against the freestone wall. To enter her home, the Spider-huntress uses a little hole left open by accident in the shut-ters. On the mouldings of the Venetian blinds, a few stray Mason-bees build their group of cells; inside the outer shutters, left ajar, a Eumenes constructs her little earthen dome, surmounted by a short, bell-mouthed neck. The common Wasp and the Polistes are my dinner-guests: they visit my table to see if the grapes served are as ripe as they look.

Here, surely—and the list is far from complete—is a company both numerous and select, whose conversation will not fail to charm my solitude, if I succeed in drawing it out. My dear beasts of former days, my old friends, and others, more recent acquaintances, all are here, hunting, foraging, building in close proximity. Besides, should we wish to vary the scene of observation, the mountain is but a few hundred steps away, with its tangle of arbutus, rock-roses and arborescent heather; with its sandy spaces dear to the Bembeces; with its marly slopes exploited by different Wasps and Bees. And that is why, foreseeing these riches, I have abandoned the town for the village and come to Sérignan to weed my turnips and water my lettuces.

Laboratories are being founded, at great expense, on our Atlantic and Mediterranean coasts, where people cut up small sea-animals, of but meagre interest to us; they spend a fortune on powerful microscopes, delicate dissecting-instruments, engines of capture, boats, fishing-crews, aquariums, to find out how the yolk of an Annelid's egg is constructed, a question whereof I have never yet been able to grasp the full importance; and they scorn the little land-animal, which lives in constant touch with us, which provides universal psychology with documents of inestimable value, which too often threatens the public wealth by destroying our crops. When shall we have an entomological laboratory for the study not of the dead insect, steeped in alcohol, but of the living insect; a laboratory having for its object the instinct, the habits, the manner of living, the work, the struggles, the propagation of that little world, with which agriculture and philosophy have most seriously to reckon?

To know thoroughly the history of the destroyer of our vines might perhaps be more important than to know how this or that nerve-fibre of a Cirriped ends; to establish by experiment the line of demarcation between intellect and instinct; to prove, by comparing facts in the zoological progression, whether human reason be an irreducible faculty or not: all this ought surely to take precedence of the number of joints in a Crustacean's antenna. These enormous questions would need an army of workers; and we have not one. The fashion is all for the Mollusc and the Zoophytes. The depths of the sea are explored with many drag-nets; the soil which we tread is consistently disregarded. While waiting for the fashion to change, I open my harmas laboratory of living entomology; and this laboratory shall not cost the ratepayers one farthing.

THE PINE PROCESSIONARY

Among Fabre's insect stories, one of the most celebrated is this record of the Pine Processionary caterpillars following their charmed circle day after day, unable to turn aside even though they faced starvation. Silk-trails are important in the lives of many other social caterpillars, including the American tent caterpillar, found in wild cherries in the spring. Since Fabre's time, the Processionaries have been given a group all their own in entomological classification and christened the Thaumatopoeidae. The species with which Fabre experimented was probably T. processionea (Lin.). The adults of these caterpillars are small nondescript brownish moths with a wing-spread of about an inch and a quarter. The material below originally appeared in Chapters I and III of THE LIFE OF THE CATERPILLAR.

In my *harmas* laboratory, now stocked with a few trees in addition to its bushes, stand some vigorous fir-trees, the Aleppo pine and the black Austrian pine, a substitute for that of the Landes. Every year the caterpillar takes possession of them and spins his great purses in their branches. In the interest of the leaves, which are horribly ravaged, as though there had been a fire, I am obliged each winter to make a strict survey and to extirpate the nests with a long forked batten.

You voracious little creatures, if I let you have your way, I should soon be robbed of the murmur of my once so leafy pines! Today I will seek compensation for all the trouble I have taken. Let us make a compact. You have a story to tell. Tell it me; and for a year, for two years or longer, until I know more or less all about it, I shall leave you undisturbed, even at the cost of lamentable suffering to the pines.

Having concluded the treaty and left the caterpillars in peace, I soon have abundant material for my observations. In return for my indulgence I get some thirty nests within a few steps of my door. If the collection were not large enough, the pine-trees in the neighbourhood would supply me with any necessary additions. But I have a preference and a decided preference for the population of my own enclosure, whose nocturnal habits are much easier to observe by lantern-light. With such treasures daily before my eyes, at any time that I wish and under natural conditions, I cannot fail to see the Processionary's story unfolded at full length. Let us try.

Drover Dingdong's Sheep followed the Ram which Panurge had maliciously thrown overboard and leapt nimbly into the sea, one after the other, "for you know," says Rabelais, "it is the nature of the sheep always to follow the first, wheresoever it goes; which makes Aristotle mark them for the most silly and foolish animals in the world."

The Pine Caterpillar is even more sheeplike, not from foolishness, but from necessity: where the first goes all the others go, in a regular string, with not an empty space between them.

They proceed in single file, in a continuous row, each touching with its head the rear of the one in front of it. The complex twists and turns described in his vagaries by the caterpillar leading the van are scrupulously described by all the others. No Greek *theoria* winding its way to the Eleusinian festivals was ever more orderly. Hence the name of Processionary given to the gnawer of the pine.

His character is complete when we add that he is a rope-dancer all his life long: he walks only on the tight-rope, a silken rail placed in position as he advances. The caterpillar who chances to be at the head of the procession dribbles his thread without ceasing and fixes it on the path which his fickle preferences cause him to take. The thread is so tiny that the eye, though armed with a magnifying-glass, suspects it rather than sees it.

But a second caterpillar steps on the slender footboard and doubles it with his thread; a third trebles it; and all the others, however many there be, add the sticky spray from their spinnerets, so much so that, when the procession has marched by, there remains, as a record of its passing, a narrow white ribbon whose dazzling whiteness shimmers in the sun. Very much more sumptuous than ours, their system of road-making consists in upholstering with silk instead of macadamizing. We sprinkle our roads with broken stones and level them by the pressure of a heavy steamroller; they lay over their paths a soft satin rail, a work of general interest to which each contributes his thread.

What is the use of all this luxury? Could they not, like other caterpillars, walk about without these costly preparations? I see two reasons for their mode of progression. It is night when the Processionaries sally forth to browse upon the pine-leaves. They leave their nest, situated at the top of a bough, in profound darkness; they go down the denuded pole till they come to the nearest branch that has not yet been gnawed, a branch which becomes lower and lower by degrees as the consumers finish stripping the upper storeys; they climb up this untouched branch and spread over the green needles.

When they have had their suppers and begin to feel the keen night air, the next thing is to return to the shelter of the house. Measured in a straight line, the distance is not great, hardly an arm's length; but it cannot be covered in this way on foot. The caterpillars have to climb down from one crossing to the next, from the needle to the twig, from the twig to the branch, from the branch to the bough and from the bough, by a no less angular path, to go back home. It is useless to rely upon sight as a guide on this long and erratic journey. The Processionary, it is true, has five ocular specks on either side of his head, but they are so infinitesimal, so difficult to make out through the magnifying-glass, that we cannot attribute to them any great power of vision. Besides, what good would those short-sighted lenses be in the absence of light, in black darkness?

It is equally useless to think of the sense of smell. Has the Processional any olfactory powers or has he not? I do not know. Without giving a positive answer to the question, I can at least declare that his sense of smell is exceedingly dull and in no way suited to help him find his way. This is proved, in my experiments, by a number of hungry caterpillars that, after a long fast, pass close beside a pine-branch without betraying any eagerness or showing a sign of stopping. It is the sense of touch that tells them where they are. So long as their lips do not chance to light upon the pasture-land, not one of them settles there, though he be ravenous. They do not hasten to food which they have scented from afar; they stop at a branch which they encounter on their way.

Apart from sight and smell, what remains to guide them in returning to the nest? The ribbon spun on the road. In the Cretan labyrinth, Theseus would have been lost but for the clue of thread with which Ariadne supplied him. The spreading maze of the pine-needles is, especially at night, as inextricable a labyrinth as that constructed for Minos.

The Processionary finds his way through it, without the possibility of a mistake, by the aid of his bit of silk. At the time for going home, each easily recovers either his own thread or one or other of the neighbouring threads, spread fanwise by the diverging herd; one by one the scattered tribe line up on the common ribbon, which started from the nest; and the sated caravan finds its way back to the manor with absolute certainty.

Longer expeditions are made in the daytime, even in winter, if the weather be fine. Our caterpillars then come down from the tree, venture on the ground, march in procession for a distance of thirty yards or so. The object of these sallies is not to look for food, for the native pine-tree is far from being exhausted: the shorn branches hardly count amid the vast leafage. Moreover, the caterpillars observe complete abstinence till nightfall. The trippers have no other object than a constitutional, a pilgrimage to the outskirts to see what these are like, possibly an inspection of the locality where, later on, they mean to bury themselves in the sand for their metamorphosis.

It goes without saying that, in these greater evolutions, the guiding cord is not neglected. It is now more necessary than ever. All contribute to it from the produce of their spinnerets, as is the invariable rule whenever there is a progression. Not one takes a step forward without fixing to the path the thread hanging from his lip.

If the series forming the procession be at all long, the ribbon is dilated sufficiently to make it easy to find; nevertheless, on the homeward journey, it is not picked up without some hesitation. For observe that the caterpillars when on the march never turn completely; to wheel round on their tight-rope is a method utterly unknown to them. In order therefore to regain the road already covered, they have to describe a zig-zag whose windings and extent are determined by the leader's fancy. Hence come gropings and roamings which are sometimes prolonged to the point of causing the herd to spend the night out of doors. It is not a serious matter. They collect into a motionless cluster. To-morrow the search will start afresh and will sooner or later be successful. Oftener still the winding curve meets the guide-thread at the first attempt. As soon as the first caterpillar has the rail between his legs, all hesitation ceases; and the band makes for the nest with hurried steps.

The use of this silk-tapestried roadway is evident from a second point of view. To protect himself against the severity of the winter which he has to face when working, the Pine Caterpillar weaves himself a shelter in which he spends his bad hours, his days of enforced idleness. Alone, with none but the meagre resources of his silk-glands, he would find difficulty in protecting himself on the top of a branch buffeted by the winds. A substantial dwelling, proof against snow, gales and icy fogs, requires the cooperation of a large number. Out of the individual's piled-up atoms, the community obtains a spacious and durable establishment.

The enterprise takes a long time to complete. Every evening, when the weather permits, the building has to be strengthened and enlarged. It is indispensable, therefore, that the corporation of workers should not be dissolved while the stormy season continues and the insects are still in the caterpillar stage. But, without special arrangements, each nocturnal expedition at grazing-time would be a cause of separation. At that moment of appetite for food there is a return to individualism. The caterpillars become more or less scattered, settling singly on the branches around; each browses his pine-needle separately. How are they to find one another afterwards and become a community again?

The several threads left on the road make this easy. With that guide, every caterpillar, however far he may be, comes back to his companions without ever missing the way. They come hurrying from a host of twigs, from here, from there, from above, from below; and soon the scattered legion reforms into a group. The silk thread is something more than a road-making expedient: it is the social bond, the system that keeps the members of the community indissolubly united.

At the head of every procession, long or short, goes a first caterpillar whom I will call the leader of the march or file, though the word leader, which I use for want of a better, is a little out of place here. Nothing, in fact, distinguishes this caterpillar from the others: it just depends upon the order in which they happen to line up; and mere chance brings him to the front. Among the Processionaries, every captain is an officer of fortune. The actual leader leads; presently he will be a subaltern, if the file should break up in consequence of some accident and be formed anew in a different order.

His temporary functions give him an attitude of his own. While the others follow passively in a close file, he, the captain, tosses himself about and with an abrupt movement flings the front of his body hither and thither. As he marches ahead he seems to be seeking his way. Does he in point of fact explore the country? Does he choose the most practicable places? Or are his hesitations merely the result of the absence of a guiding thread on ground that has not yet been covered? His subordinates follow very placidly, reassured by the cord which they hold between their legs; he, deprived of that support, is uneasy.

Why cannot I read what passes under his black, shiny skull, so like a drop of tar? To judge by actions, there is here a small dose of discernment which is able, after experimenting, to recognize excessive roughnesses, over-slippery surfaces, dusty places that offer no resistance and, above all, the threads left by other excursionists. This is all or nearly all that my long acquaintance with the Processionaries has taught me as to their mentality. Poor brains, indeed; poor creatures, whose commonwealth has its safety hanging upon a thread!

The processions vary greatly in length. The finest that I have seen manœuvring on the ground measured twelve or thirteen yards and numbered about three hundred caterpillars, drawn up with absolute precision in a wavy line. But, if there were only two in a row, the order would still be perfect: the second touches and follows the first.

By February I have processions of all lengths in the greenhouse. What tricks can I play upon them? I see only two: to do away with the leader; and to cut the thread.

The suppression of the leader of the file produces nothing striking. If the thing is done without creating a disturbance, the procession does not alter its ways at all. The second caterpillar, promoted to captain, knows the duties of his rank off-hand: he selects and leads, or rather he hesitates and gropes.

The breaking of the silk ribbon is not very important either. I remove a caterpillar from the middle of the file. With my scissors, so as not to cause a commotion in the ranks, I cut the piece of ribbon on which he stood and clear away every thread of it. As a result of this breach, the procession acquires two marching leaders, each independent of the other. It may be that the one in the rear joins the file ahead of him, from which he is separated by but a slender interval; in that case, things return to their original condition. More frequently, the two parts do not become reunited. In that case, we have two distinct processions, each of which wanders where it pleases and diverges from the other. Nevertheless, both will be able to return to the nest by discovering sooner or later, in the course of their peregrinations, the ribbon on the other side of the break.

These two experiments are only moderately interesting. I have thought out another, one more fertile in possibilities. I propose to make the caterpillars describe a close circuit, after the ribbons running from it and liable to bring about a change of direction have been destroyed. The locomotive engine pursues its invariable course so long as it is not shunted on to a branch-line. If the Processionaries find the silken rail always clear in front of them, with no switches anywhere, will they continue on the same track, will they persist in following a road that never comes to an end? What we have to do is to produce this circuit, which is unknown under ordinary conditions, by artificial means.

The first idea that suggests itself is to seize with the forceps the silk ribbon at the back of the train, to bend it without shaking it and to bring the end of it ahead of the file. If the caterpillar marching in the van steps upon it, the thing is done: the others will follow him faithfully. The operation is very simple in theory but very difficult in practice and produces no useful results. The ribbon, which is extremely slight, breaks under the weight of the grains of sand that stick to it and are lifted with it. If it does not break, the caterpillars at the back, however delicately we may go to work, feel a disturbance which makes them curl up or even let go.

There is a yet greater difficulty: the leader refuses the ribbon laid before him; the cut end makes him distrustful. Failing to see the regular, uninterrupted road, he slants off to the right or left, he escapes at a tangent. If I try to interfere and to bring him back to the path of my choosing, he persists in his refusal, shrivels up, does not budge; and soon the whole procession is in confusion. We will not insist: the method is a poor one, very wasteful of effort for at best a problematical success.

We ought to interfere as little as possible and obtain a natural closed circuit. Can it be done? Yes. It lies in our power, without the least meddling, to see a procession march along a perfect circular track. I owe this result, which is eminently deserving of our attention, to pure chance.

On the shelf with the layer of sand in which the nests are planted stand some big palm-vases measuring nearly a yard and a half in circumference at the top. The caterpillars often scale the sides and climb up to the moulding which forms a cornice around the opening. This place suits them for their processions, perhaps because of the absolute firmness of the surface, where there is no fear of landslides, as on the loose, sandy soil below; and also, perhaps, because of the horizontal position, which is favorable to repose after the fatigue of the ascent. It provides me with a circular track all ready-made. I have nothing to do but wait for an occasion propitious to my plans. This occasion is not long in coming.

On the 30th of January, 1896, a little before twelve o'clock in the day, I discover a numerous troop making their way up and gradually reaching the popular cornice. Slowly, in single file, the caterpillars climb the great vase, mount the ledge and advance in regular procession, while others are constantly arriving and continuing the series. I wait for the string to close up, that is to say, for the leader, who keeps following the circular moulding, to return to the point from which he started. My object is achieved in a quarter of an hour. The closed circuit is realized magnificently, in something very nearly approaching a circle.

The next thing is to get rid of the rest of the ascending column, which would disturb the fine order of the procession by an excess of new-comers; it is also important that we should do away with all the silken paths, both new and old, that can put the cornice into communication with the ground. With a thick hair-pencil I sweep away the surplus climbers; with a big brush, one that leaves no smell behind it—for this might afterwards prove confusing—I carefully rub down the vase and get rid of every thread which the caterpillars have laid on the march. When these preparations are finished, a curious sight awaits us.

In the uninterrupted circular procession there is no longer a leader. Each caterpillar is preceded by another on whose heels he follows, guided by the silk track, the work of the whole party; he again has a com-panion close behind him, following him in the same orderly way. And this is repeated without variation throughout the length of the chain. None commands, or rather none modifies the trail according to his fancy; all obey, trusting in the guide who ought normally to lead the march and who in reality has been abolished by my trickery.

From the first circuit of the edge of the tub the rail of silk has been laid in position and is soon turned into a narrow ribbon by the pro-cession, which never ceases dribbling its thread as it goes. The rail is simply doubled and has no branches anywhere, for my brush has de-stroyed them all. What will the caterpillars do on this deceptive, closed path? Will they walk endlessly round and round until their strength gives out entirely?

The old schoolmen were fond of quoting Buridan's Ass, that famous Donkey who, when placed between two bundles of hay, starved to death because he was unable to decide in favour of either by breaking the equilibrium between two equal but opposite attractions. They slandered the worthy animal. The Ass, who is no more foolish than any one else, would reply to the logical snare by feasting off both bundles. Will my caterpillars show a little of his mother wit? Will they, after many attempts, be able to break the equilibrium of their closed circuit, which keeps them on a road without a turning? Will they make up their minds to swerve to this side or that, which is the only method of reaching their bundle of hay, the green branch yonder, quite near, not two feet off?

I thought that they would and I was wrong. I said to myself:

"The procession will go on turning for some time, for an hour, two hours perhaps; then the caterpillars will perceive their mistake. They will abandon the deceptive road and make their descent somewhere or other."

That they should remain up there, hard pressed by hunger and the lack of cover, when nothing prevented them from going away, seemed to me inconceivable imbecility. Facts, however, forced me to accept the incredible. Let us describe them in detail.

The circular procession begins, as I have said, on the 30th of January, about midday, in splendid weather. The caterpillars march at an even pace, each touching the stern of the one in front of him. The unbroken chain eliminates the leader with his changes of direction; and all follow mechanically, as faithful to their circle as are the hands of a watch. The headless file has no liberty left, no will; it has become mere clockwork. And this continues for hours and hours. My success goes far beyond my wildest suspicions. I stand amazed at it, or rather I am stupefied.

Meanwhile, the multiplied circuits change the original rail into a superb ribbon a twelfth of an inch broad. I can easily see it glittering on the red ground of the pot. The day is drawing to a close and no alteration has yet taken place in the position of the trail. A striking proof confirms this.

The trajectory is not a plane curve, but one which, at a certain point, deviates and goes down a little way to the lower surface of the cornice, returning to the top some eight inches farther. I marked these two points of deviation in pencil on the vase at the outset. Well, all that afternoon and, more conclusive still, on the following days, right to the end of this mad dance, I see the string of caterpillars dip under the ledge at the first point and come to the top again at the second. Once the first thread is laid, the road to be pursued is permanently established.

If the road does not vary, the speed does. I measure nine centimetres a minute as the average distance covered. But there are more or less lengthy halts; the pace slackens at times, especially when the temperature falls. At ten o'clock in the evening the walk is little more than a lazy swaying of the body. I foresee an early halt, in consequence of the cold, of fatigue and doubtless also of hunger.

Grazing-time has arrived. The caterpillars have come crowding from all the nests in the greenhouse to browse upon the pine-branches planted by myself beside the silken purses. Those in the garden do the same, for the temperature is mild. The others, lined up along the earthenware cornice, would gladly take part in the feast; they are bound to have an appetite after a ten hours' walk. The branch stands green and tempting not a hand's breadth away. To reach it they need but go down; and the poor wretches, foolish slaves of their ribbon that they are, cannot make up their minds to do so. I leave the famished ones at half-past ten, persuaded that they will take counsel with their pillow and that on the morrow things will have resumed their ordinary course.

I was wrong. I was expecting too much of them when I accorded them that faint gleam of intelligence which the tribulations of a distressful stomach ought, one would think, to have aroused. I visit them at dawn. They are lined up as on the day before, but motionless. When the air grows a little warmer, they shake off their torpor, revive and start walking again. The circular procession begins anew, like that which I have already seen. There is nothing more and nothing less to be noted in their machine-like obstinacy.

This time it is a bitter night. A cold snap has supervened, was indeed foretold in the evening by the garden caterpillars, who refused to come out despite appearances which to my duller senses seemed to promise a continuation of the fine weather. At daybreak the rosemary-walks are all asparkle with rime and for the second time this year there is a sharp frost. The large pond in the garden is frozen over. What can the caterpillars in the conservatory be doing? Let us go and see.

All are ensconced in their nests, except the stubborn processionists on the edge of the vase, who, deprived of shelter as they are, seem to have spent a very bad night. I find them clustered in two heaps, without any attempt at order. They have suffered less from the cold, thus huddled together.

'Tis an ill wind that blows nobody any good. The severity of the night has caused the ring to break into two segments which will, perhaps, afford a chance of safety. Each group, as it revives and resumes its walk, will presently be headed by a leader who, not being obliged to follow a caterpillar in front of him, will possess some liberty of movement and perhaps be able to make the procession swerve to one side. Remember that, in the ordinary processions, the caterpillar walking ahead acts as a scout. While the others, if nothing occurs to create excitement, keep to their ranks, he attends to his duties as a leader and is continually turning his head to this side and that, investigating, seeking, groping, making his choice. And things happen as he decides: the band follows him faithfully. Remember also that, even on a road which has already been travelled and beribboned, the guiding caterpillar continues to explore.

There is reason to believe that the Processionaries who have lost their way on the ledge will find a chance of safety here. Let us watch them. On recovering from their torpor, the two groups line up by degrees into two distinct files. There are therefore two leaders, free to go where they please, independent of each other. Will they succeed in leaving the enchanted circle? At the sight of their large black heads swaying anxiously from side to side, I am inclined to think so for a moment. But I am soon undeceived. As the ranks fill out, the two sections of the chain meet and the circle is reconstituted. The momentary leaders once more become simple subordinates; and again the caterpillars march round and round all day.

For the second time in succession, the night, which is very calm and magnificently starry, brings a hard frost. In the morning the Processionaries on the tub, the only ones who have camped out unsheltered, are gathered into a heap which largely overflows both sides of the fatal ribbon. I am present at the awakening of the numbed ones. The first to take the road is, as luck will have it, outside the track. Hesitatingly he ventures into unknown ground. He reaches the top of the rim and descends upon the other side on the earth in the vase. He is followed by six others, no more. Perhaps the rest of the troop, who have not fully recovered from their nocturnal torpor, are too lazy to bestir themselves.

The result of this brief delay is a return to the old track. The caterpillars embark on the silken trail and the circular march is resumed, this time in the form of a ring with a gap in it. There is no attempt, however, to strike a new course on the part of the guide whom this gap has placed at the head. A chance of stepping outside the magic circle has presented itself at last; and he does not know how to avail himself of it.

As for the caterpillars who have made their way to the inside of the vase, their lot is hardly improved. They climb to the top of the palm, starving and seeking for food. Finding nothing to eat that suits them, they retrace their steps by following the thread which they have left on the way, climb the ledge of the pot, strike the procession again and, without further anxiety, slip back into the ranks. Once more the ring is complete, once more the circle turns and turns.

Then when will the deliverance come? There is a legend that tells of poor souls dragged along in an endless round until the hellish charm is broken by a drop of holy water. What drop will good fortune sprinkle on my Processionaries to dissolve their circle and bring them back to the nest? I see only two means of conjuring the spell and obtaining a release from the circuit. These two means are two painful ordeals. A strange linking of cause and effect: from sorrow and wretchedness good is to come.

And, first, shrivelling as the result of cold. The caterpillars gather together without any order, heap themselves some on the path, some, more numerous these, outside it. Among the latter there may be, sooner or later, some revolutionary who, scorning the beaten track, will trace out a new road and lead the troop back home. We have just seen an instance of it. Seven penetrated to the interior of the vase and climbed the palm. True, it was an attempt with no result, but still an attempt. For complete success, all that need be done would have been to take the opposite slope. An even chance is a great thing. Another time we shall be more successful.

In the second place, the exhaustion due to fatigue and hunger. A lame one stops, unable to go farther. In front of the defaulter the procession still continues to wend its way for a short time. The ranks close up and an empty space appears. On coming to himself and resuming the march, the caterpillar who has caused the breach becomes a leader, having nothing before him. The least desire for emancipation is all that he wants to make him launch the band into a new path which perhaps will be the saving path.

In short, when the Processionaries' train is in difficulties, what it needs, unlike ours, is to run off the rails. The side-tracking is left to the caprice of a leader who alone is capable of turning to the right or left; and this leader is absolutely non-existent so long as the ring remains unbroken. Lastly, the breaking of the circle, the one stroke of luck, is the result of a chaotic halt, caused principally by excess of fatigue or cold.

The liberating accident, especially that of fatigue, occurs fairly often. In the course of the same day, the moving circumference is cut up several times into two or three sections; but continuity soon returns and no change takes place. Things go on just the same. The bold innovator who is to save the situation has not yet had his inspiration.

There is nothing new on the fourth day, after an icy night like the previous one; nothing to tell except the following detail. Yesterday I did not remove the trace left by the few caterpillars who made their way to the inside of the vase. This trace, together with a junction connecting it with the circular road, is discovered in the course of the morning. Half the troop takes advantage of it to visit the earth in the pot and climb the palm; the other half remains on the ledge and continues to walk along the old rail. In the afternoon the band of emigrants rejoins the others, the circuit is completed and things return to their original condition.

We come to the fifth day. The night frost becomes more intense, without however as yet reaching the greenhouse. It is followed by bright sunshine in a calm and limpid sky. As soon as the sun's rays have warmed the panes a little, the caterpillars, lying in heaps, wake up and resume their evolutions on the ledge of the vase. This time the fine order of the beginning is disturbed and a certain disorder becomes manifest, apparently an omen of deliverance near at hand. The scouting-path inside the vase, which was upholstered in silk yesterday and the day before, is to-day followed to its origin on the rim by a part of the band and is then abandoned after a short loop. The other caterpillars follow the usual ribbon. The result of this bifurcation is two almost equal files, walking along the ledge in the same direction, at a short distance from each other, sometimes meeting, separating farther on, in every case with some lack of order.

Weariness increases the confusion. The crippled, who refuse to go on, are many. Breaches increase; files are split up into sections each of which has its leader, who pokes the front of his body this way and that to explore the ground. Everything seems to point to the disintegration which will bring safety. My hopes are once more disappointed. Before the night the single file is reconstituted and the invincible gyration resumed.

Heat comes, just as suddenly as the cold did. To-day, the 4th of February, is a beautiful, mild day. The greenhouse is full of life. Numerous festoons of caterpillars, issuing from the nests, meander along the sand on the shelf. Above them, at every moment, the ring on the ledge of the vase breaks up and comes together again. For the first time I see daring leaders who, drunk with heat, standing only on their hinder prolegs at the extreme edge of the earthenware rim, fling themselves forward into space, twisting about, sounding the depths. The endeavour is frequently repeated, while the whole troop stops. The caterpillars' heads give sudden jerks; their bodies wriggle.

One of the pioneers decides to take the plunge. He slips under the ledge. Four follow him. The others, still confiding in the perfidious silken path, dare not copy him and continue to go along the old road.

The short string detached from the general chain gropes about a great deal, hesitates long on the side of the vase; it goes half-way down, then climbs up again slantwise, rejoins and takes its place in the procession. This time the attempt has failed, though at the foot of the vase, not nine inches away, there lay a bunch of pine-needles which I had placed there with the object of enticing the hungry ones. Smell and sight told them nothing. Near as they were to the goal, they went up again.

No matter, the endeavour has its uses. Threads were laid on the way and will serve as a lure to further enterprise. The road of deliverance has its first landmarks. And two days later, on the eighth day of the experiment, the caterpillars—now singly, anon in small groups, then again in strings of some length—come down from the ledge by following the staked-out path. At sunset the last of the laggards is back in the nest.

Now for a little arithmetic. For seven times twenty-four hours the caterpillars have remained on the ledge of the vase. To make an ample allowance for stops due to the weariness of this one or that and above all for the rest taken during the colder hours of the night, we will deduct one-half of the time. This leaves eighty-four hours' walking. The average pace is nine centimetres a minute. The aggregate distance covered, therefore, is 453 metres, a good deal more than a quarter of a mile, which is a great walk for these little crawlers. The circumference of the vase, the perimeter of the track, is exactly 1 m. 35. Therefore the circle covered, always in the same direction and always without result, was described three hundred and thirty-five times.

These figures surprise me, though I am already familiar with the abysmal stupidity of insects as a class whenever the least accident occurs. I feel inclined to ask myself whether the Processionaries were not kept up there so long by the difficulties and dangers of the descent rather than by the lack of any gleam of intelligence in their benighted minds. The facts, however, reply that the descent is as easy as the ascent.

The caterpillar has a very supple back, well adapted for twisting round projections or slipping underneath. He can walk with the same ease vertically or horizontally, with his back down or up. Besides, he never moves forward until he has fixed his thread to the ground. With this support to his feet, he has no falls to fear, no matter what his position.

I had a proof of this before my eyes during a whole week. As I have already said, the track, instead of keeping on one level, bends twice, dips at a certain point under the ledge of the vase and reappears at the top a little farther on. At one part of the circuit, therefore, the procession walks on the lower surface of the rim; and this inverted position implies so little discomfort or danger that it is renewed at each turn for all the caterpillars from first to last.

It is out of the question then to suggest the dread of a false step on the edge of the rim which is so nimbly turned at each point of inflexion. The caterpillars in distress, starved, shelterless, chilled with cold at night, cling obstinately to the silk ribbon covered hundreds of times, because they lack the rudimentary glimmers of reason which would advise them to abandon it.

HEREDITY

« 3 »

In one of those delightful asides that add so much charm to Fabre's writings, the author here examines all his known ancestors and finds no explanation in them for his fascination by the insects. His grandparent's farm was located in the province of Guienne, of which Rodez is the capital. It was here that he felt the first awakenings of that immense curiosity about the natural world which was one of Fabre's lifelong characteristics. This chapter originally appeared in that most autobiographical of Fabre's entomological books, THE LIFE OF THE FLY.

Since Darwin bestowed upon me the title of "incomparable observer," the epithet has often come back to me, from this side and from that, without my yet understanding what particular merit I have shown. It seems to me so natural, so much within everybody's scope, so absorbing to interest one's self in everything that swarms around us! However, let us pass on and admit that the compliment is not unfounded.

My hesitation ceases if it is a question of admitting my curiosity in matters that concern the insect. Yes, I possess the gift, the instinct that impels me to frequent that singular world; yes, I know that I am capable of spending on those studies an amount of precious time which would be better employed in making provision, if possible, for the poverty of old age; yes, I confess that I am an enthusiastic observer of the animal. How was this characteristic propensity, at once the torment and delight of my life, developed? And, to begin with, how much does it owe to heredity?

The common people have no history: persecuted by the present, they cannot think of preserving the memory of the past. And yet what surpassingly instructive records, comforting too and pious, would be the family-papers that should tell us who our forebears were and speak to us of their patient struggles with harsh fate, their stubborn efforts to build up, atom by atom, what we are to-day. No story would come up with that for individual interest. But, by the very force of things, the home is abandoned; and, when the brood has flown, the nest is no longer recognized.

I, a humble journeyman in the toilers' hive, am therefore very poor in family-recollections. In the second degree of ancestry, my facts become suddenly obscured. I will linger over them a moment for two reasons: first, to enquire into the influence of heredity; and secondly, to leave my children yet one more page concerning them.

I did not know my maternal grandfather. This venerable ancestor was, I have been told, a process-server in one of the poorest parishes of the Rouergue. He used to engross on stamped paper in a primitive spelling. With his well-filled pen-case and ink-horn, he went drawing out deeds up hill and down dale, from one insolvent wretch to another more insolvent still. Amid his atmosphere of pettifoggery, this rudimentary scholar, waging battle on life's acerbities, certainly paid no attention to the insect; at most, if he met it, he would crush it under foot. The unknown animal, suspected of evil-doing, deserved no further enquiry. Grandmother, on her side, apart from her housekeeping and her beads, knew still less about anything. She looked on the alphabet as a set of hieroglyphics only fit to spoil your sight for nothing, unless you were scribbling on paper bearing the government stamp. Who in the world, in her day, among the small folk, dreamt of knowing how to read and write? That luxury was reserved for the attorney, who himself made but a sparing use of it. The insect, I need hardly say, was the least of her cares. If sometimes, when rinsing her salad at the tap, she found a Caterpillar on the lettuce-leaves, with a start of fright she would fling the loathsome thing away, thus cutting short relations reputed dangerous. In short, to both my maternal grandparents, the insect was a creature of no interest whatever and almost always a repulsive object, which one dared not touch with the tip of one's finger. Beyond a doubt, my taste for animals was not derived from them.

I have more precise information regarding my grandparents on my father's side, for their green old age allowed me to know them both. They were people of the soil, whose quarrel with the alphabet was so great that they had never opened a book in their lives; and they kept a lean farm on the cold granite ridge of the Rouergue table-land. The house, standing alone among the heath and broom, with no neighbour for many a mile around and visited at intervals by the wolves, was to them the hub of the universe. But for a few surrounding villages, whither the calves were driven on fair-days, the rest was only very vaguely known by hearsay. In this wild solitude, the mossy fens, with their quagmires oozing with iridescent pools, supplied the cows, the principal source of wealth, with rich, wet grass. In summer, on the short swards of the slopes, the sheep were penned day and night, protected from beasts of prey by a fence of hurdles propped up with pitchforks. When the grass was cropped close at one spot, the fold was shifted elsewhither. In the centre was the shepherd's rolling hut, a straw cabin. Two watch-dogs, equipped with spiked collars, were answerable for tranquillity if the thieving wolf appeared in the night from out the neighbouring woods.

Padded with a perpetual layer of cow-dung, in which I sank to my knees, broken up with shimmering puddles of dark-brown liquid manure, the farm-yard also boasted a numerous population. Here the lambs skipped, the geese trumpeted, the fowls scratched the ground and the sow grunted with her swarm of little pigs hanging to her dugs.

The harshness of the climate did not give husbandry the same chances. In a propitious season, they would set fire to a stretch of moorland bristling with gorse and send the swing-plough across the ground enriched with the cinders of the blaze. This yielded a few acres of rye, oats and potatoes. The best corners were kept for hemp, which furnished the distaffs and spindles of the house with the material for linen and was looked upon as grandmother's private crop.

Grandfather, therefore, was, before all, a herdsman versed in matters of cows and sheep, but completely ignorant of aught else. How dumbfounded he would have been to learn that, in the remote future, one of his family would become enamoured of those insignificant animals to which he had never vouchsafed a glance in his life! Had he guessed that that lunatic was myself, the scapegrace seated at the table by his side, what a smack I should have caught in the neck, what a wrathful look!

"The idea of wasting one's time with that nonsense!" he would have thundered.

For the patriarch was not given to joking. I can still see his serious face, his unclipped head of hair, often brought back behind his ears with a flick of the thumb and spreading its ancient Gallic mane over his shoulders. I see his little three-cornered hat, his small-clothes buckled at the knees, his wooden shoes, stuffed with straw, that echoed as he walked. Ah, no! Once childhood's games were past, it would never have done to rear the Grasshopper and unearth the Dung-beetle from his natural surroundings.

Grandmother, pious soul, used to wear the eccentric head-dress of the Rouergue highlanders: a large disk of black felt, stiff as a plank, adorned in the middle with a crown a finger's-breadth high and hardly wider across than a six-franc piece. A black ribbon fastened under the chin maintained the equilibrium of this elegant, but unsteady circle. Pickles, hemp, chickens, curds and whey, butter; washing the clothes, minding the children, seeing to the meals of the household: say that and you have summed up the strenuous woman's round of ideas. On her left side, the distaff, with its load of flax; in her right hand, the spindle turning under a quick twist of her thumb, moistened at intervals with her tongue: so she went through life, unweariedly, attending to the order and the welfare of the house. I see her in my mind's eye particularly on winter evenings, which were more favourable to family-talk. When the hour came for meals, all of us, big and little, would take our seats round a long table, on a couple of benches, deal planks supported by four rickety legs. Each found his wooden bowl and his tin spoon in front of him. At one end of the table always stood an enormous rye-loaf, the size of a cartwheel, wrapped in a linen cloth with a pleasant smell of washing, and remained until nothing was left of it. With a vigorous stroke, grandfather would cut off enough for the needs of the moment; then he would divide the piece among us with the one knife which he alone was entitled to wield. It was now each one's business to break up his bit with his fingers and to fill his bowl as he pleased.

Next came grandmother's turn. A capacious pot bubbled lustily and sang upon the flames in the hearth, exhaling an appetizing savour of bacon and turnips. Armed with a long metal ladle, grandmother would take from it, for each of us in turn, first the broth, wherein to soak the bread, and next the ration of turnips and bacon, partly fat and partly lean, filling the bowl to the top. At the other end of the table was the pitcher, from which the thirsty were free to drink at will. What appetites we had and what festive meals those were, especially when a cream-cheese, homemade, was there to complete the banquet!

Near us blazed the huge fire-place, in which whole tree-trunks were consumed in the extreme cold weather. From a corner of that monumental, soot-glazed chimney, projected, at a convenient height, a bracket with a slate shelf, which served to light the kitchen when we sat up late. On this we burnt chips of pine-wood, selected among the most translucent, those containing the most resin. They shed over the room a lurid red light, which saved the walnut-oil in the lamp.

When the bowls were emptied and the last crumb of cheese scraped up, grandam went back to her distaff, on a stool by the chimney-corner. We children, boys and girls, squatting on our heels and putting out our hands to the cheerful fire of furze, formed a circle round her and listened to her with eager ears. She told us stories, not greatly varied, it is true, but still wonderful, for the wolf often played a part in them. I should have very much liked to see this wolf, the hero of so many tales that made our flesh creep; but the shepherd always refused to take me into his straw hut, in the middle of the fold, at night. When we had done talking about the horrid wolf, the dragon and the serpent and when the resinous splinters had given out their last gleams, we went to sleep the sweet sleep that toil gives. As the youngest of the household, I had a right to the mattress, a sack stuffed with oat-chaff. The others had to be content with straw.

I owe a great deal to you, dear grandmother: it was in your lap that I found consolation for my first sorrows. You have handed down to me, perhaps, a little of your physical vigour, a little of your love of work; but certainly you were no more accountable than grandfather for my passion for insects.

Nor was either of my own parents. My mother, who was quite illiterate, having known no teacher than the bitter experience of a harassed life, was the exact opposite of what my tastes required for their development. My peculiarity must seek its origin elsewhere: that I will swear. But I do not find it in my father, either. The excellent man, who was hard-working and sturdily-built like grandad, had been to school as a child. He knew how to write, though he took the greatest liberties with spelling; he knew how to read and understand what he read, provided the reading presented no more serious literary difficulties than occurred in the stories in the almanack. He was the first of his line to allow himself to be tempted by the town and he lived to regret it. Badly off, having but little outlet for his industry, making God knows what shifts to pick up a livelihood, he went through all the disappointments of the country-man turned townsman. Persecuted by bad luck, borne down by the burden, for all his energy and good-will, he was far indeed from starting me in entomology. He had other cares, cares more direct and more serious. A good cuff or two when he saw me pinning an insect to a cork was all the encouragement that I received from him. Perhaps he was right.

The conclusion is positive: there is nothing in heredity to explain my taste for observation. You may say that I do not go far enough back.

Well, what should I find beyond the grandparents where my facts come to a stop? I know, partly. I should find even more uncultured ancestors: sons of the soil, ploughmen, sowers of rye, neat-herds; one and all, by the very force of things, of not the least account in the nice matters of observation.

And yet, in me, the observer, the enquirer into things began to take shape almost in infancy. Why should I not describe my first discoveries? They are ingenuous in the extreme, but will serve notwithstanding to tell us something of the way in which tendencies first show themselves. I was five or six years old. That the poor household might have one mouth less to feed, I had been placed in grandmother's care, as I have just been saying. Here, in solitude, my first gleams of intelligence were awakened amidst the geese, the calves and the sheep. Everything before that is impenetrable darkness. My real birth is at that moment when the dawn of personality rises, dispersing the mists of unconsciousness and leaving a lasting memory. I can see myself plainly, clad in a soiled frieze frock flapping against my bare heels; I remember the handkerchief hanging from my waist by a bit of string, a handkerchief often lost and replaced by the back of my sleeve.

There I stand one day, a pensive urchin, with my hands behind my back and my face turned to the sun. The dazzling splendour fascinates me. I am the Moth attracted by the light of the lamp. With what am I enjoying the glorious radiance: with my mouth or my eyes? That is the question put by my budding scientific curiosity. Reader, do not smile: the future observer is already practising and experimenting. I open my mouth wide and close my eyes: the glory disappears. I open my eyes and shut my mouth: the glory reappears. I repeat the performance, with the same result. The question's solved: I have learnt by deduction that I see the sun with my eyes. Oh, what a discovery! That evening, I told the whole house all about it. Grandmother smiled fondly at my simplicity: the others laughed at it. 'Tis the way of the world.

Another find. At nightfall, amidst the neighbouring bushes, a sort of jingle attracted my attention, sounding very faintly and softly through the evening silence. Who is making that noise? Is it a little bird chirping in his nest? We must look into the matter and that quickly. True, there is the wolf, who comes out of the woods at this time, so they tell me. Let's go all the same, but not too far: just there, behind that clump of broom. I stand on the look-out for long, but all in vain. At the faintest sound of movement in the brushwood, the jingle ceases. I try again next day and the day after. This time, my stubborn watch succeeds. Whoosh!

A grab of my hand and I hold the singer. It is not a bird; it is a kind of Grasshopper whose hind-legs my playfellows have taught me to like: a poor recompense for my prolonged ambush. The best part of the business is not the two haunches with the shrimpy flavour, but what I have just learnt. I now know, from personal observation, that the Grasshopper sings. I did not publish my discovery, for fear of the same laughter that greeted my story about the sun.

Oh, what pretty flowers, in a field close to the house! They seem to smile to me with their great violet eyes. Later on, I see, in their place, bunches of big red cherries. I taste them. They are not nice and they have no stones. What can those cherries be? At the end of the summer, grandfather comes with a spade and turns my field of observation topsy-turvy. From under ground there comes, by the basketful and sackful, a sort of round root. I know that root; it abounds in the house; time after time I have cooked it in the peat-stove. It is the potato. Its violet flower and its red fruit are pigeon-holed for good and all in my memory.

THE HUNTING WASP

« 4 »

It was among the solitary wasps that Fabre found subjects for some of his most revealing experiments. Sphex *wasps have a widespread distribution. A number of species finish the work of filling in their burrows by using a pebble as a hammer to tamp down the ground. Many are grasshopper-hunters. The Ephippiger, hunted by the Languedocian Sphex, is a long-horned green grasshopper found in the south of France. The scene of many of Fabre's wasp studies, before he retired to Sérignan, was along a deeply sunken road on the outskirts of Carpentras. This account is taken from* THE HUNTING WASPS.

When the chemist has fully prepared his plan of research, he mixes his reagent at the most convenient moment and lights a flame under his retort. He is the master of time, place and circumstances. He chooses his hour, shuts himself up in his laboratory, where nothing can come to disturb the business in hand; he produces at will this or that condition which reflection suggests to him: he is in quest of the secrets of inorganic matter, whose chemical activities science can awaken whenever it thinks fit.

The secrets of living matter—not those of anatomical structure, but really those of life in action, especially of instinct—present much more difficult and delicate conditions to the observer. Far from being able to choose his own time, he is the slave of the season, of the day, of the hour, of the very moment. When the opportunity offers, he must seize it as it comes, without hesitation, for it may be long before it presents itself again. And, as it usually arrives at the moment when he is least expecting it, nothing is in readiness for making the most of it. He must then and there improvise his little stock of experimenting-material, contrive his plans, evolve his tactics, devise his tricks; and he can think himself lucky if inspiration comes fast enough to allow him to profit by the chance offered. This chance, moreover, hardly ever comes except to those who look for it. You must watch for it patiently for days and days, now on sandy slopes exposed to the full glare of the sun, now on some path walled in by high banks, where the heat is like that of an oven, or again on some sandstone ledge which is none too steady. If it is in your power to set up your observatory under a meagre olive-tree that pretends to protect you from the rays of a pitiless sun, you may bless the fate that treats you as a sybarite: your lot is an Eden. Above all, keep your eyes open. The spot is a good one; and—who knows?—the opportunity may come at any moment.

It came, late, it is true; but still it came. Ah, if you could now observe at your ease, in the quiet of your study, with nothing to distract your mind from your subject, far from the profane wayfarer who, seeing you so busily occupied at a spot where he sees nothing, will stop, overwhelm you with queries, take you for some water-diviner, or—a graver suspicion this—regard you as some questionable character searching for buried treasure and discovering by means of incantations where the old pots full of coin lie hidden! Should you still wear a Christian aspect in his eyes, he will approach you, look to see what you are looking at and smile in a manner that leaves no doubt as to his poor opinion of people who spend their time in watching Flies. You will be lucky indeed if the troublesome visitor, with his tongue in his cheek, walks off at last without disturbing things and without repeating in his innocence the disaster brought about by my two conscripts' boots.

Should your inexplicable doings not puzzle the passer-by, they will be sure to puzzle the village keeper, that uncompromising representative of the law in the ploughed acres. He has long had his eye on you. He has so often seen you wandering about, like a lost soul, for no appreciable reason; he has so often caught you rooting in the ground, or, with infinite precautions, knocking down some strip of wall in a sunken road, that in the end he has come to look upon you with dark suspicion. You are nothing to him but a gipsy, a tramp, a poultry-thief, a shady person or, at the best, a madman. Should you be carrying your botanizing-case, it will represent to him the poacher's ferret-cage; and you would never get it out of his head that, regardless of the game-laws and the rights of landlords, you are clearing all the neighbouring warrens of their rabbits. Take care. However thirsty you may be, do not lay a finger on the nearest bunch of grapes: the man with the municipal badge will be there, delighted to have a case at last and so to receive an explanation of your highly perplexing behaviour.

I have never, I can safely say, committed any such misdemeanour; and yet, one day, lying on the sand, absorbed in the details of a Bembex' household, I suddenly heard beside me:

"In the name of the law, I arrest you! You come along with me!"

It was the keeper of Les Angles, who, after vainly waiting for an opportunity to catch me at fault and being daily more anxious for an answer to the riddle that was worrying him, at last resolved upon the brutal expedient of a summons. I had to explain things. The poor man seemed anything but convinced:

"Pooh!" he said. "Pooh! You will never make me believe that you come here and roast in the sun just to watch Flies. I shall keep an eye on you, mark you! And, the first time I . . . ! However, that'll do for the present."

And he went off. I have always believed that my red ribbon had a good deal to do with his departure. And I also put down to that red ribbon certain other little services by which I benefited during my entomological and botanical excursions. It seemed to me—or was I dreaming?—it seemed to me that, on my botanizing-expeditions up Mont Ventoux, the guide was more tractable and the donkey less obstinate.

The aforesaid bit of scarlet ribbon did not always spare me the tribulations which the entomologist must expect when experimenting on the public way. Here is a characteristic example. Ever since daybreak I have been ambushed, sitting on a stone, at the bottom of a ravine. The subject of my matutinal visit is the Languedocian Sphex. Three women, vine-pickers, pass in a group, on the way to their work. They give a glance at the man seated, apparently absorbed in reflection. At sunset, the same pickers pass again, carrying their full baskets on their heads. The man is still there, sitting on the same stone, with his eyes fixed on the same place. My motionless attitude, my long persistency in remaining at that deserted spot, must have impressed them deeply. As they passed by me, I saw one of them tap her forehead and heard her whisper to the others:

"Un paouré inoucènt, pécaïre!"

And all three made the sign of the Cross.

An innocent, she had said, un inoucènt, an idiot, a poor creature, quite harmless, but half-witted; and they had all made the sign of the Cross, an idiot being to them one with God's seal stamped upon him.

"How now!" thought I. "What a cruel mockery of fate! You, who are so laboriously seeking to discover what is instinct in the animal and what is reason, you yourself do not even possess your reason in these good women's eyes! What a humiliating reflection!"

No matter: *pécaïre,* that expression of supreme compassion, in the Provençal dialect, *pécaïre,* coming from the bottom of the heart, soon made me forget *inoucènt.*

It is in this ravine with its three grape-gathering women that I would meet the reader, if he be not discouraged by the petty annoyances of which I have given him a foretaste. The Languedocian Sphex frequents these points, not in tribes congregating at the same spot when nest-building work begins, but as solitary individuals, sparsely distributed, settling wherever the chances of their vagabondage lead them. Even as her kinswoman, the Yellow-winged Sphex, seeks the society of her kind and the animation of a yard full of workers, the Languedocian Sphex prefers isolation, quiet and solitude. Graver of gait, more formal in her manners, of a larger size and also more sombrely clad, she always lives apart, not caring what others do, disdaining company, a genuine misanthrope among the Sphegidæ. The one is sociable, the other is not: a profound difference which in itself is enough to characterize them.

This amounts to saying that, with the Languedocian Sphex, the diffi-culties of observation increase. No long-meditated experiment is pos-sible in her case; nor, when the first attempts have failed, can one hope to try them again, on the same occasion, with a second or a third sub-ject and so on. If you prepare the materials for your observation in advance, if, for instance, you have in reserve a piece of game which you propose to substitute for that of the Sphex, it is to be feared, nay, it is almost certain that the huntress will not appear; and, when she does come at last, your materials are no longer fit for use and everything has to be improvised in a hurry, that very moment, under conditions that are not always satisfactory.

Let us take heart. The site is a first-rate one. Many a time already I have surprised the Sphex here, sunning herself on a vine-leaf. The in-sect, spread out flat, is basking voluptuously in the heat and light. From time to time it has a sort of frenzied outburst of pleasure: it quivers with content; it rapidly taps its feet on its couch, producing a tattoo not unlike that of rain falling heavily on the leaf. The joyous thrum can be heard several feet away. Then immobility begins again, soon followed by a fresh nervous commotion and by the whirling of the tarsi, a symbol of supreme felicity. I have known some of these passionate sun-lovers suddenly to leave the workyard, when the larva's cave has been half-dug, and go to the nearest vine to take a bath of heat and light, after which they would come back to the burrow, as though reluctantly, just to give a perfunctory sweep and soon end by knocking off work, unable to resist the exquisite temptation of luxuriating on the vine-leaves.

It may be that the voluptuous couch is also an observatory, whence the Wasp surveys the surrounding country in order to discover and select her prey. Her exclusive game is the Ephippiger of the Vine, scattered here and there on the branches or on any brambles hard by. The joint is a substantial one, especially as the Sphex favours solely the females, whose bellies are swollen with a mighty cluster of eggs.

Let us take no notice of the repeated trips, the fruitless searches, the tedium of frequent long waiting, but rather present the Sphex suddenly to the reader as she herself appears to the observer. Here she is, at the bottom of a sunken road with high, sandy banks. She comes on foot, but gets help from her wings in dragging her heavy prize. The Ephippiger's antennæ, long and slender as threads, are the harnessing-ropes. Holding her head high, she grasps one of them in her mandibles. The antenna gripped passes between her legs; and the game follows, turned over on its back. Should the soil be too uneven and so offer resistance to this method of carting, the Wasp clasps her unwieldy burden and carries it with very short flights, interspersed, as often as possible, with journeys on foot. We never see her undertake a sustained flight, for long distances, holding the game in her legs, as is the practice of those expert aviators, the Bembeces and Cerceres, for instance, who bear through the air for more than half a mile their respective Flies or Weevils, a very light booty compared with the huge Ephippiger. The overpowering weight of her capture compels the Languedocian Sphex, to make the whole or nearly the whole journey on foot, her method of transport being consequently slow and laborious.

The same reason, the bulk and weight of the prey, have entirely reversed the usual order which the Burrowing Wasps follow in their operations. This order we know: it consists in first digging a burrow and then stocking it with provisions. As the victim is not out of proportion to the strength of the spoiler, it is quite simple to carry it flying, which means that the Wasp can choose any site that she likes for her dwelling. She does not mind how far afield she goes for her prey: once she has captured her quarry, she comes flying home at a speed which makes questions of distance quite immaterial. Hence she prefers as the site for her burrow the place where she herself was born, the place where her forbears lived; she here inherits deep galleries, the accumulated work of earlier generations; and, by repairing them a little, she makes them serve as approaches to new chambers, which are in this way better protected than they would be if they depended upon the labours of a single Wasp, who had to start boring from the surface each year. This happens, for instance, in the case of the Great Cerceris and the Bee-eating Philanthus. And, should the ancestral abode not be strong enough to withstand the rough weather from one year to the next and to be handed down to the offspring, should the burrower have each time to start her tunnelling afresh, at least the Wasp finds greater safety in places consecrated by the experience of her forerunners. Consequently she goes there to dig her galleries, each of which serves as a corridor to a group of cells, thus effecting an economy in the aggregate labour expended upon the whole business of the laying.

In this way are formed not real societies, for there are no concerted efforts towards a common object, but at least assemblies where the sight of her kinswomen and her neighbours doubtless puts heart into the labour of the individual. We can observe, in fact, between these little tribes, springing from the same stock, and the burrowers who do their work alone, a difference in activity which reminds us of the emulation prevailing in a crowded yard and the indifference of labourers who have to work in solitude. Action is contagious in animals as in men; it is fired by its own example.

To sum up: when of a moderate weight for its captor, the prey can be conveyed flying, to a great distance. The Wasp can then choose any site that she pleases for her burrow. She adopts by preference the spot where she was born and uses each passage as a common corridor giving access to several cells. The result of this meeting at a common birthplace is the formation of groups, like turning to like, which is a source of friendly rivalry. This first step towards social life comes from facilities for travelling. Do not things happen in the same way with man, if I may be permitted the comparison? When he has nothing but trackless paths, man builds a solitary hut; when supplied with good roads, he and his fellows collect in populous cities; when served by railways which, so to speak, annihilate distance, they assemble in those immense human hives called London or Paris.

The situation of the Languedocian Sphex is just the reverse. Her prey is a heavy Ephippiger, a single dish representing by itself the sum total of provisions which the other free-booters amass on numerous journeys, insect by insect. What the Cerceres and the other plunderers strong on the wing accomplish by dividing the labour she does in a single journey. The weight of the prey makes any distant flight impossible; it has to be brought home slowly and laboriously, for it is a troublesome business to cart things along the ground. This alone makes the site of the burrow dependent on the accidents of the chase: the prey comes first and the dwelling next. So there is no assembling at a common meeting-place, no association of kindred spirits, no tribes stimulating one another in their work by mutual example, but isolation in the particular spot where the chances of the day have taken the Sphex, solitary labour, carried on without animation though with unfailing diligence. First of all, the prey is sought for, attacked, reduced to helplessness. Not until after that does the digger trouble about the burrow. A favourable place is chosen, as near as possible to the spot where the victim lies, so as to cut short the tedious work of transport; and the chamber of the future larva is rapidly hollowed out and at once receives the egg and the victuals. There you have an example of the inverted method of the Languedocian Sphex, a method, as all my observations go to prove, diametrically opposite to that of the other Hymenoptera. I will give some of the more striking of these observations.

When caught digging, the Languedocian Sphex is always alone, sometimes at the bottom of a dusty recess left by a stone that has dropped out of an old wall, sometimes ensconced in the shelter formed by a flat, projecting bit of sandstone, a shelter much sought after by the fierce Eyed Lizard to serve as an entrance-hall to his lair. The sun beats full upon it; it is an oven. The soil, consisting of old dust that has fallen little by little from the roof, is very easy to dig. The cell is soon scooped out with the mandibles, those pincers which are also used for digging, and the tarsi, which serve as rubbish-rakes. Then the miner flies off, but with a slow flight and no sudden display of wing-power, a manifest sign that the insect is not contemplating a distant expedition. We can easily follow it with our eyes and perceive the spot where it alights, usually ten or twelve yards away. At other times, it decides to walk. It goes off and makes hurriedly for a spot where we will have the indiscretion to follow it, for our presence does not trouble it at all. On reaching its destination, either on foot or on the wing, it looks round for some time, as we gather from its undecided attitude and its journeys hither and thither. It looks round; at last it finds or rather retrieves something. The object recovered is an Ephippiger, half-paralysed, but still moving her tarsi, antennæ and ovipositor. She is a victim which the Sphex certainly stabbed not long ago with a few stings. After the operation, the Wasp left her prey, an embarrassing burden amid the suspense of house-hunting; she abandoned it perhaps on the very spot where she captured it, contenting herself with making it more or less conspicuous by placing it on some grass-tuft, in order to find it more easily later; and, trusting to her good memory to return presently to the spot where the booty lies, she set out to explore the neighbourhood with the object of finding a suitable site and there digging a burrow. Once the home was ready, she came back to her prize, which she found again without much hesitation, and she now prepares to lug it home. She bestrides the victim, seizes one or both of the antennæ and off she goes, tugging and dragging with all the strength of her loins and jaws.

Sometimes, she has only to make one journey; at other times and more often, the carter suddenly plumps down her load and quickly runs home. Perhaps it occurs to her that the entrance-door is not wide enough to admit so substantial a morsel; perhaps she remembers some lack of finish that might hamper the storing. And, in point of fact, the worker does touch up her work: she enlarges the doorway, smooths the threshold, strengthens the ceiling. It is all done with a few strokes of the tarsi. Then she returns to the Ephippiger, lying yonder, on her back, a few steps away. The hauling begins again. On the road, the Sphex seems struck with a new idea, which flashes through her quick brain. She has inspected the door, but has not looked inside. Who knows if all is well in there? She hastens to see, dropping the Ephippiger before she goes. The interior is inspected; and apparently a few pats of the trowel are administered with the tarsi, giving a last polish to the walls. Without lingering too long over these delicate aftertouches, the Wasp goes back to her booty and harnesses herself to its antennæ. Forward! Will the journey be completed this time? I would not answer for it. I have known a Sphex, more suspicious than the others, perhaps, or more neglectful of the minor architectural details, to repair her omissions, to dispel her doubts, by abandoning her prize on the way five or six times running, in order to hurry to the burrow, which each time was touched up a little or merely inspected within. It is true that others make straight for their destination, without even stopping to rest. I must also add that, when the Wasp goes home to improve the dwelling, she does not fail to give a glance from a distance every now and then at the Ephippiger over there, to make sure that nothing has happened to her. This solicitude recalls that of the Sacred Beetle when he leaves the hall which he is excavating in order to come and feel his beloved pellet and bring it a little nearer to him.

The inference to be drawn from the details which I have related is manifest. The fact that every Languedocian Sphex surprised in her mining-operations, even though it be at the very beginning of the digging, at the first stroke of the tarsus in the dust, afterwards, when the home is prepared, makes a short excursion, now on foot, anon flying, and invariably finds herself in possession of a victim already stabbed, already paralysed, compels us to conclude, in all certainty, that this Wasp does her work as a huntress first and as a burrower after, so that the place of the capture decides the place of the home.

This reversal of procedure, which causes the food to be prepared before the larder, whereas hitherto we have seen the larder come before the food, I attribute to the weight of the Sphex' prey, a prey which it is not possible to carry far through the air. It is not that the Languedocian Sphex is ill-built for flight: on the contrary, she can soar magnificently; but the prey which she hunts would weigh her down if she had no other support than her wings. She needs the support of the ground for her hauling-work, in which she displays wonderful strength. When laden with her prey, she always goes afoot, or takes but very short flights, even under conditions when flight would save her time and trouble. I will quote an instance taken from my latest observations on this curious Wasp.

A Sphex appears unexpectedly, coming I know not whence. She is on foot, dragging her Ephippiger, a capture which apparently she has made that moment in the neighbourhood. In the circumstances, it behoves her to dig herself a burrow. The site is as bad as bad can be. It is a well-beaten path, hard as stone. The Sphex, who has no time to make laborious excavations, because the already captured prize must be stored as quickly as possible, the Sphex wants soft ground, wherein the larva's chamber can be contrived in one short spell of work. I have described her favourite soil, namely, the dust of years which has accumulated at the bottom of some hole in a wall or of some little shelter under the rocks. Well, the Sphex whom I am now observing stops at the foot of a house with a newly-whitewashed front some twenty to twenty-five feet high. Her instinct tells her that up there, under the red tiles of the roof, she will find nooks rich in old dust. She leaves her prey at the foot of the house and flies up to the roof. For some time, I see her looking here, there and everywhere. After finding a proper site, she begins to work under the curve of a pantile. In ten minutes or fifteen at most, the home is ready. The insect now flies down again. The Ephippiger is promptly found. She has to be taken up. Will this be done on the wing, as circumstances seem to demand? Not at all. The Sphex adopts the toilsome method of scaling a perpendicular wall,

with a surface smoothed by the mason's trowel and measuring twenty
to twenty-five feet in height. Seeing her take this road, dragging the
game between her legs, I at first think the feat impossible; but I am
soon reassured as to the outcome of the bold attempt. Getting a foothold on the little roughnesses in the mortar, the plucky insect, despite
the hindrance of her heavy load, walks up this vertical plane with the
same assured gait and the same speed as on level ground. The top is
reached without the least accident; and the prey is laid temporarily
on the edge of the roof, upon the rounded back of a tile. While the digger
gives a finishing touch to the burrow, the badly-balanced prey slips and
drops to the foot of the wall. The thing must be done all over again and
once more by laboriously climbing the height. The same mistake is repeated. Again the prey is incautiously left on the curved tile, again it
slips and again it falls to the ground. With a composure which accidents such as these cannot disturb, the Sphex for the third time hoists
up the Ephippiger by scaling the wall and, better-advised, drags her
forthwith right into the home.

As even under these conditions no attempt has been made to carry
the prey on the wing, it is clear that the Wasp is incapable of long
flight with so heavy a load. To this incapacity we owe the few characteristics that form the subject of this chapter. A quarry that is not too big
to permit the effort of flying makes of the Yellow-winged Sphex a semi-
social species, that is to say, one seeking the company of her fellows;
a quarry too heavy to carry through the air makes of the Languedocian
Sphex a species vowed to solitary labour, a sort of savage disdainful of
the pleasures that come from the proximity of one's kind. The lighter
or heavier weight of the game selected here determines the fundamental
character of the huntress.

THE WISDOM OF INSTINCT

« 5 »

Continuing his questioning of the Sphex *"through the language of experiment," Fabre is seen in this chapter in the midst of classic researches into the nature of instinct. It was in this field that he made his greatest contributions to science.* THE HUNTING WASPS *is also the source of this selection.*

To paralyse her prey, the Languedocian Sphex, I have no doubt, pursues the method of the Cricket-huntress and drives her lancet repeatedly into the Ephippiger's breast in order to strike the ganglia of the thorax. The process of wounding the nerve-centres must be familiar to her; and I am convinced beforehand of her consummate skill in that scientific operation. This is an art thoroughly known to all the Hunting Wasps, who carry a poisoned dart that has not been given them in vain. At the same time, I must confess that I have never yet succeeded in witnessing the deadly performance. This omission is due to the solitary life led by the Languedocian Sphex.

When a number of burrows are dug on a common site and then provisioned, one has but to wait on the spot to see how one huntress and now another arrive with the game which they have caught. It is easy in these circumstances to try upon the new arrivals the substitution of a live prey for the doomed victim and to repeat the experiment as often as we wish. Besides, the certainty that we shall not lack subjects of observation, as and when wanted, enables us to arrange everything in advance. With the Languedocian Sphex, these conditions of success do not exist. To set out expressly to look for her, with one's material prepared, is almost useless, as the solitary insect is scattered one by one over vast expanses of ground. Moreover, if you do come upon her, it will most often be in an idle hour and you will get nothing out of her.

As I said before, it is nearly always unexpectedly, when your thoughts are elsewhere engaged, that the Sphex appears, dragging her Ephippiger after her.

This is the moment, the only propitious moment to attempt a substitution of prey and invite the huntress to let you witness her lancet-thrusts. Quick, let us procure an alternative morsel, a live Ephippiger! Hurry, time presses: in a few minutes, the burrow will have received the victuals and the glorious occasion will be lost! Must I speak of my mortification at these moments of good fortune, the mocking bait held out by chance? Here, before my eyes, is matter for interesting observations; and I cannot profit by it! I cannot surprise the Sphex' secret for the lack of something to offer her in the place of her prize! Try it for yourself, try setting out in quest of an alternative piece with only a few minutes at your disposal, when it took me three days of wild running about before I found Weevils for my Cerceres! And yet I made the desperate experiment twice over. Ah, if the keeper had caught me this time, tearing like mad through the vineyards, what a good opportunity it would have been for crediting me with robbery and having me up before the magistrate! Vine-branches and clusters of grapes: not a thing did I respect in my mad rush, hampered by the trailing shoots. I must have an Ephippiger at all costs, I must have him that moment. And once I did get my Ephippiger during one of these frenzied expeditions. I was radiant with joy, never suspecting the bitter disappointment in store for me.

If only I arrive in time, if only the Sphex be still engaged in transport work! Thank heaven, everything is in my favour! The Wasp is still some distance away from her burrow and still dragging her prize along. With my forceps I pull gently at it from behind. The huntress resists, stubbornly clutches the antennæ of her victim and refuses to let go. I pull harder, even drawing the carter back as well; it makes no difference: the Sphex does not loose her hold. I have with me a pair of sharp scissors, belonging to my little entomological case. I use them and promptly cut the harness-ropes, the Ephippiger's long antennæ. The Sphex continues to move ahead, but soon stops, astonished at the sudden decrease in the weight of the burden which she is trailing, for this burden is now reduced merely to the two antennæ, snipped off by my mischievous wiles. The real load, the heavy, pot-bellied insect, remains behind and is instantly replaced by my live specimen. The Wasp turns round, lets go the ropes that now draw nothing after them and re-traces her steps. She comes face to face with the prey substituted for her own. She examines it, walks round it gingerly, then stops, moistens her foot with saliva and begins to wash her eyes. In this attitude of meditation, can some such thought as the following pass through her mind:

"Come now! Am I awake or am I asleep? Do I know what I am about or do I not? That thing's not mine. Who or what is trying to humbug me?"

At any rate, the Sphex shows no great hurry to attack my prey with her mandibles. She keeps away from it and shows not the smallest wish to seize it. To excite her, I offer the insect to her in my fingers, I almost thrust the antennæ under her teeth. I know that she does not suffer from shyness; I know that she will come and take from your fingers, without hesitation, the prey which you have snatched from her and afterwards present to her. But what is this? Scorning my offers, the Sphex retreats instead of snapping up what I place within her reach. I put down the Ephippiger, who, obeying a thoughtless impulse, unconscious of danger, goes straight to his assassin. Now we shall see! Alas, no: the Sphex continues to recoil, like a regular coward, and ends by flying away. I never saw her again. Thus ended, to my confusion, an experiment that had filled me with such enthusiasm.

Later and by degrees, as I inspected an increasing number of burrows, I came to understand my failure and the obstinate refusal of the Sphex. I always found the provisions to consist, without a single exception, of a female Ephippiger, harbouring in her belly a copious and succulent cluster of eggs. This appears to be the favourite food of the grubs. Well, in my hurried rush through the vines, I had laid my hands on an Ephippiger of the other sex. I was offering the Sphex a male. More far-seeing than I in this important question of provender, the Wasp would have nothing to say to my game:

"A male, indeed! Is that a dinner for my larvæ? What do you take them for?"

What nice discrimination they have, these dainty epicures, who are able to differentiate between the tender flesh of the female and the comparatively dry flesh of the males! What an unerring glance, which can distinguish at once between the two sexes, so much alike in shape and colour! The female carries a sword at the tip of her abdomen, the ovipositor wherewith the eggs are buried in the ground; and that is about the only external difference between her and the male. This distinguishing feature never escapes the perspicacious Sphex; and that is why, in my experiment, the Wasp rubbed her eyes, hugely puzzled at beholding swordless a prey which she well knew carried a sword when she caught it. What must not have passed through her little Sphex brain at the sight of this transformation?

Let us now watch the Wasp when, having prepared the burrow, she goes back for her victim, which, after its capture and the operation that paralysed it, she has left at no great distance. The Ephippiger is in a condition similar to that of the Cricket sacrificed by the Yellow-winged Sphex, a condition proving for certain that stings have been driven into her thoracic ganglia. Nevertheless, a good many movements still continue; but they are disconnected, though endowed with a certain vigour. Incapable of standing on its legs, the insect lies on its side or on its back. It flutters its long antennæ and also its palpi; it opens and closes its mandibles and bites as hard as in the normal state. The abdomen heaves rapidly and deeply. The ovipositor is brought back sharply under the belly, against which it almost lies flat. The legs stir, but languidly and irregularly; the middle legs seem more torpid than the others. If pricked with a needle, the whole body shudders convulsively; efforts are made to get up and walk, but without success. In short, the insect would be full of life, but for its inability to move about or even to stand upon its legs. We have here therefore a wholly local paralysis, a paralysis of the legs, or rather a partial abolition and ataxy of their movements. Can this very incomplete inertia be caused by some special arrangement of the victim's nervous system, or does it come from this, that the Wasp perhaps administers only a single prick, instead of stinging each ganglion of the thorax, as the Cricket-huntress does? I cannot tell.

Still, for all its shivering, its convulsions, its disconnected movements, the victim is none the less incapable of hurting the larva that is meant to devour it. I have taken from the burrow of the Sphex Ephippigers struggling just as lustily as when they were first half-paralysed; and nevertheless the feeble grub, hatched but a few hours since, was digging its teeth into the gigantic victim in all security; the dwarf was biting into the colossus without danger to itself. This striking result is due to the spot selected by the mother for laying her egg. I have already said how the Yellow-winged Sphex glues her egg to the Cricket's breast, a little to one side, between the first and second pair of legs. Exactly the same place is chosen by the White-edged Sphex; and a similar place, a little farther back, towards the root of one of the large hind-thighs, is adopted by the Languedocian Sphex, all three thus giving proof, by this uniformity, of wonderful discernment in picking out the spot where the egg is bound to be safe.

Consider the Ephippiger pent in the burrow. She lies stretched upon her back, absolutely incapable of turning. In vain, she struggles, in vain she writhes: the disordered movements of her legs are lost in space, the room being too wide to afford them the support of its walls. The grub cares nothing for the victim's convulsions: it is at a spot where naught can reach it, not tarsi, nor mandibles, nor ovipositor, nor antennæ; a spot absolutely stationary, devoid of so much as a surface tremor. It is in perfect safety, on the sole condition that the Ephippiger cannot shift her position, turn over, get upon her feet; and this one condition is admirably fulfilled.

But, with several heads of game, all in the same stage of paralysis, the larva's danger would be great. Though it would have nothing to fear from the insect first attacked, because of its position out of the reach of its victim, it would have every occasion to dread the proximity of the others, which, stretching their legs at random, might strike it and rip it open with their spurs. This is perhaps the reason why the Yellow-winged Sphex, who heaps up three or four Crickets in the same cell, practically annihilates all movement in its victims, whereas the Languedocian Sphex, victualling each burrow with a single piece of game, leaves her Ephippigers the best part of their power of motion and contents herself with making it impossible for them to change their position or stand upon their legs. She may thus, though I cannot say so positively, economize her dagger-thrusts.

While the only half-paralysed Ephippiger cannot imperil the larva, fixed on a part of the body where resistance is impossible, the case is different with the Sphex, who has to cart her prize home. First, having still, to a great extent, preserved the use of its tarsi, the victim clutches with these at any blade of grass encountered on the road along which it is being dragged; and this produces an obstacle to the hauling-process which is difficult to overcome. The Sphex, already heavily burdened by the weight of her load, is liable to exhaust herself with her efforts to make the other insect relax its desperate grip in grassy places. But this is the least serious drawback. The Ephippiger preserves the complete use of her mandibles, which snap and bite with their customary vigour. Now what these terrible nippers have in front of them is just the slender body of the enemy, at a time when she is in her hauling-attitude. The antennæ, in fact, are grasped not far from their roots, so that the mouth of the victim dragged along on its back faces either the thorax or the abdomen of the Sphex, who, standing high on her long legs, takes good care, I am convinced, not to be caught in the mandibles yawning underneath her. At all events, a moment of forgetfulness, a slip, the merest trifle can bring her within the reach of two powerful nippers, which would not neglect the opportunity of taking a pitiless vengeance. In the more difficult cases at any rate, if not always, the action of those formidable pincers must be done away with; and the fish-hooks of the legs must be rendered incapable of increasing their resistance to the process of transport.

How will the Sphex go to work to obtain this result? Here man, even the man of science, would hesitate, would waste his time in barren efforts and would perhaps abandon all hope of success. He can come and take one lesson from the Sphex. She, without ever being taught it, without ever seeing it practised by others, understands her surgery through and through. She knows the most delicate mysteries of the physiology of the nerves, or rather she behaves as if she did. She knows that under her victim's skull there is a circlet of nervous nuclei, something similar to the brain of the higher animals. She knows that this main centre of innervation controls the action of the mouth-parts and moreover is the seat of the will, without whose orders not a single muscle acts; lastly, she knows that, by injuring this sort of brain, she will cause all resistance to cease, the insect no longer possessing any will to resist. As for the mode of operating, this is the easiest matter in the world to her; and, when we have been taught in her school, we are free to try her process in our turn. The instrument employed is no longer the sting; the insect, in its wisdom, has deemed compression preferable to a poisoned thrust. Let us accept its decision, for we shall see presently how prudent it is to be convinced of our own ignorance in the presence of the animal's knowledge. Lest by editing my account I should fail to give a true impression of the sublime talent of this masterly operator, I here copy out my note as I pencilled it on the spot, immediately after the stirring spectacle.

The Sphex finds that her victim is offering too much resistance, hooking itself here and there to blades of grass. She then stops to perform upon it the following curious operation, a sort of *coup de grâce*. The Wasp, still astride her prey, forces open the articulation of the neck, high up, at the nape. Then she seizes the neck with her mandibles and, without making any external wound, probes as far forward as possible under the skull, so as to seize and chew up the ganglia of the head. When this operation is done, the victim is utterly motionless, incapable of the least resistance, whereas previously the legs, though deprived of the power of connected movement needed for walking, vigorously opposed the process of traction.

There is the fact in all its eloquence. With the points of its mandibles, the insect, while leaving uninjured the thin and supple membrane of the neck, goes rummaging into the skull and munching the brain. There is no effusion of blood, no wound, but simply an external pressure. Of course, I kept for my own purposes the Ephippiger paralysed before my eyes, in order to ascertain the effects of the operation at my leisure; also of course, I hastened to repeat in my turn, upon live Ephippigers, what the Sphex had just taught me. I will here compare my results with the Wasp's.

Two Ephippigers whose cervical ganglia I squeeze and compress with a forceps fall rapidly into a state resembling that of the victims of the Sphex. Only, they grate their cymbals if I tease them with a needle; and the legs still retain a few disordered and languid movements. The difference no doubt is due to the fact that my patients were not previously injured in their thoracic ganglia, as were those of the Sphex, who were first stung on the breast. Allowing for this important condition, we see that I was none too bad a pupil and that I imitated pretty closely my teacher of physiology, the Sphex. I confess, it was not without a certain satisfaction that I succeeded in doing almost as well as the insect.

As well? What am I talking about? Wait a bit and you shall see that I still have much to learn from the Sphex. For what happens is that my two patients very soon die: I mean, they really die; and, in four or five days, I have nothing but putrid corpses before my eyes. And the Wasp's Ephippiger? I need hardly say that the Wasp's Ephippiger, even ten days after the operation, is perfectly fresh, just as she will be required by the larva for which she has been destined. Nay, more: only a few hours after the operation under the skull, there reappeared, as though nothing had occurred, the disorderly movements of the legs, antennæ, palpi, ovipositor and mandibles; in a word, the insect returned to the condition wherein it was before the Sphex bit its brain. And these movements were kept up after, though they became feebler every day. The Sphex had merely reduced her victim to a passing state of torpor, lasting amply long enough to enable her to bring it home without resistance; and I, who thought myself her rival, was but a clumsy and barbarous butcher: I killed my prize. She, with her inimitable dexterity, shrewdly compressed the brain to produce a lethargy of a few hours; I, brutal through ignorance, perhaps crushed under my forceps that delicate organ, the main seat of life. If anything could prevent me from blushing at my defeat, it would be the conviction that very few, if any, could vie with these clever ones in cleverness.

Ah, I now understand why the Sphex does not use her sting to injure the cervical ganglia! A drop of poison injected here, at the centre of vital force, would destroy the whole nervous system; and death would follow soon after. But it is not death that the huntress wishes to obtain; the larvæ have not the least use for dead game, for a corpse, in short, smelling of corruption; and all that she wants to bring about is a lethargy, a passing torpor, which will put a stop to the victim's resistance during the carting process, this resistance being difficult to overcome and moreover dangerous for the Sphex. The torpor is obtained by a method known in laboratories of experimental physiology: compression of the brain. The Sphex acts like a Flourens, who, laying bare an animal's brain and bearing upon the cerebral mass, forthwith suppresses intelligence, will, sensibility and movement. The pressure is removed; and everything reappears. Even so do the remains of the Ephippiger's life reappear, as the lethargic effects of a skilfully-directed pressure pass off. The ganglia of the skull, squeezed between the mandibles but without fatal contusions, gradually recover their activity and put an end to the general torpor. Admit that it is all alarmingly scientific.

Fortune has her entomological whims: you run after her and catch no glimpse of her; you forget about her and behold, she comes tapping at your door! How vainly I watched and waited, how many useless journeys I made to see the Languedocian Sphex sacrifice her Ephippigers! Twenty years pass; these pages are in the printer's hands; and, one day early this month, on the 8th of August 1878, my son Émile comes rushing into my study:

"Quick!" he shouts. "Come quick: there's a Sphex dragging her prey under the plane-trees, outside the door of the yard!"

Émile knew all about the business, from what I had told him, to amuse him when we used to sit up late, and better still from similar incidents which he had witnessed in our life out of doors. He is right. I run out and see a magnificent Languedocian Sphex dragging a paralysed Ephippiger by the antennæ. She is making for the hen-house close by and seems anxious to scale the wall, with the object of fixing her burrow under some tile on the roof; for, a few years ago, in the same place, I saw a Sphex of the same species accomplish the ascent with her game and make her home under the arch of a badly-joined tile. Perhaps the present Wasp is descended from the one who performed that arduous climb.

A like feat seems about to be repeated; and this time before nu-

merous witnesses, for all the family, working under the shade of the
plane-trees, come and form a circle around the Sphex. They wonder at
the unceremonious boldness of the insect, which is not diverted from
its work by a gallery of onlookers; all are struck by its proud and lusty
bearing, as, with raised head and the victim's antennæ firmly gripped
in its mandibles, it drags the enormous burden after it. I, alone among
the spectators, feel a twinge of regret at the sight:

"Ah, if only I had some live Ephippigers!" I cannot help saying, with
not the least hope of seeing my wish realized.

"Live Ephippigers?" replies Émile. "Why, I have some perfectly fresh
ones, caught this morning!"

He dashes upstairs, four steps at a time, and runs to his little den,
where a fence of dictionaries encloses a park for the rearing of some
fine caterpillars of the Spurge Hawkmoth. He brings me three Ephippigers, the best that I could wish for, two females and a male.

How did these insects come to be at hand, at the moment when they
were wanted, for an experiment tried in vain twenty years ago? That
is another story. A Lesser Grey Shrike had nested in one of the tall plane-
trees of the avenue. Now a few days earlier, the mistral, the brutal north-
west wind of our parts, blew with such violence as to bend the branches
as well as the reeds; and the nest, turned upside down by the swaying of
its support, had dropped its contents, four small birds. Next morning, I
found the brood upon the ground; three were killed by the fall, the
fourth was still alive. The survivor was entrusted to the cares of Émile,
who went Cricket-hunting twice a day on the neighbouring grass-plots
for the benefit of his young charge. But Crickets are small and the nurseling's appetite called for many of them. Another dish was preferred,
the Ephippiger, of whom a stock was collected from time to time
among the stalks and prickly leaves of the eryngo. The three insects
which Émile brought me came from the Shrike's larder. My pity for
the fallen nestling had procured me this unhoped-for success.

After making the circle of spectators stand back so as to leave the
field clear for the Sphex, I take away her prey with a pair of pincers and
at once give her in exchange one of my Ephippigers, carrying a sword
at the end of her belly, like the game which I have abstracted. The
dispossessed Wasp stamps her feet two or three times; and that is the
only sign of impatience which she gives. She goes for her new prey,
which is too stout, too obese even to try to avoid pursuit, grips it with
her mandibles by the saddle-shaped corselet, gets astride and, curving
her abdomen, slips the end of it under the Ephippiger's thorax. Here,

no doubt, some stings are administered, though I am unable to state the number exactly, because of the difficulty of observation. The Ephippiger, a peaceable victim, suffers herself to be operated on without resistance; she is like the silly Sheep of our slaughter-houses. The Sphex takes her time and wields her lancet with a deliberation which favours accuracy of aim. So far, the observer has nothing to complain of; but the prey touches the ground with its breast and belly and exactly what happens underneath escapes his eye. As for interfering and lifting the Ephippiger a little, so as to see better, that must not be thought of: the murderess would resheathe her weapon and retire. The act that follows is easy to observe. After stabbing the thorax, the tip of the abdomen appears under the victim's neck, which the operator forces open by pressing the nape. At this point, the sting probes with marked persistency, as if the prick administered here were more effective than elsewhere. One would be inclined to think that the nerve-centre attacked is the lower part of the œsophageal chain; but the continuance of movement in the mouth-parts—the mandibles, jaws and palpi—controlled by this seat of innervation shows that such is not the case. Through the neck, the Sphex reaches simply the ganglia of the thorax, or at any rate the first of them, which is more easily accessible through the thin skin of the neck than through the integuments of the chest.

And in a moment it is all over. Without the least shiver denoting pain, the Ephippiger becomes henceforth an inert mass. I remove the Sphex' patient for the second time and replace it by the other female at my disposal. The same proceedings are repeated, followed by the same result. The Sphex has performed her skilful surgery thrice over, almost in immediate succession, first with her own prey and then with my substitutes. Will she do so a fourth time with the male Ephippiger whom I still have left? I have my doubts, not because the Wasp is tired, but because the game does not suit her. I have never seen her with any prey but females, who, crammed with eggs, are the food which the larvæ appreciate above all others. My suspicion is well-founded; deprived of her capture, the Sphex stubbornly refuses the male whom I offer to her. She runs hither and thither, with hurried steps, in search of the vanished game; three or four times, she goes up to the Ephippiger, walks round him, casts a scornful glance at him; and at last she flies away. He is not what her larvæ want; experiment demonstrates this once again after an interval of twenty years.

The three females stabbed, two of them before my eyes, remain in my possession. In each case, all the legs are completely paralysed.

Whether lying naturally, on its belly or on its back or side, the insect retains indefinitely whatever position we give it. A continued fluttering of the antennæ, a few intermittent pulsations of the belly and the play of the mouth-parts are the only signs of life. Movement is destroyed but not susceptibility; for, at the least prick administered to a thin-skinned spot, the whole body gives a slight shudder. Perhaps, some day, physiology will find in such victims the material for valuable work on the functions of the nervous system. The Wasp's sting, so incomparably skilful at striking a particular point and administering a wound which affects that point alone, will supplement, with immense advantage, the experimenter's brutal scalpel, which rips open where it ought to give merely a light touch. Meanwhile, here are the results which I have obtained from the three victims, but in another direction.

As only the movement of the legs has been destroyed, without any wound save that of the nerve-centres, which are the seat of that movement, the insect must die of inanition and not of its injuries. The experiment was conducted as follows: two sound and healthy Ephippigers, just as I picked them up in the fields, were imprisoned without food, one in the dark, the other in the light. The second died in four days, the first in five. This difference of a day is easily explained. In the light, the insect made greater exertions to recover its liberty; and, as every movement of the animal machine is accompanied by a corresponding expenditure of energy, a greater sum total of activity has involved a more rapid consumption of the reserve force of the organism. In the light, there is more restlessness and a shorter life; in the dark, less restlessness and a longer life, while no food at all was taken in either case.

One of my three stabbed Ephippigers was kept in the dark, fasting. In her case, there were not only the conditions of complete abstinence and darkness, but also the serious wounds inflicted by the Sphex; and nevertheless for seventeen days I saw her continually waving her antennæ. As long as this sort of pendulum keeps on swinging, the clock of life does not stop. On the eighteenth day, the creature ceased its antennary movements and died. The badly-wounded insect therefore lived, under the same conditions, four times as long as the insect that was untouched. What seemed as though it should be a cause of death was really a cause of life.

However paradoxical it may seem at first sight, this result is exceedingly simple. When untouched, the insect exerts itself and consequently uses up its reserves. When paralysed, it has merely the feeble, internal movements which are inseparable from any organism; and its substance is economized in proportion to the weakness of the action displayed. In the first case, the animal machine is at work and wears itself out; in the second, it is at rest and saves itself. There being no nourishment now to repair the waste, the moving insect spends its nutritive reserves in four days and dies; the motionless insect does not spend them and lives for eighteen days. Life is a continual dissolution, the physiologists tell us; and the Sphex' victims give us the neatest possible demonstration of the fact.

One remark more. Fresh food is absolutely necessary for the Wasp's larvæ. If the prey were warehoused in the burrow intact, in four or five days it would be a corpse abandoned to corruption; and the scarce-hatched grub would find nothing to live upon but a putrid mass. Pricked with the sting, however, it can keep alive for two or three weeks, a period more than long enough to allow the egg to hatch and the larva to grow. The paralysing of the victim therefore has a twofold result: first, the living dish remains motionless and the safety of the delicate grub is not endangered; secondly, the meat keeps good a long time and thus ensures wholesome food for the larva. Man's logic, enlightened by science, could discover nothing better.

My two other Ephippigers stung by the Sphex were kept in the dark with food. To feed inert insects, hardly differing from corpses except by the perpetual waving of their long antennæ, seems at first an impossibility; still, the play of the mouth-parts gave me some hope and I tried. My success exceeded my anticipations. There was no question here, of course, of giving them a lettuce-leaf or any other piece of green stuff on which they might have browsed in their normal state; they were feeble valetudinarians, who needed spoon-feeding, so to speak, and supporting with liquid nourishment. I used sugar-and-water.

Laying the insect on its back, I place a drop of the sugary fluid on its mouth with a straw. The palpi at once begin to stir; the mandibles and jaws move. The drop is swallowed with evident satisfaction, especially after a somewhat prolonged fast. I repeat the dose until it is refused. The meal takes place once a day, sometimes twice, at irregular intervals, lest I should become too much of a slave to my patients. Well, one of the Ephippigers lived for twenty-one days on this meagre fare. It was not much, compared with the eighteen days of the one whom I had left to die of starvation. True, the insect had twice had a bad fall, having dropped from the experimenting-table to the floor owing to some piece of awkwardness on my part. The bruises which it received must have hastened its end. The other, which suffered no accidents, lived for forty days. As the nourishment employed, sugar-and-water, could not indefinitely take the place of the natural green food, it is very likely that the insect would have lived longer still if the usual diet had been possible. And so the point which I had in view is proved: the victims stung by the Digger-wasps die of starvation and not of their wounds.

THE IGNORANCE OF INSTINCT

« 6 »

As was often the case with Fabre's major researches, the field-notes of many successive summers were combined to give the final picture of the wisdom and the ignorance of the instinctive acts of the Sphex. The Bee-Eating Philanthi, mentioned in this chapter are also hunting wasps. Several species are found in America, all of them preying on bees and in some instances becoming serious problems around apiaries. This section originally formed Chapter X of THE HUNTING WASPS.

The Sphex has shown us how infallibly and with what transcendental art she acts when guided by the unconscious inspiration of her instinct; she is now going to show us how poor she is in resource, how limited in intelligence, how illogical even, in circumstances outside of her regular routine. By a strange inconsistency, characteristic of the instinctive faculties, profound wisdom is accompanied by an ignorance no less profound. To instinct nothing is impossible, however great the difficulty may be. In building her hexagonal cells, with their floors consisting of three lozenges, the Bee solves with absolute precision the arduous problem of how to achieve the maximum result at a minimum cost, a problem whose solution by man would demand a powerful mathematical mind. The Wasps whose larvæ live on prey display in their murderous art methods hardly rivalled by those of a man versed in the intricacies of anatomy and physiology. Nothing is difficult to instinct, so long as the act is not outside the unvarying cycle of animal existence; on the other hand, nothing is easy to instinct, if the act is at all removed from the course usually pursued. The insect which astounds us, which terrifies us with its extraordinary intelligence surprises us, the next moment, with its stupidity, when confronted with some simple fact that happens to lie outside its ordinary practice. The Sphex will supply us with a few instances.

Let us follow her dragging her Ephippiger home. If fortune smile upon us, we may witness some such little scene as that which I will now describe. When entering her shelter under the rock, where she has made her burrow, the Sphex finds, perched on a blade of grass, a Praying Mantis, a carnivorous insect which hides cannibal habits under a pious appearance. The danger threatened by this robber ambushed on her path must be known to the Sphex, for she lets go her game and pluckily rushes upon the Mantis, to inflict some heavy blows and dislodge her, or at all events to frighten her and inspire her with respect. The robber does not move, but closes her lethal machinery, the two terrible saws of the arm and fore-arm. The Sphex goes back to her capture, harnesses herself to the antennæ and boldly passes under the blade of grass whereon the other sits perched. By the direction of her head we can see that she is on her guard and that she holds the enemy rooted, motionless, under the menace of her eyes. Her courage meets with the reward which it deserves: the prey is stored away without further mishap.

A word more on the Praying Mantis, or, as they say in Provence, *lou Prégo Diéou,* the Pray-to-God. Her long, pale-green wings, like spreading veils, her head raised heavenwards, her folded arms, crossed upon her breast, are in fact a sort of travesty of a nun in ecstasy. And yet she is a ferocious creature, loving carnage. Though not her favourite spots, the work-yards of the various Digger-wasps receive her visits pretty frequently. Posted near the burrows, on some bramble or other, she waits for chance to bring within her reach some of the arrivals, forming a double capture for her, as she seizes both the huntress and her prey. Her patience is long put to the test: the Wasp suspects something and is on her guard; still, from time to time, a rash one gets caught. With a sudden rustle of wings half-unfurled as by the violent release of a clutch, the Mantis terrifies the newcomer, who hesitates for a moment, in her fright. Then, with the sharpness of a spring, the toothed fore-arm folds back on the toothed upper arm; and the insect is caught between the blades of the double saw. It is as though the jaws of a Wolf-trap were closing on the animal that had nibbled at its bait. Thereupon, without unloosing the cruel machine, the Mantis gnaws her victim by small mouthfuls. Such are the ecstasies, the prayers, the mystic meditations of the *Prégo Diéou.*

Of the scenes of carnage which the Praying Mantis has left in my memory, let me relate one. The thing happens in front of a work-yard of Bee-eating Philanthi. These diggers feed their larvæ on Hive-bees, whom they catch on the flowers while gathering pollen and honey. If the Philanthus who has made a capture feels that her Bee is swollen with honey, she never fails, before storing her, to squeeze her crop, either on the way or at the entrance of the dwelling, so as to make her disgorge the delicious syrup, which she drinks by licking the tongue which her unfortunate victim, in her death-agony, sticks out of her mouth at full length. This profanation of a dying creature, whose enemy squeezes its belly to empty it and feast on the contents, has something so hideous about it that I should denounce the Philanthus as a brutal murderess, if animals were capable of wrongdoing. At the moment of some such horrible banquet, I have seen the Wasp, with her prey, seized by the Mantis: the bandit was rifled by another bandit. And here is an awful detail: while the Mantis held her transfixed under the points of the double saw and was already munching her belly, the Wasp continued to lick the honey of her Bee, unable to relinquish the delicious food even amid the terrors of death. Let us hasten to cast a veil over these horrors.

We will return to the Sphex, with whose burrow we must make ourselves acquainted before we go further. This burrow is a hole made in fine sand, or rather in a sort of dust at the bottom of a natural shelter. Its entrance-passage is very short, merely an inch or two, without a bend, and leads to a single, roomy, oval chamber. The whole thing is a rough den, hastily dug out, rather than a leisurely and artistically excavated dwelling. I have explained that the reason for this simplicity is that the game is captured first and set down for a moment on the hunting-field while the Wasp hurriedly makes a burrow in the vicinity, a method of procedure which allows of but one chamber or cell to each retreat. For who can tell whither the chances of the day will lead the huntress for her second capture? The prisoner is heavy and the burrow must therefore be near; so to-day's home, which is too far away for the next Ephippiger to be conveyed to it, cannot be utilized to-morrow. Thus, as each prey is caught, there is a fresh excavation, a fresh burrow, with its single chamber, now here, now there. Having said this, we will try a few experiments to see how the insect behaves when we create circumstances new to it.

Experiment I

A Sphex, dragging her prey along, is a few inches from the burrow. Without disturbing her, I cut with a pair of scissors the Ephippiger's antennæ, which the Wasp, as we know, uses for harness-ropes. On recovering from the surprise caused by the sudden lightening of her load, the Sphex goes back to her victim and, without hesitation, now seizes the root of the antenna, the short stump left by the scissors. It is very short indeed, hardly a millimetre; no matter: it is enough for the Sphex, who grips this fag-end of a rope and resumes her hauling. With the greatest precaution, so as not to injure the Wasp, I now cut the two antennary stumps level with the skull. Finding nothing left to catch hold of at the familiar points, the insect seizes, close by, one of the victim's long palpi and continues its hauling-work, without appearing at all perturbed by this change in the harness. I leave it alone. The prey is brought home and placed so that its head faces the entrance to the burrow; and the Wasp goes in by herself, to make a brief inspection of the inside of the cell before proceeding to warehouse the provisions. Her behaviour reminds us of that of the Yellow-winged Sphex in similar circumstances. I take advantage of this short moment to seize the abandoned prey, remove all its palpi and place it a little farther off, about half a yard from the burrow. The Sphex reappears and goes straight to her captive, whom she has seen from her threshold. She looks at the top of the head, she looks underneath, on either side and finds nothing to take hold of. A desperate attempt is made: the Wasp, opening wide her mandibles, tries to grab the Ephippiger by the head; but the pincers have not a sufficient compass to take in so large a bulk and they slip off the round, polished skull. She makes several fresh endeavours, each time without result. She is at length convinced of the uselessness of her efforts. She draws back a little to one side and appears to be renouncing further attempts. One would say that she was discouraged; at least, she smoothes her wings with her hind-legs, while with her front tarsi, which she first puts into her mouth, she washes her eyes. This, so it has always seemed to me, is a sign in Hymenoptera of giving up a job.

Nevertheless there is no lack of parts by which the Ephippiger might be seized and dragged along as easily as by the antennæ and the palpi. There are the six legs, there is the ovipositor: all organs slender enough to be gripped boldly and to serve as hauling-ropes. I agree that the easiest way to effect the storing is to introduce the prey head first, drawn down by the antennæ; but it would enter almost as readily if drawn by a leg, especially one of the front legs, for the orifice is wide and the passage short or sometimes even non-existent. Then how is it that the Sphex did not once try to seize one of the six tarsi or the tip of the ovipositor, whereas she attempted the impossible, the absurd, in striving

to grip, with her much too short mandibles, the huge skull of her prey?
Can it be that the idea did not occur to her? Then we will try to suggest it.

I offer her, right under her mandibles, first a leg, next the end of the
abdominal rapier. The insect obstinately refuses to bite; my repeated
blandishments lead to nothing. A singular huntress, to be embarrassed
by her game, not knowing how to seize it by a leg when she is not able
to take it by the horns! Perhaps my prolonged presence and the unusual
events that have just occurred have disturbed her faculties. Then let us
leave the Sphex to herself, between her Ephippiger and her burrow;
let us give her time to collect herself and, in the calm of solitude, to
think out some way of managing her business. I leave her therefore
and continue my walk; and, two hours later, I return to the same place.
The Sphex is gone, the burrow is still open and the Ephippiger is lying
just where I placed her. Conclusion: the Wasp has tried nothing; she
went away, abandoning everything, her home and her game, when, to
utilize them both, all that she had to do was to take her prey by one leg.
And so this rival of Flourens, who but now was startling us with her
cleverness as she dexterously squeezed her victim's brain to produce
lethargy, becomes incredibly helpless in the simplest case outside her
usual habits. She, who so well knows how to attack a victim's thoracic
ganglia with her sting and its cervical ganglia with her mandibles; she,
who makes such a judicious difference between a poisoned prick annihi-
lating the vital influence of the nerves forever and a pressure causing
only momentary torpor, cannot grip her prey by this part when it is
made impossible for her to grip it by any other. To understand that she
can take a leg instead of an antenna is utterly beyond her powers. She
must have the antenna, or some other string attached to the head, such
as one of the palpi. If these cords did not exist, her race would perish,
for lack of the capacity to solve this trivial problem.

Experiment II

The Wasp is engaged in closing her burrow, where the prey has been
stored and the egg laid upon it. With her front tarsi, she brushes her
doorstep, working backwards and sweeping into the entrance a stream
of dust which passes under her belly and spurts behind in a parabolic
spray as continuous as a liquid spray, so nimble is the sweeper in her
actions. From time to time, the Sphex picks out with her mandibles a
few grains of sand, so many solid blocks which she inserts one by one

into the mass of dust, causing it all to cake together by beating and com-
pressing it with her forehead and mandibles. Walled up by this masonry,
the entrance-door soon disappears from sight.

I intervene in the middle of the work. Pushing the Sphex aside, I
carefully clear the short gallery with the blade of a knife, take away
the materials that close it and restore full communication between the
cell and the outside. Then, with my forceps, without damaging the edifice, I take the Ephippiger from the cell, where she lies with her head
at the back and her ovipositor towards the entrance. The Wasp's egg
is on the victim's breast, at the usual place, the root of one of the hinder
thighs: a proof that the Sphex was giving the finishing touch to the
burrow, with the intention of never returning.

Having done this and put the stolen prey safely away in a box, I yield
my place to the Sphex, who has been on the watch beside me while I
was rifling her home. Finding the door open, she goes in and stays for
a few moments. Then she comes out and resumes her work where I
interrupted it, that is to say, she starts conscientiously stopping the entrance to the cell by sweeping dust backwards and carrying grains of
sand, which she continues to heap up with scrupulous care, as though
she were doing useful work. When the door is once again thoroughly
walled up, the insect brushes itself, seems to give a glance of satisfaction
at the task accomplished and finally flies away.

The Sphex must have known that the burrow contained nothing, because she went inside and even stayed there for some time; and yet,
after this inspection of the pillaged abode, she once more proceeds to
close up the cell with the same care as though nothing out of the way
had happened. Can she be proposing to use this burrow later, to return
to it with a fresh victim and lay a new egg there? If so, her work of
closing would be intended to prevent the access of intruders to the dwelling during her absence; it would be a measure of prudence against the
attempts of other diggers who might covet the ready-made chamber;
it might also be a wise precaution against internal dilapidations. And,
as a matter of fact, some Hunting Wasps do take care to protect the
entrance to the burrow by closing it temporarily, when the work has to
be suspended for a time. Thus I have seen certain Ammophilæ, whose
burrow is a perpendicular shaft, block the entrance to the home with
a small flat stone when the insect goes off hunting or ceases its mining-
operations at sunset, the hour for striking work. But this is a slight affair,
a mere slab laid over the mouth of the shaft. When the insect comes,

it only takes a moment to remove the little flat stone; and the entrance is free.

On the other hand, the obstruction which we have just seen built by the Sphex is a solid barrier, a stout piece of masonry, where dust and gravel form alternate layers all the way down the passage. It is a definite performance and not a provisional defence, as is proved by the care with which it is constructed. Besides, as I think I have shown pretty clearly, it is very doubtful, considering the way in which she acts, whether the Sphex will ever return to make use of the home which she has prepared. The next Ephippiger will be caught elsewhere; and the warehouse destined to receive her will be dug elsewhere too. But these, after all, are only arguments: let us rather have recourse to experiment, which is more conclusive here than logic.

I allowed nearly a week to elapse, in order to give the Sphex time to return to the burrow which she had so methodically closed and to make use of it for her next laying if such were her intention. Events corresponded with the logical inferences: the burrow was in the condition wherein I left it, still firmly closed, but without provisions, egg or larva. The proof was decisive: the Wasp had not been back.

So the plundered Sphex enters her house, makes a leisurely inspection of the empty chamber and, a moment afterwards, behaves as though she had not perceived the disappearance of the bulky prey which but now filled the cell. Did she, in fact, fail to notice the absence of the provisions and the egg? Is she, who is so clear-sighted in her murderous proceedings, dense enough not to realize that the cell is empty? I dare not accuse her of such stupidity. She is aware of it. But then why that other piece of stupidity which makes her close—and very conscientiously close—an empty burrow, one which she does not purpose to victual later? Here the work of closing is useless, is supremely absurd; no matter: the insect performs it with the same ardour as though the larva's future depended on it. The insect's various instinctive actions are then fatally linked together. Because one thing has been done, a second thing must inevitably be done to complete the first or to prepare the way for its completion; and the two acts depend so closely upon each other that the performing of the first entails that of the second, even when, owing to casual circumstances, the second has become not only inopportune but sometimes actually opposed to the insect's interests. What object can the Sphex have in blocking up a burrow which has become useless, now that it no longer contains the victim and the egg, and which will always remain useless, since the insect will not return to it? The only way to explain this inconsequent action is to look upon it as the inevitable complement of the actions that went before. In the normal order of things, the Sphex hunts down her prey, lays an egg and closes her burrow. The hunting has been done; the game, it is true, has been withdrawn by me from the cell; never mind: the hunting has been done, the egg has been laid; and now comes the business of closing up the home. This is what the insect does, without another thought, without in the least suspecting the futility of her present labours.

Experiment III

To know everything and to know nothing, according as it acts under normal or exceptional conditions: that is the strange antithesis presented by the insect race. Other examples, also drawn from the Sphex tribe, will confirm this conclusion. The White-edged Sphex (*S. albisecta*) attacks medium-sized Locusts, whereof the different species to be found in the neighbourhood of the burrow all furnish her with their tribute of victims. Because of the abundance of these Acridians, there is no need to go hunting far afield. When the burrow, which takes the form of a perpendicular shaft, is ready, the Sphex merely explores the purlieus of her lair, within a small radius, and is not long in finding some Locust browsing in the sunshine. To pounce upon her and sting her, despite her kicking, is to the Sphex the matter of a moment. After some fluttering of its wings, which unfurl their carmine or azure fan, after some drowsy stretching of its legs, the victim ceases to move. It has now to be brought home, on foot. For this laborious operation, the Sphex employs the same method as her kinswomen, that is to say, she drags her prize along between her legs, holding one of its antennæ in her mandibles. If she encounters some grassy jungle, she goes hopping and flitting from blade to blade, without ever letting slip her prey. When at last she comes within a few feet of her dwelling, she performs a manœuvre which is also practised by the Languedocian Sphex; but she does not attach as much importance to it, for she frequently neglects it. Leaving her captive on the road, the Wasp hurries home, though no apparent danger threatens her abode, and puts her head through the entrance several times, even going part of the way down the burrow. She next returns to the Locust and, after bringing her nearer the goal, leaves her a second time to revisit the burrow. This performance is repeated over and over again, always with the same anxious haste.

These visits are sometimes followed by grievous accidents. The victim, rashly abandoned on hilly ground, rolls to the bottom of the slope; and the Sphex on her return, no longer finding it where she left it, is obliged to seek for it, sometimes fruitlessly. If she find it, she must renew a toilsome climb, which does not prevent her from once more abandoning her booty on the same unlucky declivity. Of these repeated visits to the mouth of the shaft, the first can be very logically explained. The Wasp, before arriving with her heavy burden, enquires whether the entrance to the home be really clear, whether nothing will hinder her from bringing in her game. But, once this first reconnaissance is made, what can be the use of the rest, following one after the other, at close intervals? Is the Sphex so volatile in her ideas that she forgets the visit which she has just paid and runs afresh to the burrow a moment later, only to forget this new inspection also and to start doing the same thing over and over again? That would be a memory with very fleeting recollections, whence the impression vanished almost as soon as it was produced. Let us not linger too long on this obscure point.

At last the game is brought to the brink of the shaft, with its antennæ hanging down the hole. We now again see, faithfully imitated, the method employed in the like case by the Yellow-winged Sphex and also, but under less striking conditions, by the Languedocian Sphex. The Wasp enters alone, inspects the interior, reappears at the entrance, lays hold of the antennæ and drags the Locust down. While the Locust-huntress was making her examination of the home, I have pushed her prize a little farther back; and I obtained results similar in all respects to those which the Cricket-huntress gave me. Each Sphex displays the same obstinacy in diving down her burrow before dragging in the prey. Let us recall here that the Yellow-winged Sphex does not always allow herself to be caught by this trick of pulling away her Cricket. There are picked tribes, strong-minded families which, after a few disappointments, see through the experimenter's wiles and know how to baffle them. But these revolutionaries, fit subjects for progress, are the minority; the remainder, mulish conservatives clinging to the old manners and customs, are the majority, the crowd. I am unable to say whether the Locust-huntress also varies in ingenuity according to the district which she hails from.

But here is something more remarkable; and it is this with which I wanted to conclude the present experiment. After repeatedly withdrawing the White-edged Sphex' prize from the mouth of the pit and compelling her to come and fetch it again, I take advantage of her descent to the bottom of the shaft to seize the prey and put it in a place of safety where she cannot find it. The Sphex comes up, looks about for a long time and, when she is convinced that the prey is really lost, goes down into her home again. A few moments after, she reappears. Is it with the intention of resuming the chase? Not the least in the world: the Sphex begins to stop up the burrow. And what we see is not a temporary clos-ing, effected with a small flat stone, a slab covering the mouth of the well; it is a final closing, carefully done with dust and gravel swept into the passage until it is filled up. The White-edged Sphex makes only one cell at the bottom of her shaft and puts one head of game into this cell. That single Locust has been caught and dragged to the edge of the hole. If she was not stored away, it was not the huntress' fault, but mine. The Wasp performed her task according to the inflexible rule; and, also ac-cording to the inflexible rule, she completes her work by stopping up the dwelling, empty though it be. We have here an exact repetition of the useless exertions made by the Languedocian Sphex whose home has just been plundered.

Experiment IV

It is almost impossible to make certain whether the Yellow-winged Sphex, who constructs several cells at the end of the same passage and stacks several Crickets in each, is equally illogical when accidentally disturbed in her proceedings. A cell can be closed though empty or im-perfectly victualled and the Wasp will none the less continue to come to the same burrow in order to work at the others. Nevertheless, I have reason to believe that this Sphex is subject to the same aberrations as her two kinswomen. My conviction is based on the following facts: the number of Crickets found in the cells, when all the work is done, is usually four to each cell, although it is not uncommon to find only three, or even two. Four appears to me to be the normal number, first, because it is the most frequent and, secondly, because, when rearing young larvæ dug up while they were still engaged on their first joint, I found that all of them, those actually provided with only two or three pieces of game as well as those which had four, easily managed the vari-ous Crickets wherewith I served them one by one, up to and including the fourth, but that after this they refused all nourishment, or barely touched the fifth ration. If four Crickets are necessary to the larva to acquire the full development called for by its organization, why are some-times only three, sometimes only two provided for it? Why this enormous difference in the quantity of the victuals, some larvæ having twice as much as the others? It cannot be because of any difference in the size of the dishes provided to satisfy the grub's appetite, for all have very much the same dimensions; and it can therefore be due only to the wastage of game on the way. We find, in fact, at the foot of the banks whose upper stages are occupied by the Sphex-wasps, Crickets that have been paralysed but lost, owing to the slope of the ground, down which they have slipped when the huntresses have momentarily left them, for some reason or other. These Crickets fall a prey to the Ants and Flies; and the Sphex-wasps who come across them take good care not to pick them up, for, if they did, they would themselves be admitting enemies into the house.

These facts seem to me to prove that, while the Yellow-winged Sphex' arithmetical powers enable her to calculate exactly how many victims to capture, she cannot achieve a census of those which have safely reached their destination. It is as though the insect had no mathematical guide beyond an irresistible impulse that prompts her to hunt for game a definite number of times. When the Sphex has made the requisite number of journeys, when she has done her utmost to store the captures that result from these, her work is ended; and she closes the cell whether completely or incompletely provisioned. Nature has endowed her with only those faculties called for in ordinary circumstances by the interests of her larvæ; and, as these blind faculties, which cannot be modified by experience, are sufficient for the preservation of the race, the insect is unable to go beyond them.

I conclude therefore as I began: instinct knows everything, in the undeviating paths marked out for it; it knows nothing, outside those paths. The sublime inspirations of science and the astounding inconsistencies of stupidity are both its portion, according as the insect acts under normal or accidental conditions.

OF STUDY AND OBSERVATION

« 7 »

Fabre was well past eighty when he wrote these words. They were among the last from his pen, forming the conclusion of the tenth and final volume of his great SOUVENIRS ENTOMOLOGIQUES. *In the English translation, they form the beginning of Chapter XVII of* THE SACRED BEETLE AND OTHERS.

Begun to-day and dropped to-morrow, taken up again later and again abandoned, according to the chances of the day, the study of instinct makes but halting progress. The changing seasons brings unwelcome delays, forcing the observer to wait till the following year or even longer for the answer to his eager questions. Moreover, the problem often crops up unexpectedly, as the result of some casual incident of slight interest in itself, and it comes in a form so vague that it gives little basis for precise investigation. How can one investigate what has not yet been suspected? We have no facts to go upon and are consequently unable to tackle the problem frankly.

To collect these facts by fragments, to subject those fragments to varied tests in order to try their value, to make them into a sheaf of rays lighting up the darkness of the unknown and gradually causing it to emerge: all this demands a long space of time, especially as the favourable periods are brief. Years elapse; and then very often the perfect solution has not appeared. There are always gaps in our sheaf of light; and always behind the mysteries which the rays have penetrated stand others, still shrouded in darkness.

I am perfectly aware that it would be preferable to avoid repetitions and to give a complete story every time; but, in the domain of instinct, who can claim a harvest that leaves no grain for other gleaners? Sometimes the handful of corn left on the field is of more importance than

66

the reaper's sheaves. If we had to wait until we knew every detail of the question studied, no one would venture to write the little that he knows. From time to time, a few truths are revealed, tiny pieces of the vast mosaic of things. Better to divulge the discovery, however humble it be. Others will come who, also gathering a few fragments, will assemble the whole into a picture ever growing larger but ever notched by the unknown.

And then the burden of years forbids me to entertain long hopes. Distrustful of the morrow, I write from day to day, as I make my observations. This method, one of necessity rather than choice, sometimes results in the reopening of old subjects, when new investigations throw light within and enable me to complete or it may be to modify the first text.

INSTINCTS OF A CATERPILLAR

« 8 »

The Great Peacock Moth, of whose caterpillar Fabre here writes, is Saturnia pyri *(Schiff).* It is related to the largest and showiest of American moths, the Polyphemus, the Cecropia and the Luna. At one time the scientific name, Bombyx, embraced a wide variety of moths. It is now confined to the silkworm insects. The Mulberry Bombyx is, of course, Bombyx mori, *the commercial silkworm.* THE MASON-WASPS, *Chapter V, is the source of this selection.*

The Great Peacock is the largest Moth of our district. Her caterpillar, which is yellow-hued, with turquoise-blue spots surrounded by black hairs, spins itself, at the foot of the almond-trees, a robust cocoon whose ingenious construction has long been celebrated. At the moment of her deliverance, the Mulberry Bombyx has in her stomach a particular solvent which the new-born Moth disgorges against the wall of the cocoon to soften it, to dissolve the gum that sticks the threads together and in this way to force an exit by the mere pressure of her head. With the aid of this reagent, the recluse is able triumphantly to attack her silken prison at the fore-end, the rear-end or the side, as I discover by turning the chrysalis in its cocoon, which I slit with a pair of scissors and then sew up again. Whatever the spot to be perforated for the emergence, a spot which my intervention varies at will, the liquid disgorged promptly soaks into and softens the wall, whereupon the captive, struggling with her fore-limbs and pushing her forehead against the tangle of unstuck threads, makes herself a passage with the same ease as in her natural liberation.

The Great Peacock is not endowed with this method of delivery by means of a solvent; her stomach is incapable of preparing the corrosive calculated to destroy, at any point, the defensive enclosure which is now a prison-wall. Indeed, if I reverse the chrysalis in its cocoon, opened and then closed with a few stitches, the Moth always dies, being powerless to free herself. When the point to be forced is changed, the release becomes impossible. To emerge from this shell, a genuine strong-box, a special method is therefore necessary, one having no relation to the chemical method of the Mulberry Bombyx. Let me describe, as others have done before me, how things happen.

At the fore-end of the cocoon, a conical end, whereas the other is rounded, the threads are not glued together; every elsewhere, the silken web is cemented with a gummy product that turns it into a stout, waterproof parchment. Those front threads, which are almost straight, converge at their free end and form a cone-shaped series of palisades, having as their common base the circle where the use of the gummy cement is suddenly discontinued. The arrangement can best be compared with the mouth of an Eel-pot, which the fish readily enters by following the funnel of osier-switches, but from which the imprudent one cannot get out again, because the narrow passage closes its palisade at the least effort to push through.

Another very accurate comparison is provided by the Mouse-traps with an entrance consisting of a bunch of wires arranged in a truncated cone. Attracted by the bait, the rodent enters the orifice of the trap, enlarging it with a gentle thrust; but, when it becomes a question of departure, the wires, at first so tractable, become an insuperable barrier of halberds. Both devices permit entrance and forbid exit. If we invert the arrangement of the conical palisade, making it point outwards from within, its action is reversed: exit is permitted and entrance forbidden.

This is the case with the Great Peacock's cocoon, which has a slight improvement to its credit: its mouth, shaped like the Eel-pot or Mousetrap aforesaid, is formed of a numerous series of cones, fitting one within the other and overlapping. In order to emerge, the Moth has only to push her head in front of her; the several rows of uncemented threads yield without difficulty. Once the recluse is liberated, these threads resume their position, so that there is nothing outside to show whether the cocoon is empty or inhabited.

Easy exit is not enough: there must also be an inviolable refuge during the labour of metamorphosis. The cell whose door is open for exit must have the same door closed against entrance, so that no evil-minded one may make his way inside. The mechanism of the Eel-pot's mouth admirably fulfils this condition, which is as necessary to the safety of the Great Peacock as the first. To enter through the multiple fences of converging threads, which constitute a more effectual obstacle the harder they are pushed, would be impossible to any creature that might bethink itself of attempting to violate the dwelling. I am well-acquainted with the secrets of this lock, which contrives, like any fine piece of workmanship, to combine simple means with important results; and yet I always stand amazed when, with an open cocoon in my fingers, I try to pass a pencil through the entrance. When pushed outwards from within, it passes immediately; when pushed inwards from without, it is invincibly checked.

I am lingering over these details to show the importance which the good construction of her palisade of threads possesses for the Great Peacock. If ill-ordered, entangled and therefore intractable when pushed, the series of boxed cones will offer an insurmountable resistance and the Moth will perish, a victim of the caterpillar's imperfect art. If constructed with mathematical accuracy, but with sparse rows of threads in insufficient numbers, it will leave the retreat exposed to dangers from without and the chrysalis will become the prey of some intruder, of whom there are many in search of somnolent nymphs, forming easy victims. For the caterpillar, therefore, this double-acting mouth is a work of the highest importance. It has to expend upon it all that it possesses in foresight, in gleams of reason and in art capable of modification when circumstances require; it must in short give proof of the best of which its talents are capable. Let us follow it in its labours; let us interpose the experimental test; and we shall learn some curious facts.

The cocoon and its opening are constructed simultaneously. When it has woven this or that part of the general wall, the caterpillar turns about, if need be, and with its unbroken thread proceeds to continue the palisade of converging filaments. To this end it pokes its head to the end of the roughly-defined funnel and then withdraws it, doubling the thread as it goes. This alternation of thrusts and withdrawals results in a circle of doubled filaments, which do not adhere to one another. The shift is not a long one; when the palisade is a row the richer, the caterpillar resumes its work upon the shell, a task which it again abandons to busy itself with the funnel; and so on, over and over again, the emission of the gummy product being suspended when the threads are to be left free and copiously effected when they have to be stuck together in order to obtain a solid texture.

The exit-funnel is not, as we see, a piece of work executed continuously; the caterpillar works at it intermittently, as the general shell progresses. From the beginning to the end of its spinning-period, so long as the reservoirs of silk are not exhausted, it multiplies the tiers without neglecting the rest of the cocoon. These tiers take the form of cones enclosed one within the other and of increasingly obtuse angles, until the last to be spun are so flat as to become almost level surfaces.

If nothing happens to disturb the worker, the work is performed with a perfection that would do credit to a discerning industry capable of realizing the why and wherefore of things. Can the caterpillar be said to have any conception, however slight, of the importance of its task, of the future function of its overlapping conical palisades? This is what we are about to learn.

I take a pair of scissors and remove the conical extremity while the spinner is working at the other end. The cocoon is now wide open. The caterpillar soon turns about. It thrusts its head into the wide breach which I have just made; it seems to be exploring the outside and enquiring into the accident that has occurred. I expect to see it repair the disaster and entirely reconstruct the cone destroyed by my scissors. It does, in fact, work at it for a time; it erects a row of converging threads; then, without paying further heed to the disaster, it applies its spinnerets elsewhere and continues to thicken the cocoon.

Grave doubts come to my mind: the cone built upon the breach consists of sparse filaments; it is, moreover, very flat and does not project anything like so much as the original cone. What I took at first to be a work of repair is merely a work of continuation. The caterpillar, put to the test by my tricks, has not modified the course of its work; despite the imminence of the danger, it has confined itself to the tier of threads which it would have fitted inside the preceding tier but for the snip of my scissors.

I let things go on for a while; and, when the mouth has once again acquired a certain solidity, I cut it off for the second time. The insect displays the same lack of perspicacity as before, replacing the absent cone by one with an even more obtuse angle, that is to say, continuing its usual task, without any attempt at a thorough restoration, despite the extreme urgency. If the store of silk were nearly at an end, I should sympathize with the troubles of the sorely-tried caterpillar doing its best to repair its house with the scanty materials that remain at its disposal; but I see it foolishly squandering its product on the additional upholstering of a shell which may be strong enough as it is, while economizing to the point of stinginess in the matter of the fence, which, if neglected, will leave the cell and its inhabitant at the mercy of the first thief that comes along. There is no lack of silk: the spinner applies layer upon layer to the points that are unhurt; but at the breach it employs only the quantity required under ordinary conditions. This is not economy imposed by shortage; it is blind clinging to custom. And so my commiseration changes to amazement in the presence of such profound stupidity, which applies itself to the superfluous work of upholstery in a dwelling henceforth uninhabitable, instead of attending, while there is yet time, to the business of repairing the ruins.

I make my cut a third time. When the moment has come to resume the series of boxed cones, the caterpillar arms the breach with bristles arranged in a disk, as they appear in the last courses of the undisturbed structure. This configuration shows that the end of the task is at hand. The cocoon is strengthened for a little longer; then rest ensues and the metamorphosis begins in a dwelling with a niggardly fence to it, one which would not strike terror into the puniest invader.

To sum up, the caterpillar, incapable of perceiving the dangers attendant upon an incomplete palisade, resumes its work, after each amputation of the cocoon, at the point where it had left it before the accident. Instead of thoroughly restoring the ruined exit, which its very abundant store of silk would allow it to do; instead of reerecting on the breach a projecting cone of many layers, to replace the one removed by my scissors, it runs up layers of threads that become gradually flatter and flatter and form a continuation and not a reconstruction of the missing layers. Moreover, this work of fence-building, the need for which would seem imperious to any reasoning creature, does not appear to preoccupy the caterpillar more than usual, for it keeps on alternating this work with that of the cocoon, which is much less urgent. Everything goes by rote, as though the serious incident of the housebreaking had not occurred. In a word, the caterpillar does not begin all over again a thing once made and then destroyed; it continues it. The early stages of the work are lacking; no matter: the sequel follows without any modification in the plans.

It would be easy for me, if my argument were not already quite clear, to give a host of similar examples showing plainly that the intelligence of the insect is absolutely deficient in rational discernment, even when the great perfection of the work would seem to allow the artisan a certain perspicacity. We will confine ourselves for the moment to the three cases which I have mentioned. The Pelopæus goes on storing Spiders for an egg that has been removed; she perseveres in making hunting-trips that are henceforth useless; she hoards victuals that are destined to nourish nothing; she multiplies her battues to fill with game a larder which is forthwith emptied by my tweezers; lastly, she closes, with every customary care, a cell that no longer contains anything whatever: she sets her seal on emptiness. She does even absurder things: she plasters the site of her vanished nest, covering an imaginary structure and putting a roof to a house which at the moment is tucked away in my pocket. In the case of the Great Peacock, the caterpillar, despite the certain loss of the coming Moth, instead of beginning all over again the mouth of the Eel-pot cut down by my scissors, quietly continues its spinning, without in any way modifying the regular course of the work; and, when the time comes for making the last tiers of defensive filaments, it erects them upon the dangerous breach, but neglects to rebuild the ruined portion of the barricade. Indifferent to the indispensable, it occupies itself with the superfluous.

What are we to conclude from these facts? I would fain believe, for the sake of my insects' reputation, in some distraction on their part, in some individual giddiness which would not taint the general perspicacity; I should like to regard their aberrations merely as isolated and exceptional actions, which would not affect their judgment as a whole. Alas, a long series of glaring facts would impose silence on my attempts at rehabilitation! Any species, no matter which, when subjected to experimental tests, is guilty of similar inconsistencies in the course of its disturbed industry. Constrained by the inexorable logic of the facts, I therefore state the deductions suggested by observations as follows: the insect is neither free nor conscious in its industry, which in its case is an external function with phases regulated almost as strictly as the phases of an internal function, such as digestion. It builds, weaves, hunts, stabs and paralyses, even as it digests, even as it secretes the poison of its sting, the silk of its cocoon or the wax of its combs, always without the least understanding of the means or the end. It is ignorant of its wonderful talents just as the stomach is ignorant of its skilful chemistry. It can add nothing essential to them nor subtract anything from them, any more than it is able to increase or diminish the pulsations of its dorsal vessel.

Test it with an accident and you affect it not at all: such as it is in the undisturbed exercise of its calling, such it will remain should circumstances arise demanding some modification in the conduct of its task. Experience does not teach it; time does not awaken a glimmer in the darkness of its unconsciousness. Its art, perfect in its speciality, but inept in the face of the slightest new difficulty, is handed down immutably, as the art of the suction-pump is handed down to the babe at the breast. To expect the insect to alter the essential points of its industry is to hope that the babe will change its manner of sucking. Both equally ignorant of what they are doing, they persevere in the method prescribed for the safeguarding of the species, precisely because their ignorance forbids them to make any sort of essay or attempt.

The insect, then, lacks the aptitude for reflection, the aptitude that harks back and reverts to the antecedent, without which the consequent would lose all its value. In the phases of its industry, each action accomplished counts as valid by the mere fact that it has been accomplished; the insect does not go back to it, should some accident demand; the consequent follows without troubling about the missing antecedent. A blind impulse urges it from one act to a second, from this second to a third and so on until the task is completed; but it is impossible for the insect to reascend the current of its activity should accidental conditions arise and call for this, however imperatively. Having passed through the complete cycle, the work is considered to be most logically performed by a worker devoid of all logic.

THE GREAT PEACOCK MOTH

« 9 »

Printed first in English in THE LIFE OF THE CATERPILLAR, *this account of the assembling of the male moths is one of the best known of Fabre's experiments. It has even been mentioned in a Hollywood movie. However, before it got on the screen, it was given a super-colossal twist. According to the dialogue: "This scientist, Fabre, was in the heart of Africa. He caught a female moth. When he returned home, a male followed him all the way from Africa to Paris!" In several of his experiments, including this one, Fabre tended to underrate the importance of smell. The smelling equipment of many insects is highly specialized; it is concentrated on detecting one or a very few odors; it is like a radio permanently set to a certain wavelength. In laboratory tests, since Fabre's time, scientists have removed the scent-producing organs from female moths and have found that the males fly direct to this fragment of the moth's body, ignoring the female entirely. Phil Rau, an amateur experimenter near St. Louis, Missouri, found that a number of American moths tend to arrive at certain hours of the night; the males of one moth, for instance, are more likely to appear before midnight, those of another moth after midnight.*

It was a memorable evening. I shall call it the Great Peacock evening. Who does not know the magnificent Moth, the largest in Europe, clad in maroon velvet with a necktie of white fur? The wings, with their sprinkling of grey and brown, crossed by a faint zig-zag and edged with smoky white, have in the centre a round patch, a great eye with a black pupil and a variegated iris containing successive black, white, chestnut and purple arcs.

Well, on the morning of the 6th of May, a female emerges from her cocoon in my presence, on the table of my insect-laboratory. I forthwith cloister her, still damp with the humours of the hatching, under a wire-gauze bell-jar. For the rest, I cherish no particular plans. I incarcerate her from mere habit, the habit of the observer always on the look-out for what may happen.

It was a lucky thought. At nine o'clock in the evening, just as the household is going to bed, there is a great stir in the room next to mine. Little Paul, half-undressed, is rushing about, jumping and stamping, knocking the chairs over like a mad thing. I hear him call me:

"Come quick!" he screams. "Come and see these Moths, big as birds! The room is full of them!"

I hurry in. There is enough to justify the child's enthusiastic and hyperbolical exclamations, an invasion as yet unprecedented in our house, a raid of giant Moths. Four are already caught and lodged in a bird-cage. Others, more numerous, are fluttering on the ceiling.

At this sight, the prisoner of the morning is recalled to my mind.

"Put on your things, laddie," I say to my son. "Leave your cage and come with me. We shall see something interesting."

We run downstairs to go to my study, which occupies the right wing of the house. In the kitchen I find the servant, who is also bewildered by what is happening and stands flicking her apron at great Moths whom she took at first for Bats.

The Great Peacock, it would seem, has taken possession of pretty well every part of the house. What will it be around my prisoner, the cause of this incursion? Luckily, one of the two windows of the study had been left open. The approach is not blocked.

We enter the room, candle in hand. What we see is unforgettable. With a soft flick-flack the great Moths fly around the bell-jar, alight, set off again, come back, fly up to the ceiling and down. They rush at the candle, putting it out with a stroke of their wings; they descend on our shoulders, clinging to our clothes, grazing our faces. The scene suggests a wizard's cave, with its whirl of Bats. Little Paul holds my hand tighter than usual, to keep up his courage.

How many of them are there? About a score. Add to these the number that have strayed into the kitchen, the nursery and the other rooms of the house; and the total of those who have arrived from the outside cannot fall far short of forty. As I said, it was a memorable evening, this Great Peacock evening. Coming from every direction and apprised I know not how, here are forty lovers eager to pay their respects to the marriageable bride born that morning amid the mysteries of my study.

For the moment let us disturb the swarm of wooers no further. The flame of the candle is a danger to the visitors, who fling themselves into it madly and singe their wings. We will resume the observation to-morrow with an experimental interrogatory thought out beforehand.

But first let us clear the ground and speak of what happens every night during the week that my observation lasts. Each time it is pitch dark, between eight and ten o'clock, when the Moths arrive one by one. It is stormy weather, the sky is very much overcast and the darkness is so profound that even in the open air, in the garden, far from the shadow of the trees, it is hardly possible to see one's hand before one's face.

In addition to this darkness there is the difficulty of access. The house is hidden by tall plane-trees; it is approached by a walk thickly bordered with lilac- and rose-trees, forming a sort of outer vestibule; it is protected against the mistral by clumps of pines and screens of cypresses. Clusters of bushy shrubs make a rampart a few steps away from the door. It is through this tangle of branches, in complete darkness, that the Great Peacock has to tack about to reach the object of his pilgrimage.

Under such conditions, the Brown Owl would not dare leave the hole in his olive-tree. The Moth, better-endowed with his faceted optical organs than the night-bird with its great eyes, goes forward without hesitating and passes through without knocking against things. He directs his tortuous flight so skilfully that, despite the obstacles overcome, he arrives in a state of perfect freshness, with his big wings intact, with not a scratch upon him. The darkness is light enough for him.

Even if we grant that it perceives certain rays unknown to common retinæ, this extraordinary power of sight cannot be what warns the Moth from afar and brings him hurrying to the spot. The distance and the screens interposed make this quite impossible.

Besides, apart from deceptive refractions, of which there is no question in this case, the indications provided by light are so precise that we go straight to the thing seen. Now the Moth sometimes blunders, not as to the general direction which he is to take, but as to the exact spot where the interesting events are happening. I have said that the children's nursery, which is at the side of the house opposite my study, the real goal of my visitors at the present moment, was occupied by the Moths before I went there with a light in my hand. These certainly were ill-informed. There was the same throng of hesitating visitors in the kitchen; but here the light of a lamp, that irresistible lure to nocturnal insects, may have beguiled the eager ones.

Let us consider only the places that were in the dark. In these there are several stray Moths. I find them more or less everywhere around the actual spot aimed at. For instance, when the captive is in my study, the visitors do not all enter by the open window, the safe and direct road, only two or three yards away from the caged prisoner. Several of them come in downstairs, wander about the hall and at most reach the staircase, a blind alley barred at the top by a closed door.

These data tell us that the guests at this nuptial feast do not make straight for their object, as they would if they derived their information from some kind of luminous radiation, whether known or unknown to our physical science. It is something else that apprises them from afar, leads them to the proximity of the exact spot and then leaves the final discovery to the airy uncertainty of random searching. It is very much like the way in which we ourselves are informed by hearing and smell, guides which are far from accurate when we want to decide the precise point of origin of the sound or the smell.

What are the organs of information that direct the rutting Moth on his nightly pilgrimage? One suspects the antennæ, which, in the males, do in fact seem to be questioning space with their spreading tufts of feathers. Are those glorious plumes mere ornaments, or do they at the same time play a part in the perception of the effluvia that guide the enamoured swain? A conclusive experiment seems to present no difficulty. Let us try it.

On the day after the invasion, I find in the study eight of my visitors of the day before. They are perched motionless on the transoms of the second window, which is kept closed. The others, when their dance was over, about ten o'clock in the evening, went out as they came in, that is to say, through the first window, which is left open day and night. Those eight persevering ones are just what I want for my schemes.

With a sharp pair of scissors, without otherwise touching the Moths, I cut off their antennæ, near the base. The patients take hardly any notice of the operation. Not one moves; there is scarcely a flutter of the wings. These are excellent conditions: the wound does not seem at all serious. Undistraught by pain, the Moths bereft of their horns will adapt themselves all the better to my plans. The rest of the day is spent in placid immobility on the cross-bars of the window.

There are still a few arrangements to be made. It is important in particular to shift the scene of operations and not to leave the female before the eyes of the maimed ones at the moment when they resume their nocturnal flight, else the merit of their quest would disappear. I therefore move the bell-jar with its captives and place it under a porch at the other end of the house, some fifty yards from my study.

When night comes, I go to make a last inspection of my eight victims. Six have flown out through the open window; two remain behind, but these have dropped to the floor and no longer have the strength to turn over if I lay them on their backs. They are exhausted, dying. Pray do not blame my surgical work. This quick decreptitude occurs invariably, even without the intervention of my scissors.

Six, in better condition, have gone off. Will they return to the bait that attracted them yesterday? Though deprived of their antennæ, will they be able to find the cage, now put in another place, at a considerable distance from its original position?

The cage is standing in the dark, almost in the open air. From time to time, I go out with a lantern and a Butterfly-net. Each visitor is captured, examined, catalogued and forthwith let loose in an adjoining room, of which I close the door. This gradual elimination will enable me to tell the exact number, with no risk of counting the same Moth more than once. Moreover, the temporary gaol, which is spacious and bare, will in no way endanger the prisoners, who will find a quiet retreat there and plenty of room. I shall take similar precautions during my subsequent investigations.

At half past ten no more arrive. The sitting is over. In all, twenty-five males have been caught, of whom only one was without antennæ. Therefore, of the six on whom I operated yesterday and who were hale enough to leave my study and go back to the fields, one alone has returned to the bell-jar. It is a poor result, on which I dare not rely when it comes to asserting or denying that the antennæ play a guiding part. We must begin all over again, on a larger scale.

Next morning I pay a visit to the prisoners of the day before. What I see is not encouraging. Many are spread out on the floor, almost lifeless. Several of them give hardly a sign of life when I take them in my fingers. What can I hope from these cripples? Still, let us try. Perhaps they will recover their vigour when the time comes to dance the lovers' round.

The twenty-four new ones undergo amputation of the antennæ. The old, hornless one is left out of count, as dying or close to it. Lastly, the prison-door is left open for the remainder of the day. He who will may leave the room, he who can shall join in the evening festival. In order to put such as go out to the test of searching for the bride, the cage,

which they would be sure to notice on the threshold, is once more removed. I shift it to a room in the opposite wing, on the ground-floor.
The access to this room is of course left free.

Of the twenty-four deprived of their antennæ, only sixteen go outside. Eight remain, powerless to move. They will soon die where they
are. Out of the sixteen who have left, how many are there that return
to the cage in the evening? Not one! I sit up to capture just seven, all
newcomers, all sporting feathers. This result would seem to show that
the amputation of the antennæ is a rather serious matter. Let us not
draw conclusions yet: a doubt remains and an important one.

"A nice state I'm in!" said Mouflard, the Bull-pup, when his pitiless
breeder has docked his ears. "How dare I show my face before the other
Dogs?"

Can it be that my Moths entertain Master Mouflard's apprehensions?
Once deprived of their fine plumes, dare they no longer appear amidst
their rivals and a-wooing go? Is it bashfulness on their part or lack of
guidance? Or might it not rather be exhaustion after a wait that exceeds the duration of an ephemeral flame? Experiment shall tell us.

On the fourth evening, I take fourteen Moths, all new ones, and imprison them, as they arrive, in a room where I intend them to pass the
night. Next morning, taking advantage of their daytime immobility, I
remove a little of the fur from the centre of their corselet. The silky fleece
comes off so easily that this slight tonsure does not inconvenience the
insects at all; it deprives them of no organ which may be necessary to
them later, when the time comes to find the cage. It means nothing
to the shorn ones; to me it means the unmistakable sign that the callers
have repeated their visit.

This time there are no weaklings incapable of flight. At night, the
fourteen shaven Moths escape into the open. Of course the place of the
cage is once more changed. In two hours, I capture twenty Moths, including two tonsured ones, no more. Of those who lost their antennæ
two days ago, not one puts in an appearance. Their nuptial time is over
for good and all.

Only two return out of the fourteen marked with a bald patch. Why
do the twelve others hang back, although supplied with what we have
assumed to be their guides, their antennary plumes? Why again that
formidable list of defaulters, which we find nearly always after a night
of sequestration? I perceive but one reply: the Great Peacock is quickly
worn out by the ardours of pairing-time.

With a view to his wedding, the one and only object of his life, the

Moth is gifted with a wonderful prerogative. He is able to discover the object of his desire in spite of distance, obstacles and darkness. For two or three evenings, he is allowed a few hours wherein to indulge his search and his amorous exploits. If he cannot avail himself of them, all is over: the most exact of compasses fails, the brightest of lamps expires. What is the use of living after that? Stoically we withdraw into a corner and sleep our last sleep, which is the end of our illusions and of our woes alike.

The Great Peacock becomes a Moth only in order to perpetuate his species. He knows nothing of eating. While so many others, jolly companions one and all, flit from flower to flower, unrolling the spiral of their proboscis and dipping it into the honeyed cups, he, the incomparable faster, wholly freed from the bondage of the belly, has no thought of refreshment. His mouth-parts are mere rudiments, vain simulacra, not real organs capable of performing their functions. Not a sup enters his stomach: a glorious privilege, save that it involves a brief existence. The lamp needs its drop of oil, if it is not to be extinguished. The Great Peacock renounces that drop, but at the same time he renounces long life. Two or three evenings, just time enough to allow the couple to meet, and that is all: the big Moth has lived.

Then what is the meaning of the staying away of those who have lost their antennæ? Does it show that the absence of these organs has made them incapable of finding the wire bell in which the prisoner awaits them? Not at all. Like the shorn ones, whose operation has left them uninjured, they prove only that their time is up. Whether maimed or intact, they are unfit for duty because of their age; and their non-return is valueless as evidence. For lack of the time necessary for experimenting, the part played by the antennæ escapes us. Doubtful it was and doubtful it remains.

My caged prisoner lives for eight days. Every evening she draws for my benefit a swarm of visitors, in varying numbers, now to one part of the house, now to another, as I please. I catch them, as they come, with the net and transfer them, the moment they are captured, to a closed room, in which they spend the night. Next morning, I mark them with a tonsure on the thorax.

The aggregate of the visitors during those eight evenings amounts to a hundred and fifty, an astounding number when I consider how hard I had to seek during the following two years to collect the materials necessary for continuing these observations. Though not impossible to find in my near neighbourhood, the cocoons of the Great Peacock are at least very rare, for old almond-trees, on which the caterpillars live, are scarce in these parts. For two winters I visited every one of those decayed trees at the lower part of the trunk, under the tangle of hard grasses in which they are clad, and time after time I returned empty-handed. Therefore my hundred and fifty Moths came from afar, from very far, within a radius of perhaps a mile and a half or more. How did they know of what was happening in my study?

The perceptive faculties can receive information from a distance by means of three agents: light, sound and smell. Is it permissible to speak of vision in this instance? I will readily admit that sight guides the visitors once they have passed through the open window. But before that, in the mystery out of doors! It would not be enough to grant them the fabulous eye of the Lynx, which was supposed to see through walls; we should have to admit a keenness of sight which could be exercised miles away. It is useless to discuss anything so outrageous; let us pass on.

Sound is likewise out of the question. The great fat Moth, capable of sending a summons to such a distance, is mute even to the most acute hearing. It is just possible that she possesses delicate vibrations, passionate quivers, which might perhaps be perceptible with the aid of an extremely sensitive microphone; but remember that the visitors have to be informed at considerable distances, thousands of yards away. Under these conditions, we cannot waste time thinking of acoustics. That would be to set silence the task of waking the surrounding air.

There remains the sense of smell. In the domain of our senses, scent, better than anything else, would more or less explain the onrush of the Moths, even though they do not find the bait that allures them until after a certain amount of hesitation. Are there, in point of fact, effluvia similar to what we call odour, effluvia of extreme subtlety, absolutely imperceptible to ourselves and yet capable of impressing a sense of smell better-endowed than ours? There is a very simple experiment to be made. It is a question of masking those effluvia, of stifling them under a powerful and persistent odour, which masters the olfactory sense entirely. The too-strong scent will neutralize the very faint one.

I begin by sprinkling naphthaline in the room where the males will be received this evening. Also, in the bell-jar, beside the female, I lay a big capsule full of the same stuff. When the visiting-hour comes, I have only to stand in the doorway of the room to get a distinct smell of gasworks. My artifice fails. The Moths arrive as usual, they enter the room, pass through its tarry atmosphere and make for the cage with as much certainty of direction as though in unscented surroundings.

My confidence in the olfactory explanation is shaken. Besides, I am now unable to go on. Worn out by her sterile wait, my prisoner dies on the ninth day, after laying her unfertilized eggs on the wirework of the cage. In the absence of a subject of experiment, there is no more to be done until next year.

This time I shall take my precautions, I shall lay in a stock so as to be able to repeat as often as I wish the experiments which I have already tried and those which I am contemplating. To work, then; and that without delay.

In the summer, I proclaim myself a buyer of caterpillars at a sou apiece. The offer appeals to some urchins in the neighbourhood, my usual purveyors. On Thursdays, emancipated from the horrors of parsing, they scour the fields, find the fat caterpillar from time to time and bring him to me clinging to the end of a stick. They dare not touch him, poor mites; they are staggered at my audacity when I take him in my fingers as they might take the familiar Silk-worm.

Reared on almond-tree branches, my menagerie in a few days supplies me with magnificent cocoons. In the winter, assiduous searches at the foot of the fostering tree complete my collection. Friends interested in my enquiries come to my assistance. In short, by dint of trouble, much running about, commercial bargains and not a few scratches from brambles, I am the possessor of an assortment of cocoons, of which twelve, bulkier and heavier than the others, tell me that they belong to females.

A disappointment awaits me, for May arrives, a fickle month which brings to naught my preparations, the cause of so much anxiety. We have winter back again. The mistral howls, tears the budding leaves from the plane-trees and strews the ground with them. It is as cold as in December. We have to light the fires again at night and resume the thick clothes which we were beginning to leave off.

My Moths are sorely tried. They hatch late and are torpid. Around my wire cages, in which the females wait, one to-day, another to-morrow, according to the order of their birth, few males or none come from the outside. And yet there are some close at hand, for the plumed gallants resulting from my harvest were placed out in the garden as soon as they were hatched and recognized. Whether near neighbours or strangers from afar, very few arrive; and these are only half-hearted. They enter for a moment, then disappear and do not return. The lovers have grown cold.

It is also possible that the low temperature is unfavourable to the tell-tale effluvia, which might well be enhanced by the warmth and decreased by the cold, as happens with scents. My year is lost. Oh, what laborious work is this experimenting at the mercy of the sudden changes and deceptions of a short season!

I begin all over again, for the third time. I rear caterpillars, I scour the country in search of cocoons. When May returns, I am suitably provided. The weather is fine and responds to my hopes. I once more see the incursions which had struck me so powerfully at the beginning, at the time of the historic invasion which first led to my researches.

Nightly the visitors turn up, in squads of twelve, twenty or more. The female, a lusty, big-bellied matron, clings firmly to the trellis-work of the cage. She makes no movement, gives not so much as a flutter of the wings, seems indifferent to what is going on. Nor is there any odour, so far as the most sensitive nostrils in the household can judge, nor any rustle perceptible to the most delicate hearing among my family, all of whom are called in to bear evidence. In motionless contemplation she waits.

The others, in twos or threes or more, flop down upon the dome of the cage, run about it briskly in every direction, lash it with the tips of their wings in continual movement. There are no affrays between rivals. With not a sign of jealousy in regard to the other suitors, each does his utmost to enter the enclosure. Tiring of their vain attempts, they fly away and join the whirling throng of dancers. Some, giving up all hope, escape through the open window; fresh arrivals take their places; and, on the top of the cage, until ten o'clock in the evening, attempts to approach are incessantly renewed, soon to be abandoned and as soon resumed.

Every evening the cage is moved to a different place. I put it on the north side and the south, on the ground-floor and the first floor, in the right wing and fifty yards away in the left, in the open air or hidden in a distant room. All these sudden displacements, contrived if possible to put the seekers off the scent, do not trouble the Moths in the least. I waste my time and ingenuity in trying to deceive them.

Recollection of places plays no part here. Yesterday, for instance, the female was installed in a certain room. The feathered males came fluttering thither for a couple of hours; several even spent the night there. Next day, at sunset, when I move the cage, all are out of doors. Ephemeral though they be, the newest comers are ready to repeat their nocturnal expeditions a second time and a third. Where will they go first, these veterans of a day?

They know all about the meeting-place of yesterday. One is inclined to think that they will go back to it, guided by memory, and that, finding nothing left, they will proceed elsewhither to continue their investigations. But no: contrary to my expectations, they do nothing of the sort. Not one reappears in the place which was so thickly crowded last night; not one pays even a short visit. The room is recognized as deserted, without the preliminary enquiry which recollection would seem to demand. A more positive guide than memory summons them elsewhere.

Until now the female has been left exposed, under the meshes of a wire gauze. The visitors, whose eyes are used to piercing the blackest gloom, can see her by the vague light of what to us is darkness. What will happen if I imprison her under an opaque cover? According to its nature, will not this cover either set free or arrest the tell-tale effluvia?

Physical science is to-day preparing to give us wireless telegraphy, by means of the Hertzian waves. Can the Great Peacock have anticipated our efforts in this direction? In order to set the surrounding air in motion and to inform pretenders miles away, can the newly-hatched bride have at her disposal electric or magnetic waves, which one sort of screen would arrest and another let through? In a word, does she, in her own manner, employ a kind of wireless telegraphy? I see nothing impossible in this: insects are accustomed to invent things quite as wonderful.

I therefore lodge the female in boxes of various characters. Some are made of tin, some of cardboard, some of wood. All are hermetically closed, are even sealed with stout putty. I also use a glass bell-jar standing on the insulating support of a pane of glass.

Well, under these conditions of strict closing, never a male arrives, not one, however favourable the mildness and quiet of the evening. No matter its nature, whether of metal or glass, of wood or cardboard, the closed receptacle forms an insuperable obstacle to the effluvia that betray the captive's whereabouts.

A layer of cotton two fingers thick gives the same result. I place the female in a large jar, tying a sheet of wadding over the mouth by way of a lid. This is enough to keep the neighbourhood in ignorance of the secrets of my laboratory. No male puts in an appearance.

On the other hand, make use of ill-closed, cracked boxes, or even hide them in a drawer, in a cupboard; and, notwithstanding this added mystery, the Moths will arrive in numbers as great as when they come thronging to the trellised cage standing in full view on a table. I have retained a vivid recollection of an evening when the recluse was waiting in a hat-box at the bottom of a closed wall-cupboard. The Moths arrived, went to the door, struck it with their wings, knocked at it to express their wish to enter. Passing wayfarers, coming no one knows whence across the fields, they well knew what was inside there, behind those boards.

We must therefore reject the idea of any means of information similar to that of wireless telegraphy, for the first screen set up, whether a good conductor or a bad, stops the female's signals completely. To give these a free passage and carry them to a distance, one condition is indispensable: the receptacle in which the female is contained must be imperfectly closed, so as to establish a communication between the inner and the outer air. This brings us back to the probability of an odour, though that was contradicted by my experiment with naphthaline.

My stock of cocoons is exhausted and the problem is still obscure. Shall I try again another year, the fourth? I abandon the thought for the following reasons: Moths that mate at night are difficult to observe if I want to watch their intimate actions. The gallant certainly needs no illuminant to attain his ends; but my feeble human powers of vision cannot dispense with one at night. I must have at least a candle, which is often extinguished by the whirling swarm. A lantern saves us from these sudden eclipses; but its dim light, streaked with broad shadows, does not suit a conscientious observer like myself, who wants to see and to see clearly.

Nor is this all. The light of a lamp diverts the Moths from their object, distracts them from their business and, if persistent, gravely compromises the success of the evening. The visitors no sooner enter the room than they make a wild rush for the flame, singe their fluff in it and thenceforth, frightened by the scorching received, cease to be trustworthy witnesses. When they are not burnt, when they are kept at a distance by a glass chimney, they perch as close as they can to the light and there stay, hypnotized.

One evening, the female was in the dining-room, on a table facing the open window. A lighted paraffin-lamp, with a large white-enamel shade, was hanging from the ceiling. Two of the arrivals alighted on the dome of the cage and fussed around the prisoner; seven others, after greeting her as they passed, made for the lamp, circled about it a little and then, fascinated by the radiant glory of the opal cone, perched on it, motionless, under the shade. Already the children's hands were raised to seize them.

"Don't," I said. "Leave them alone. Let us be hospitable and not disturb these pilgrims to the tabernacle of light."

All that evening, not one of the seven budged. Next morning, they were still there. The intoxication of light had made them forget the intoxication of love.

With creatures so madly enamoured of the radiant flame, precise and prolonged experiment becomes unfeasible the moment the observer requires an artificial illuminant. I give up the Great Peacock and his nocturnal nuptials. I want a Moth with different habits, equally skilled in keeping conjugal appointments, but performing in the daytime.

Before continuing with a subject that fulfils these conditions, let us drop chronological order for a moment and say a few words about a late-comer who arrived after I had completed my enquiries, I mean the Lesser Peacock (*Attacus pavonia minor,* LIN.). Somebody brought me, I don't know where from, a magnificent cocoon loosely wrapped in an ample white-silk envelope. Out of this covering, with its thick, irregular folds, it was easy to extract a case similar in shape to the Great Peacock's, but a good deal smaller. The fore-end, worked into the fashion of an eel-trap by means of free and converging fibres, which prevent access to the dwelling while permitting egress without a breach of the walls, indicated a kinswoman of the big nocturnal Moth; the silk bore the spinner's mark.

And, in point of fact, towards the end of March, on the morning of Palm Sunday, the cocoon with the eel-trap formation provides me with a female of the Lesser Peacock, whom I at once seclude under a wire-gauze bell in my study. I open the window to allow the event to be made known all over the district; I want the visitors, if any come, to find free entrance. The captive grips the wires and does not move for a week.

A gorgeous creature is my prisoner, in her brown velvet streaked with wavy lines. She has white fur around her neck; a speck of carmine at the tip of the upper wings; and four large, eye-shaped spots, in which black, white, red and yellow-ochre are grouped in concentric crescents. The dress is very like that of the Great Peacock, but less dark in colouring. I have seen this Moth, so remarkable for size and costume, three or four times in my life. It was only the other day that I first saw the cocoon. The male I have never seen. I only know that, according to the books, he is half the size of the female and of a brighter and more florid colour, with orange-yellow on the lower wings.

Will he come, the unknown spark, the plume-wearer on whom I

have never set eyes, so rare does he appear to be in my part of the country? In his distant hedges will he receive news of the bride that awaits him on my study table? I venture to feel sure of it; and I am right. Here he comes, even sooner than I expected.

On the stroke of noon, as we were sitting down to table, little Paul who is late owing to his eager interest in what is likely to happen, suddenly runs up to us, his cheeks aglow. In his fingers flutters a pretty Moth, a Moth caught that moment hovering in front of my study. Paul shows me his prize; his eyes ask an unspoken question.

"Hullo!" I say. "This is the very pilgrim we were expecting. Let's fold up our napkins and go and see what's happening. We can dine later."

Dinner is forgotten in the presence of the wonders that are taking place. With inconceivable punctuality, the plume-wearers hasten to answer the captive's magic call. They arrive one by one, with a tortuous flight. All of them come from the north. This detail has its significance. As a matter of fact, during the past week we have experienced a fierce return of winter. The north wind has been blowing a gale, killing the imprudent almond-blossoms. It was one of those ferocious storms which, as a rule, usher in the spring in our part of the world. Today the temperature has suddenly grown milder, but the wind is still blowing from the north.

Now at this first visit all the Moths hurrying to the prisoner enter the enclosure from the north; they follow the movement of the air; not one beats against it. If their compass were a sense of smell similar to our own, if they were guided by odoriferous particles dissolved in the air, they ought to arrive from the opposite direction. If they came from the south, we might believe them to be informed by effluvia carried by the wind; coming as they do from the north, through the mistral, that mighty sweeper of the atmosphere, how can we suppose them to have perceived, at a great distance, what we call a smell? This reflux of scented atoms in a direction contrary to the aerial current seems to me inadmissible.

For a couple of hours, in radiant sunshine, the visitors come and go outside the front of the study. Most of them search for a long while, exploring the wall, flitting along the ground. To see their hesitation, one would think that they were at a loss to discover the exact place of the bait that attracts them. Though they have come from very far without a mistake, they seem uncertain of their bearings once they are on the spot. Nevertheless, sooner or later they enter the room and pay their respects to the captive, without much importunity. At two o'clock all is over. Ten Moths have been here.

All through the week, each time at noon-day, when the light is at its brightest, Moths arrive, but in decreasing numbers. The total is nearly forty. I see no reason to repeat experiments which could add nothing to what I already know; and I confine myself to stating two facts. In the first place, the Lesser Peacock is a day insect, that is to say, he celebrates his wedding in the brilliant light of the middle of the day. He needs radiant sunshine. The Great Peacock, on the contrary, whom he so closely resembles in his adult form and in the work which he does as a caterpillar, requires the dusk of the early hours of the night. Let him who can explain this strange contrast of habits.

THE SNAIL-SHELL BEE

« 10 »

With metallic blue, green or purple bodies, the Osmia *bees of America choose burrows in the earth or in decayed wood rather than snailshells for their tiny earthen cells. Sometimes they form them inside dry plant galls after they have served as the larval homes of other insects. At other times, they place them on or under stones. They are commonly known as Mason-Bees. Fabre will have more to say of these wild bees in a later chapter. This section originally appeared in* THE MASON-BEES.

Near Sérignan are some great quarries of coarse limestone, characteristic of the miocene formation of the Rhone valley. These have been worked for many generations. The ancient public buildings of Orange, notably the colossal frontage of the theatre whither all the intellectual world once flocked to hear Sophocles' *Œdipus Tyrannus,* derive most of their material from these quarries. Other evidence confirms what the similarity of the hewn stone tells us. Among the rubbish that fills up the spaces between the tiers of seats, they occasionally discover the Marseilles obol, a bit of silver stamped with the four-spoked wheel, or a few bronze coins bearing the effigy of Augustus or Tiberius. Scattered also here and there among the monuments of antiquity are heaps of refuse, accumulations of broken stones in which various Hymenoptera, including the Three-horned Osmia in particular, take possession of the dead Snail-shell.

The quarries form part of an extensive plateau which is so arid as to be nearly deserted. In these conditions, the Osmia, at all times faithful to her birth-place, has little or no need to emigrate from her heap of stones and leave the shell for another dwelling which she would have to go and seek at a distance. Since there are heaps of stone there, she probably has no other dwelling than the Snail-shell. Nothing tells us that the present-day generations are not descended in the direct line from the generations contemporary with the quarryman who lost his as or his obol at this spot. All the circumstances seem to point to it: the Osmia of the quarries is an inveterate user of Snail-shells; so far as heredity is concerned, she knows nothing whatever of reeds. Well, we must place her in the presence of these new lodgings.

I collect during the winter about two dozen well-stocked Snail-shells and instal them in a quiet corner of my study, as I did at the time of my enquiries into the distribution of the sexes. The little hive with its front pierced with forty holes has bits of reed fitted to it. At the foot of the five rows of cylinders I place the inhabited shells and with these I mix a few small stones, the better to imitate the natural conditions. I add an assortment of empty Snail-shells, after carefully cleaning the interior so as to make the Osmia's stay more pleasant. When the time comes for nest-building, the stay-at-home insect will have, close beside the house of its birth, a choice of two habitations: the cylinder, a novelty unknown to its race; and the spiral staircase, the ancient ancestral home.

The nests were finished at the end of May and the Osmiæ began to answer my list of questions. Some, the great majority, settled exclusively in the reeds; the others remained faithful to the Snail-shell or else entrusted their eggs partly to the spirals and partly to the cylinders. With the first, who were the pioneers of cylindrical architecture, there was no hesitation that I could perceive: after exploring the stump of reed for a time and recognizing it as serviceable, the insect instals itself there and, an expert from the first touch, without apprenticeship, without groping, without any tendencies bequeathed by the long practice of its predecessors, builds its straight row of cells on a very different plan from that demanded by the spiral cavity of the shell, which increases in size as it goes on.

The slow school of the ages, the gradual acquisitions of the past, the legacies of heredity count for nothing, therefore, in the Osmia's education. Without any novitiate on its own part or that of its forebears, the insect is versed straight away in the calling which it has to pursue; it possesses, inseparable from its nature, the qualities demanded by its craft: some which are invariable and belong to the province of instinct; others flexible, belonging to the domain of discernment. To divide a free lodging into chambers by means of mud partitions; to fill those chambers with a heap of pollen-flour, with a few sups of honey in the central part where the egg is to lie; in short, to prepare board and lodging for the unknown, for a family which the mothers have never seen in the past and will never see in the future: this, in its essential features, is the function of the Osmia's instinct. Here, everything is harmoniously, inflexibly, permanently preordained; the insect has but to follow its blind impulse to attain the goal. But the free lodging offered by chance varies exceedingly in hygienic conditions, in shape and in capacity. Instinct, which does not choose, which does not contrive, would, if it were alone, leave the insect's existence in peril. To help her out of her predicament, in these complex circumstances, the Osmia possesses her little stock of discernment, which distinguishes between the dry and the wet, the solid and the fragile, the sheltered and the exposed; which recognizes the worth or the worthlessness of a site and knows how to sprinkle it with cells according to the size and shape of the space at disposal. Here, slight industrial variations are necessary and inevitable; and the insect excels in them without any apprenticeship, as the experiment with the native Osmia of the quarries has just proved.

Animal resources have a certain elasticity, within narrow limits. What we learn from the animals' industry at a given moment is not always the full measure of their skill. They possess latent powers held in reserve for certain emergencies. Long generations can succeed one another without employing them; but, should some circumstance require it, suddenly those powers burst forth, free of any previous attempts, even as the spark potentially contained in the flint flashes forth independently of all preceding gleams. Could one who knew nothing of the Sparrow but her nest under the eaves suspect the ball-shaped nest at the top of a tree? Would one who knew nothing of the Osmia save her home in the Snail-shell expect to see her accept as her dwelling a stump of reed, a paper funnel, a glass tube? My neighbour the Sparrow, impulsively taking it into her head to leave the roof for the plane-tree, the Osmia of the quarries, rejecting the natal cabin, the spiral of the shell, for my cylinder, alike show us how sudden and spontaneous are the industrial variations of animals.

THE SACRED BEETLE

« 11 »

Fabre spent nearly half a century studying the life of the Sacred Beetle. He began his experiments with the Scarab in the middle of the Nineteenth Century; he wrote the final chapter of its life-history almost fifty years later. There are more than 20,000 species in the widespread Scarab family. The most famous is the sacred Scarab of the Nile. Probably the most familiar relative in America is the June-Bug or May Beetle. Shakespeare's "shard-borne beetle with his drowsy hum" was a member of the Scarab family. The Gymnopleurus, mentioned by Fabre in this chapter, is a smaller dung beetle. The Hydrophilis is a predaceous water beetle. The source of the following selection is THE SACRED BEETLE AND OTHERS.

It happened like this. There were five or six of us: myself, the oldest, officially their master but even more their friend and comrade; they, lads with warm hearts and joyous imaginations, overflowing with that youthful vitality which makes us so enthusiastic and so eager for knowledge. We started off one morning down a path fringed with dwarf elder and hawthorn, whose clustering blossoms were already a paradise for the Rosechafer ecstatically drinking in their bitter perfumes. We talked as we went. We were going to see whether the Sacred Beetle had yet made his appearance on the sandy plateau of Les Angles, whether he was rolling that pellet of dung in which ancient Egypt beheld an image of the world; we were going to find out whether the stream at the foot of the hill was not hiding under its mantle of duckweed young Newts with gills like tiny branches of coral; whether that pretty little fish of our rivulets, the Stickleback, had donned his wedding scarf of purple and blue; whether the newly arrived Swallow was skimming the meadows on pointed wing, chasing the Craneflies, who scatter their eggs as

93

they dance through the air; if the Eyed Lizard was sunning his blue-speckled body on the threshold of a burrow dug in the sandstone; if the Laughing Gull, travelling from the sea in the wake of the legions of fish that ascend the Rhône to milt in its waters, was hovering in his hundreds over the river, ever and anon uttering his cry so like a maniac's laughter; if . . . but that will do. To be brief, let us say that, like good simple folk who find pleasure in all living things, we were off to spend a morning at the most wonderful of festivals, life's springtime awakening.

Our expectations were fulfilled. The Stickleback was dressed in his best: his scales would have paled the lustre of silver; his throat was flashing with the brightest vermilion. On the approach of the great black Horse-leech, the spines on his back and sides started up, as though worked by a spring. In the face of this resolute attitude, the bandit turns tail and slips ignominiously down among the water-weeds. The placid mollusc tribe—Planorbes, Limnaei and other Water-snails—were sucking in the air on the surface of the water. The Hydrophilus and her hideous larva, those pirates of the ponds, darted amongst them, wringing a neck or two as they passed. The stupid crowd did not seem even to notice it. But let us leave the plain and its waters and clamber up the bluff to the plateau above us. Up there, Sheep are grazing and Horses being exercised for the approaching races, while all are distributing manna to the enraptured Dung-beetles.

What excitement over a single patch of Cow-dung! Never did adventurers hurrying from the four corners of the earth display such eagerness in working a Californian claim. Before the sun becomes too hot, they are there in their hundreds, large and small, of every sort, shape and size, hastening to carve themselves a slice of the common cake. There are some that labour in the open air and scrape the surface; there are others that dig themselves galleries in the thick of the heap, in search of choice veins; some work the lower stratum and bury their spoil without delay in the ground just below; others again, the smallest, keep on one side and crumble a morsel that has slipped their way during the mighty excavations of their more powerful fellows. Some, newcomers and doubtless the hungriest, consume their meal on the spot; but the greater number dream of accumulating stocks that will allow them to spend long days in affluence, down in some safe retreat. A nice, fresh patch of dung is not found just when you want it, in the barren plains overgrown with thyme; a windfall of this sort is as manna from the sky; only fortune's favourites receive so fair a portion. Wherefore the riches of to-day are prudently hoarded for the morrow. The stercoraceous scent has carried the glad tidings half a mile around; and all have hastened up to get a store of provisions. A few laggards are still arriving, on the wing or on foot.

Who is this that comes trotting towards the heap, fearing lest he reach it too late? His long legs move with awkward jerks, as though driven by some mechanism within his belly; his little red antennae unfurl their fan, a sign of anxious greed. He is coming, he has come, not without sending a few banqueters sprawling. It is the Sacred Beetle, clad all in black, the biggest and most famous of our Dung-beetles. Behold him at table, beside his fellow-guests, each of whom is giving the last touches to his ball with the flat of his broad fore-legs or else enriching it with yet one more layer before retiring to enjoy the fruit of his labours in peace. Let us follow the construction of the famous ball in all its phases.

The clypeus, or shield, that is, the edge of the broad, flat head, is notched with six angular teeth arranged in a semicircle. This constitutes the tool for digging and cutting up, the rake that lifts and casts aside the unnutritious vegetable fibres, goes for something better, scrapes and collects it. A choice is thus made, for these connoisseurs differentiate between one thing and another, making a rough selection when the Beetle is occupied with his own provender, but an extremely scrupulous one when it is a case of constructing the maternal ball, which has a central cavity in which the egg will hatch. Then every scrap of fibre is conscientiously rejected and only the stercoral quintessence is gathered as the material for building the inner layer of the cell. The young larva, on issuing from the egg, thus finds in the very walls of its lodging a food of special delicacy which strengthens its digestion and enables it afterwards to attack the coarse outer layers.

Where his own needs are concerned, the Beetle is less particular and contents himself with a very general sorting. The notched shield then does its scooping and digging, its casting aside and scraping together more or less at random. The fore-legs play a mighty part in the work. They are flat, bow-shaped, supplied with powerful nervures and armed on the outside with five strong teeth. If a vigorous effort be needed to remove an obstacle or to force a way through the thickest part of the heap, the Dung-beetle makes use of his elbows, that is to say, he flings his toothed legs to right and left and clears a semicircular space with an energetic sweep. Room once made, a different kind of work is found for these same limbs: they collect armfuls of the stuff raked together by the shield and push it under the insect's belly, between the four hinder legs. These are formed for the turner's trade. They are long and slender, especially the last pair, slightly bowed and finished with a very sharp claw. They are at once recognized as compasses, capable of embracing a globular body in their curved branches and of verifying and correcting its shape. Their function is, in fact, to fashion the ball.

Armful by armful, the material is heaped up under the belly, between the four legs, which, by a slight pressure, impart their own curve to it and give it a preliminary outline. Then, every now and again, the rough-hewn pill is set spinning between the four branches of the double pair of spherical compasses; it turns under the Dung-beetle's belly until it is rolled into a perfect ball. Should the surface layer lack plasticity and threaten to peel off, should some too-stringy part refuse to yield to the action of the lathe, the fore-legs touch up the faulty places; their broad paddles pat the ball to give consistency to the new layer and to work the recalcitrant bits into the mass.

Under a hot sun, when time presses, one stands amazed at the turner's feverish activity. And so the work proceeds apace: what a moment ago was a tiny pellet is now a ball the size of a walnut; soon it will be the size of an apple. I have seen some gluttons manufacture a ball the size of a man's fist. This indeed means food in the larder for days to come!

The Beetle has his provisions. The next thing is to withdraw from the fray and transport the victuals to a suitable place. Here the Scarab's most striking characteristics begin to show themselves. Straightway he begins his journey; he clasps his sphere with his two long hind-legs, whose terminal claws, planted in the mass, serve as pivots; he obtains a purchase with the middle pair of legs; and, with his toothed fore-arms, pressing in turn upon the ground, to do duty as levers, he proceeds with his load, he himself moving backwards, body bent, head down and hind-quarters in the air. The rear legs, the principal factor in the mechanism, are in continual movement backwards and forwards, shifting the claws to change the axis of rotation, to keep the load balanced and to push it along by alternate thrusts to right and left. In this way, the ball finds itself touching the ground by turns with every point of its surface, a process which perfects its shape and gives an even consistency to its outer layer by means of pressure uniformly distributed.

And now to work with a will! The thing moves, it begins to roll; we shall get there, though not without difficulty. Here is a first awkward place: the Beetle is wending his way athwart a slope and the heavy mass tends to follow the incline; the insect, however, for reasons best known to itself, prefers to cut across this natural road, a bold project which may be brought to naught by a false step or by a grain of sand which disturbs the balance of the load. The false step is made: down goes the ball to the bottom of the valley; and the insect, toppled over by the shock, is lying on its back, kicking. It is soon up again and hastens to harness itself once more to its load. The machine works better than ever. But look out, you dunderhead! Follow the dip of the valley: that will save labour and mishaps; the road is good and level; your ball will roll quite easily. Not a bit of it! The Beetle prepares once again to mount the slope that has already been his undoing. Perhaps it suits him to return to the heights. Against that I have nothing to say: the Scarab's judgment is better than mine as to the advisability of keeping to lofty regions; he can see farther than I can in these matters. But at least take this path, which will lead you up by a gentle incline! Certainly not! Let him find himself near some very steep slope, impossible to climb, and that is the very path which the obstinate fellow will choose. Now begins a Sisyphean labour. The ball, that enormous burden, is painfully hoisted, step by step, with infinite precautions, to a certain height, always backwards. We wonder by what miracle of statics a mass of this size can be kept upon the slope. Oh! An all-advised movement frustrates all this toil: the ball rolls down, dragging the Beetle with it. Once more the heights are scaled and another fall is the sequel. The attempt is renewed, with greater skill this time at the difficult points; a wretched grass-root, the cause of the previous falls, is carefully got over. We are almost there; but steady now, steady! It is a dangerous ascent and the merest trifle may yet ruin everything. For see, a leg slips on a smooth bit of gravel! Down come ball and Beetle, all mixed up together. And the insect begins over again, with indefatigable obstinacy. Ten times, twenty times, he will attempt the hopeless ascent, until his persistence vanquishes all obstacles, or until, wisely recognizing the futility of his efforts, he adopts the level road.

The Scarab does not always push his precious ball alone: sometimes he takes a partner; or, to be accurate, the partner takes him. This is the way in which things usually happen: once his ball is ready, a Dung-beetle issues from the crowd and leaves the workyard, pushing his prize backwards. A neighbour, a newcomer, whose own task is hardly begun, abruptly drops his work and runs to the moving ball, to lend a hand to the lucky owner, who seems to accept the proffered aid kindly. Henceforth the two work in partnership. Each does his best to push the pellet to a place of safety. Was a compact really concluded in the workyard, a tacit agreement to share the cake between them? While one was kneading and moulding the ball, was the other tapping rich veins whence to extract choice materials and add them to the common store? I have never observed any such collaboration; I have always seen each Dung-beetle occupied solely with his own affairs in the works. The last-comer, therefore, has no acquired rights.

Can it then be a partnership between the two sexes, a couple intending to set up house? I thought so for a time.

The evidence of the scalpel compelled me to abandon my belief in this domestic idyll. There is no outward difference between the two sexes in the Scarabaei. I therefore dissected the pair of Dung-beetles engaged in trundling one and the same ball; and they very often proved to be of the same sex.

Neither community of family nor community of labour! Then what is the motive for this apparent partnership? It is purely and simply an attempt at robbery. The zealous fellow-worker, on the false plea of lending a helping hand, cherishes a plan to purloin the ball at the first opportunity. To make one's own ball at the heap means hard work and patience; to steal one ready-made, or at least to foist one's self as a guest, is a much easier matter. Should the owner's vigilance slacken, you can run away with his property; should you be too closely watched, you can sit down to table uninvited, pleading services rendered. It is "Heads I win, tails you lose" in these tactics, so that pillage is practised as one of the most lucrative of trades. Some go to work craftily, in the way which I have described: they come to the aid of a comrade who has not the least need of them and hide the most barefaced greed under the cloak of charitable assistance. Others, bolder perhaps, more confident in their strength, go straight to their goal and commit robbery with violence.

Scenes are constantly happening such as this: a Scarab goes off, peacefully, by himself, rolling his ball, his lawful property, acquired by conscientious work. Another comes flying up, I know not whence, drops down heavily, folds his dingy wings under their cases and, with the back of his toothed fore-arms, knocks over the owner, who is powerless to ward off the attack in his awkward position, harnessed as he is to his property. While the victim struggles to his feet, the other perches himself atop the ball, the best position from which to repel an assailant. With his fore-arms crossed over his breast, ready to hit back, he awaits events. The dispossessed one moves round the ball, seeking a favourable spot at which to make the assault; the usurper spins round on the roof of the citadel, facing his opponent all the time. If the latter raise himself in order to scale the wall, the robber gives him a blow that stretches him on his back. Safe at the top of his fortress, the besieged Beetle could foil his adversary's attempts indefinitely if the latter did not change his tactics. He turns sapper so as to reduce the citadel with the garrison. The ball, shaken from below, totters and begins rolling, carrying with it the thieving Dung-beetle, who makes violent efforts to maintain his position on the top. This he succeeds in doing—though not invariably—thanks to hurried gymnastic feats which land him higher on the ball and make up for the ground which he loses by its rotation. Should a false movement bring him to earth, the chances become equal and the struggle turns into a wrestling-match. Robber and robbed grapple with each other, breast to breast. Their legs lock and unlock, their joints intertwine, their horny armour clashes and grates with the rasping sound of metal under the file. Then the one who succeeds in throwing his opponent and releasing himself scrambles to the top of the ball and there takes up his position. The siege is renewed, now by the robber, now by the robbed, as the chances of the hand-to-hand conflict may decree. The former, a brawny desperado, no novice at the game, often has the best of the fight. Then, after two or three unsuccessful attempts, the defeated Beetle wearies and returns philosophically to the heap, to make himself a new pellet. As for the other, with all fear of a surprise attack at an end, he harnesses himself to the conquered ball and pushes it whither he pleases. I have sometimes seen a third thief appear upon the scene and rob the robber. Nor can I honestly say that I was sorry.

I ask myself in vain what Proudhon introduced into Scarabæan morality the daring paradox that "property means plunder," or what diplomatist taught the Dung-beetle the savage maxim that "might is right." I have no data that would enable me to trace the origin of these spoliations, which have become a custom, of this abuse of strength to capture a lump of ordure. All that I can say is that theft is a general practice among the Scarabs. These dung-rollers rob one another with a calm effrontery which, to my knowledge, is without a parallel. I leave it to future observers to elucidate this curious problem in animal psychology and I go back to the two partners rolling their ball in concert.

But first let me dispel a current error in the text-books. I find in M. Émile Blanchard's magnificent work, *Métamorphoses, mœurs et instincts des insectes*, the following passage:

"Sometimes our insect is stopped by an insurmountable obstacle; the ball has fallen into a hole. At such moments the Ateuchus gives evidence of a really astonishing grasp of the situation as well as of a system of ready communication between individuals of the same species which is evern more remarkable. Recognizing the impossibility of coaxing the ball out of the hole, the Ateuchus seems to abandon it and flies away. If you are sufficiently endowed with that great and noble virtue called patience, stay by the forsaken ball: after a while, the Ateuchus will return to the same spot and will not return alone; he will be accompanied by two, three, four or five companions, who will all alight at the place indicated and will combine their efforts to raise the load. The Ateuchus has been to fetch reinforcements; and this explains why it is such a common sight, in the dry fields, to see several Ateuchi joining in the removal of a single ball."

Finally, I read in Illiger's *Entomological Magazine:*

"A *Gymnopleurus pilularius,* while constructing the ball of dung destined to contain her eggs, let it roll into a hole, whence she strove for a long time to extract it unaided. Finding that she was wasting her time in vain efforts, she ran to a neighbouring heap of manure to fetch three individuals of her own species, who, uniting their strength to hers, succeeded in withdrawing the ball from the cavity into which it had fallen and then returned to their manure to continue their work."

I crave a thousand pardons of my illustrious master, M. Blanchard, but things certainly do not happen as he says. To begin with, the two accounts are so much alike that they must have had a common origin. Illiger, on the strength of observations not continuous enough to deserve blind confidence, put forward the case of his Gymnopleurus; and the same story was repeated about the Scarabaei because it is, in fact, quite usual to see two of these insects occupied together either in rolling a ball or in getting it out of a troublesome place. But this cooperation in no way proves that the Dung-beetle who found himself in difficulties went to requisition the aid of his mates. I have had no small measure of the patience recommended by M. Blanchard; I have lived laborious days in close intimacy, if I may say so, with the Sacred Beetle; I have done everything that I could think of in order to enter into his ways and habits as thoroughly as possible and to study them from life; and I have never seen anything that suggested either nearly or remotely the idea of companions summoned to lend assistance. As I shall presently relate, I have subjected the Dung-beetle to far more serious trials than that of getting his ball into a hole; I have confronted him with much graver difficulties than that of mounting a slope, which is sheer sport to the obstinate Sisyphus, who seems to delight in the rough gymnastics involved in climbing steep places, as if the ball thereby grew firmer and accordingly increased in value; I have created artificial situations in which the insect had the uttermost need of help; and never did my eyes detect any evidence of friendly services being rendered by comrade to comrade. I have seen Beetles robbed and Beetles robbing and nothing more. If a number of them were gathered around the same pill, it meant that a battle was taking place. My humble opinion, therefore, is that the incident of a number of Scarabaei collected around the same ball with thieving intentions has given rise to these stories of comrades called in to lend a hand. Imperfect observations are responsible for this transformation of the bold highwayman into a helpful companion who has left his work to do another a friendly turn.

It is no light matter to attribute to an insect a really astonishing grasp of a situation, combined with an even more amazing power of communication between individuals of the same species. Such an admission involves more than one imagines. That is why I insist on my point. What! Are we to believe that a Beetle in distress will conceive the idea of going in quest of help? We are to imagine him flying off and scouring the country to find fellow-workers on some patch of dung; when he has found them, we are to suppose that he addresses them, in some sort of pantomime, by gestures with his antennae more particularly, in some such words as these:

"I say, you fellows, my load's upset in a hole over there; come and help me get it out. I'll do as much for you one day!"

And we are to believe that his comrades understand! And, more incredible still, that they straightway leave their work, the pellet which they have just begun, the beloved pill exposed to the cupidity of others and certain to be filched in their absence, and go to the help of the suppliant! I am profoundly incredulous of such unselfishness; and my incredulity is confirmed by what I have witnessed for years and years, not in glass-cases but in the very places where the Scarab works. Apart from its maternal solicitude, in which respect it is nearly always admirable, the insect cares for nothing but itself, unless it lives in societies, like the Hive-bees, the Ants and the rest.

But let me end this digression, which is excused by the importance of the subject. I was saying that a Sacred Beetle, in possession of a ball which he is pushing backwards, is often joined by another, who comes hurrying up to lend an assistance which is anything but disinterested, his intention being to rob his companion if the opportunity present itself. Let us call the two workers partners, though that is not the proper name for them, seeing that the one forces himself upon the other, who probably accepts outside help only for fear of a worse evil. The meeting, by the way, is absolutely peaceful. The owner of the ball does not cease work for an instant on the arrival of the newcomer; and his uninvited assistant seems animated by the best intentions and sets to work on the spot. The way in which the two partners harness themselves differs. The proprietor occupies the chief position, the place of honour: he pushes at the rear, with his hind-legs in the air and his head down. His subordinate is in front, in the reverse posture, head up, toothed arms on the ball, long hind-legs on the ground. Between the two, the ball rolls along, one driving it before him, the other pulling it towards him.

The efforts of the couple are not always very harmonious, the more so as the assistant has his back to the road to be traversed, while the owner's view is impeded by the load. The result is that they are constantly having accidents, absurd tumbles, taken cheerfully and in good part: each picks himself up quickly and resumes the same position as before. On level ground, this system of traction does not correspond with the dynamic force expended, through lack of precision in the combined movements: the Scarab at the back would do as well and better if left to himself. And so the helper, having given a proof of his good-will at the risk of throwing the machinery out of gear, now decides to keep still, without letting go of the precious ball, of course. He already looks upon that as his: a ball touched is a ball gained. He won't be so silly as not to stick to it: the other might give him the slip!

So he gathers his legs flat under his belly, encrusting himself, so to speak, on the ball and becoming one with it. Henceforth, the whole concern—the ball and the Beetle clinging to its surface—is rolled along by the efforts of the lawful owner. The intruder sits tight and lies low, heedless whether the load pass over his body, whether he be at the top, bottom or side of the rolling ball. A queer sort of assistant, who gets a free ride so as to make sure of his share of the victuals!

But a steep ascent heaves in sight and gives him a fine part to play. He takes the lead now, holding up the heavy mass with his toothed arms, while his mate seeks a purchase in order to hoist the load a little higher. In this way, by a combination of well-directed efforts, the Beetle above gripping, the one below pushing, I have seen a couple mount hills which would have been too much for a single porter, however persevering. But in times of difficulty not all show the same zeal: there are some who, on awkward slopes where their assistance is most needed, seem blissfully unaware of the trouble. While the unhappy Sisyphus exhausts himself in attempts to get over the bad part, the other quietly leaves him to it: imbedded in the ball, he rolls down with it if it comes to grief and is hoisted up with it when they start afresh.

I have often tried the following experiment on the two partners in order to judge their inventive faculties when placed in a serious predicament. Suppose them to be on level ground, number two seated motionless on the ball, number one busy pushing. Without disturbing the latter, I nail the ball to the ground with a long, strong pin. It stops suddenly. The Beetle, unaware of my perfidy, doubtless believes that some natural obstacle, a rut, a tuft of couch-grass, a pebble, bars the way. He redoubles his efforts, struggles his hardest; nothing happens.

"What can the matter be? Let's go and see."

The Beetle walks two or three times round his pellet. Discovering nothing to account for its immobility, he returns to the rear and starts pushing again. The ball remains stationary.

"Let's look up above."

The Beetle goes up to find nothing but his motionless colleague, for I had taken care to drive in the pin so deep that the head disappeared in the ball. He explores the whole upper surface and comes down again. Fresh thrusts are vigorously applied in front and at the sides, with the same absence of success. There is not a doubt about it: never before was Dung-beetle confronted with such a problem in inertia.

Now is the time, the very time, to claim assistance, which is all the easier as his mate is there, close at hand, squatting on the summit of the ball. Will the Scarab rouse him? Will he talk to him like this:

"What are you doing there, lazybones? Come and look at the thing: it's broken down!"

Nothing proves that he does anything of the kind, for I see him steadily shaking the unshakable, inspecting his stationary machine on every side, while all this time his companion sits resting. At long last, however, the latter becomes aware that something unusual is happening; he is apprised of it by his mate's restless tramping and by the immobility of the ball. He comes down, therefore, and in his turn examines the machine. Double harness does no better than single harness. This is beginning to look serious. The little fans of the Beetles' antennae open and shut, open again, betraying by their agitation acute anxiety. Then a stroke of genius ends the perplexity:

"Who knows what's underneath?"

They now start exploring below the ball; and a little digging soon reveals the presence of the pin. They recognize at once that the trouble is there.

If I had had a voice in their deliberations, I should have said:

"We must make a hole in the ball and pull out that skewer which is holding it down."

This most elementary of all proceedings and one so easy to such expert diggers was not adopted, was not even tried. The Dung-beetle was shrewder than man. The two colleagues, one on this side, one on that, slip under the ball, which begins to slide up the pin, getting higher and higher in proportion as the living wedges make their way underneath. The clever operation is made possible by the softness of the material, which gives easily and makes a channel under the head of the immovable stake. Soon the pellet is suspended at a height equal to the thickness of the Scarabs' bodies. The rest is not such plain sailing. The Dung-beetles, who at first were lying flat, rise gradually to their feet, still pushing with their backs. The work becomes harder and harder as the legs, in straightening out, lose their strength; but none the less they do it. Then comes a time when they can no longer push with their backs, the limit of their height having been reached. A last resource remains, but one much less favourable to the development of motive power. This is for the insect to adopt one or other of its postures when harnessed to the ball, head down or up, and to push with its hind- or fore-legs, as the case may be. Finally the ball drops to the ground, unless we have used too long a pin. The gash made by our stake is repaired more or less and the carting of the precious pellet is at once resumed.

But, should the pin really be too long, then the ball, which remains firmly fixed, ends by being suspended at a height above that of the insect's full stature. In that case, after vain evolutions around the unconquerable greased pole, the Dung-beetles throw up the sponge, unless we are sufficiently kind-hearted to finish the work ourselves and restore their treasure to them. Or again we can help them by raising the floor with a small flat stone, a pedestal from the top of which it is possible for the Beetle to continue his labours. Its use does not appear to be immediately understood, for neither of the two is in any hurry to take advantage of it. Nevertheless, by accident or design, one or other at last finds himself on the stone. Oh, joy! As he passed, he felt the ball touch his back. At that contact, courage returns; and his efforts begin once more. Standing on his helpful platform, the Scarab stretches his joints, rounds his shoulders, as one might say, and shoves the pellet upwards. When his shoulders no longer avail, he works with his legs, now upright, now head downwards. There is a fresh pause, accompanied by fresh signs of uneasiness, when the limit of extension is reached. Thereupon, without disturbing the creature, we place a second little stone on the top of the first. With the aid of this new step, which provides a fulcrum for its levers, the insect pursues its task. Thus adding story upon story as required, I have seen the Scarab, hoisted to the summit of a tottering pile three or four fingers'-breadth in height, persevere in his work until the ball was completely detached.

Had he some vague consciousness of the service performed by the gradual raising of the pedestal? I venture to doubt it, though he cleverly took advantage of my platform of little stones. As a matter of fact, if the very elementary idea of using a higher support in order to reach something placed above one's grasp were not beyond the Beetle's comprehension, how is it that, when there are two of them, neither thinks of lending the other his back so as to raise him by that much and make it possible for him to go on working? If one helped the other in this way, they could reach twice as high. They are very far, however, from any such cooperation. Each pushes the ball, with all his might, I admit, but he pushes as if he were alone and seems to have no notion of the happy result that would follow a combined effort. In this instance, when the ball is nailed to the ground by a pin, they do exactly what they do in corresponding circumstances, as, for example, when the load is brought to a standstill by some obstacle, caught in a loop of couch-grass or transfixed by some spiky bit of stalk that has run into the soft, rolling mass. I produced artificially a stoppage which is not really very different from those occurring naturally when the ball is being rolled amid the thousand and one irregularities of the ground; and the Beetle behaves, in my experimental tests, as he would have behaved in any other circumstances in which I had no part. He uses his back as a wedge and a lever and pushes with his feet, without introducing anything new into his methods, even when he has a companion and can avail himself of his assistance.

When he is all alone in face of the difficulty, when he has no assistant, his dynamic operations remain absolutely the same; and his efforts to move his transfixed ball end in success, provided that we give him the indispensable support of a platform, built up little by little. If we deny him this succour, then, no longer encouraged by the contact of his beloved ball, he loses heart and sooner or later flies away, doubtless with many regrets, and disappears. Where to? I do not know. What I do know is that he does not return with a gang of fellow-labourers whom he has begged to help him. What would he do with them, he who cannot make use of even one comrade?

But perhaps my experiment, which leaves the ball suspended at an inaccessible height and the insect with its means of action exhausted, is a little too far removed from ordinary conditions. Let us try instead a miniature pit, deep enough and steep enough to prevent the Dung-beetle, when placed at the bottom, from rolling his load up the side. These are exactly the conditions stated by Messrs. Blanchard and Illiger. Well, what happens? When dogged but utterly fruitless efforts have convinced him of his helplessness, the Beetle takes wing and disappears. Relying upon what these learned writers said, I have waited long hours for the insect to return reinforced by a few friends. I have always waited in vain. Many a time also I have found the pellet several days later just where I left it, stuck at the top of a pin or in a hole, proving that nothing fresh had happened in my absence. A ball abandoned from necessity is a ball abandoned for good, with no attempt at salvage with the aid of others. A dexterous use of wedge and lever to set the ball rolling again is therefore, when all is said, the greatest intellectual effort which I have observed in the Sacred Beetle. To make up for what the experiment refutes, namely an appeal for help among fellow-workers, I gladly chronicle this feat of mechanical prowess for the Dung-beetles' greater glory.

Directing their steps at random, over sandy plains thick with thyme, over cart-ruts and steep places, the two Beetle brethren roll the ball along for some time, thus giving its substance a certain consistency which may be to their liking. While still on the road, they select a favourable spot. The rightful owner, the Beetle who throughout has kept the place of honour, behind the ball, the one in short who has done almost all the carting by himself, sets to work to dig the dining-room. Beside him is the ball, with number two clinging to it, shamming dead. Number one attacks the sand with his sharp-edged forehead and his toothed legs; he flings armfuls of it behind him; and the work of excavating proceeds apace. Soon the Beetle has disappeared from view in the half-dug cavern. Whenever he returns to the upper air with a load, he invariably glances at his ball to see if all is well. From time to time, he brings it nearer the threshold of the burrow; he feels it and seems to acquire new vigour from the contact. The other, lying demure and motionless on the ball, continues to inspire confidence. Meanwhile the underground hall grows larger and deeper; and the digger's field of operations is now too vast for any but very occasional appearances. Now is the time. The crafty sleeper awakens and hurriedly decamps with the ball, which he pushes behind him with the speed of a pickpocket anxious not to be caught in the act. This breach of trust rouses my indignation, but the historian triumphs for the moment over the moralist and I leave him alone: I shall have time enough to intervene on the side of law and order if things threaten to turn out badly.

The thief is already some yards away. His victim comes out of the burrow, looks around and finds nothing. Doubtless an old hand himself, he knows what this means. Scent and sight soon put him on the track. He makes haste and catches up the robber; but the artful dodger, when he feels his pursuer close on his heels, promptly changes his posture, gets on his hind-legs and clasps the ball with his toothed arms, as he does when acting as an assistant.

You rogue, you! I see through your tricks: you mean to plead as an excuse that the pellet rolled down the slope and that you are only trying to stop it and bring it back home. I, however, an impartial witness, declare that the ball was quite steady at the entrance to the burrow and did not roll of its own accord. Besides, the ground is level. I declare that I saw you set the thing in motion and make off with unmistakable intentions. It was an attempt at larceny, or I've never seen one!

My evidence is not admitted. The owner cheerfully accepts the other's excuses; and the two bring the ball back to the burrow as though nothing had happened.

If the thief, however, has time to get far enough away, or if he manages to cover his trail by adroitly doubling back, the injury is irreparable. To collect provisions under a blazing sun, to cart them a long distance, to dig a comfortable banqueting-hall in the sand and then—just when everything is ready and your appetite, whetted by exercise, lends an added charm to the approaching feast—suddenly to find yourself cheated by a crafty partner is, it must be admitted, a reverse of fortune that would dishearten most of us. The Dung-beetle does not allow himself to be cast down by this piece of ill-luck; he rubs his cheeks, spreads his antennae, sniffs the air and flies to the nearest heap to begin all over again.

A BEETLE PATRIARCH

« 12 »

The Geotrupes, like the Scarab, is a dung beetle, a squarish, black, heavy-bodied insect often encountered in pastures. Many species are able to produce a faint squeaking sound by rubbing their legs against their abdomens. In America, some species live in the stems of decaying toadstools. Various other insects besides the Geotrupes exhibit what appears to be a "weather sense." Of the various beetles mentioned in this chapter, the Saperdae are long-horned beetles, the Capricorns are wood-boring beetles, the Buprestes are also wood-borers in their larval state, the Aphodii and the Anthophagi are dung-feeders. This section was first published in THE SACRED BEETLE AND OTHERS.

To complete the cycle of the year in the adult form, to see one's self surrounded by one's sons at the spring festival, to double and treble one's family: that surely is a most exceptional privilege in the insect world. The Bees, the aristocracy of instinct, perish once the honey-pot is filled; the Butterflies, the aristocracy not of instinct but of dress, die when they have fastened their packet of eggs in a propitious spot; the richly-armoured Ground-beetles succumb when the germs of a posterity are scattered beneath the stones.

So with the others, except among the social insects, where the mother survives, either alone or accompanied by her attendants. It is a general law: the insect is born orphaned of both its parents. And lo, by an unexpected turn of fate, the humble scavenger escapes the catastrophes that devour the mighty! The Dung-beetle, sated with days, becomes a patriarch.

This longevity explains first of all a fact that struck me long ago, when, to learn a little about the tribes whose history attracted me so greatly, I used to stick rows of Beetles on pins in my boxes. Ground-beetles, Rose-chafers, Buprestes, Capricorns, Saperdae and the rest were collected one by one, after prolonged search. Now and again a lucky find would make my cheeks glow with excitement. Exclamations broke from our prentice band when one of these rarities was captured. A touch of jealousy accompanied our congratulations of the proud possessor. It was bound to be so; for think: there were not enough to go round.

A Scalary Saperda, the denizen of dead cherry-trees, clad in deep yellow with ladder-like markings of black velvet; a purply Ground-beetle, edged with amethyst along his ebony wing-cases; a brilliant Buprestis, wedding the sheen of gold and copper to the gorgeous green of malachite: these were great events, far too infrequent to satisfy us all.

With the Dung-beetles you can sing a different song! These are the ones if you want to fill the greediest of asphyxiating-phials to the neck. They, especially the smaller ones, are a numberless multitude when the others are few and far between. I remember Anthophagi and Aphodii swarming by the thousand under one shelter. You could have shovelled them up if you wished.

To this day I am still astonished when I see these crowds again; as of old, the abundance of the Dung-beetle family forms a striking contrast with the comparative scarcity of the others. If it occurred to me to go a-hunting once more and renew the quest to which I owe moments of such sheer delight, I should be certain of filling my flasks with Scarabæi, Copres, Geotrupes, Anthophagi and other members of the same corporation before achieving any measure of success with the rest of the series. By the time that May comes, the distiller of ordure is there in numbers; and in July and August, those months of blazing heat which see the suspension of labour in the fields, the dealer in unsavoury matter is still at work while the others have taken to earth and are lying in motionless torpor. He and his contemporary, the Cicada, represent almost by themselves such activity as prevails during the torrid days.

May not this greater frequency of the Dung-beetles, at least in my part of the world, be due to the longevity of the adult form? I think so. Whereas the other insects are summoned to enjoy the fine weather only in successive generations, these receive a general invitation, father and sons together, daughters and mother together. Being equally prolific, they are therefore represented twice over.

And they really deserve it, in consideration of the services which they render. There is a general hygienic law which requires that every putrid thing shall disappear in the shortest possible time. Paris has not yet solved the formidable problem of her sewage, which sooner or later will become a question of life or death for the monstrous city. One asks one's self whether the centre of light is not doomed to be extinguished some day in the reeking exhalations of a soil saturated with putrescence. What this agglomeration of millions of men cannot obtain, with all its treasures of wealth and talent, the smallest hamlet possesses without going to any expense or even troubling to think about it.

Nature, so lavish of her cares in respect of rural health, is indifferent to the welfare of cities, if not actively hostile to it. She has created for the fields two classes of scavengers, whom nothing wearies, whom nothing repels. One of these, consisting of Flies, Silphae, Dermestes, Necrophori, Histers is charged with the dissection of corpses. They cut and hash, they elaborate the waste matter of death in their stomachs in order to restore it to life.

A Mole ripped open by the ploughshare soils the path with its entrails, which soon turn purple; a Snake lies on the grass, crushed by the foot of a wayfarer who thought, the fool, that he was performing a good work; an unfledged bird, fallen from its nest, lies, a crushed and pathetic heap, at the foot of the tree that carried it; thousands of other similar remains, of every sort and kind, are scattered here and there, threatening danger through their effluvia, if no steps be taken to put things right. Have no fear: no sooner is a corpse signalled in any direction than the little undertakers come trotting along. They work away at it, empty it, consume it to the bone, or at least reduce it to the dryness of a mummy. In less than twenty-four hours, Mole, Snake, bird have disappeared and the requirements of health are satisfied.

The same zeal for their task exists in the second class of scavengers. The village hardly knows those ammonia-scented refuges to which the townsman repairs to relieve his wretched needs. A little bit of a wall, a hedge, a bush is all that the peasant asks as a retreat at the moment when he would fain be alone. I need say no more to suggest the encounters to which such free and easy manners expose you! Enticed by the patches of lichen, the cushions of moss, the tufts of houseleek and other pretty things that adorn old stones, you go up to a sort of wall that supports a vineyard. Faugh! At the foot of the daintily-decked shelter, what an unconcealed abomination! You flee: lichens, mosses and houseleek tempt you no more. But come back on the morrow. The thing has disappeared, the place is clean: the Dung-beetles have been that way.

A whole world is benefited by the agricultural industry of the Dung-beetle, that burier of manure: first the plant and then all that live upon the plant. A small world, a very small world, as small as you please, but after all not a negligible world. It is of such trifles that the great integral of life is composed, even as the integral of the mathematicians is composed of quantities neighbouring on o.

Agricultural chemistry teaches us that, to employ the stable-dung to the best purpose, we should put it into the ground, so far as possible, while fresh. When diluted by the rain and dissipated by the air, it becomes lifeless and devoid of fertilizing elements. The agronomic truth of such high importance is quite familiar to the Geotrupes and his colleagues. In their burying-work, they invariably aim at materials of recent date. Just as they are eager to put away the produce of the moment, all saturated with its potassium, its nitrates and its phosphates, even so do they scorn the stuff hardened into brick by the sun or rendered infertile by long exposure to the air. The valueless residue does not interest them; they leave this barren rubbish to others.

We now know about the Geotrupes as a sanitary expert and as a collector of manure. We are going to see him in a third aspect, that of the sagacious weather-prophet. It is popularly believed, in the countryside, that a swarm of agitated Geotrupes, skimming the ground with an air of great business in the evening, is a sign of fine weather on the morrow. Is this rustic prognostication worth anything? My cages shall tell us. I watch my boarders closely all through the autumn, the period when they build their nests; I note the state of the sky on the day before and register the weather of the next day. I use no thermometer, no barometer, none of the scientific implements employed in the meteorological observatories. I confine myself to the summary information derived from my personal impressions.

The Geotrupes do not leave their burrows until after sundown. With the last glimmer of daylight, if the air be calm and the temperature mild, they roam about, flying low with a humming noise, seeking the materials which have accumulated for them in the course of the day. If they come upon something that suits them, they drop down heavily, tumbling over in their clumsy eagerness, thrust themselves into their new treasure and spend the best part of the night in burying it. In this way the dirt of the fields is made to disappear in a single night.

There is one condition indispensable to this purging-process: the atmosphere must be still and warm. Should it rain, the Geotrupes will not stir out of doors. They have sufficient resources underground for a prolonged holiday. Should it be cold, should the northwind blow, they will not sally forth either. In both cases, my cages remain deserted on the surface. We will leave out of the question these periods of enforced leisure and consider only those evenings on which the atmospheric conditions are favourable to foraging-expeditions or at least seem to me as though they ought to be. I will summarize the details in my note-book in three general cases.

First case. A glorious evening. The Geotrupes fuss about the cages, impatient to hasten to their nocturnal task. Next day, magnificent weather. The prophecy, of course, is of the simplest. To-day's fine weather is only the continuation of yesterday's. If the Geotrupes know nothing more than this, they hardly deserve their reputation. However, let us pursue the experiment before drawing any conclusions.

Second case. Again a fine evening. My experience seems to say that the condition of the sky forebodes a fine morning. The Geotrupes think otherwise. They do not come out. Which of the two will be right, man or Dung-beetle? The Dung-beetle: thanks to the keenness of his perceptions, he foresees, he scents a downpour. Rain comes during the night and lasts for part of the day.

Third case. The sky is overcast. Will the south-wind, gathering its clouds, bring us rain? I am of that opinion, appearances seem so much to point that way. The Geotrupes, however, fly and buzz around their cages. Their prophecy is correct and I am wrong. The threat of rain is dispelled and the sun next morning rises radiantly.

They seem to be influenced above all by the electric tension of the atmosphere. On hot and sultry evenings, when a storm is brewing, I see them moving about even more than usual. The morrow is always marked by violent claps of thunder.

There you have the upshot of my observations, which were continued for three months. Whatever the condition of the sky, whether clear or clouded, the Geotrupes announce fair weather or storm by their excited movements at twilight. They are living barometers, more worthy of belief perhaps, in such contingencies, than the barometer of our scientists. The exquisite sensitiveness of life is mightier than the brute weight of a column of mercury.

I will end by mentioning a fact that well deserves further investiga-

tion when circumstances permit. On the twelfth, thirteenth and fourteenth of November 1894, the Geotrupes in my cages are in an extraor-
dinarily agitated condition. Never before and never since I have seen
such animation. They clamber wildly up the wires; at every moment,
they take wing and at once bump against the walls and are flung to
the ground. Their restlessness continues until a late hour of the night,
a very unusual thing with them. Out of doors, a few free neighbours
run up and complete the riot in front of my house. What can be happening to bring these strangers here and especially to throw my cages
into such a state of excitement?

After a few hot days, which are most exceptional at this time of the
year, the southwind prevails, foretelling that rain is at hand. On the
evening of the fourteenth, an endless procession of broken clouds passes
before the face of the moon. It is a magnificent sight. During the night,
the wind drops. There is not a breath of air. The sky is a uniform grey.
The rain pours straight down, monotonously, continuously, depress-
ingly. It looks as though it would never stop. And it goes on, in fact,
until the eighteenth of the month.

Did the Geotrupes, who were so restless on the twelfth, foresee this
deluge? They did. But as a rule they do not quit their burrows at the
approach of rain. Something very extraordinary must have happened,
therefore, to upset them in this way.

The newspapers explained the riddle. On the twelfth, a storm of
unprecedented violence burst over the north of France. The great
barometrical depression which caused it was echoed in my district; and
the Geotrupes marked this profound disturbance by their exceptional
display of emotion. They told me of the hurricane before the papers did,
had I but been able to understand them. Was this simply a chance coincidence, or was it a case of cause and effect? In the absence of sufficient
evidence, I will end on this note of interrogation.

THE BEMBEX WASP

« 13 »

A lover of the summer heat, the Bembex often excavates its burrow in the most exposed portions of a sandbank or open space. In the Indiana dunes, I have seen such wasps mining into sand that was almost too hot to touch. Unlike most digger wasps, the Bembex—which specializes in capturing flies—keeps opening its tunnel and adding fresh food to the larder of its larva. The Eristalis fly that Fabre mentions is one of the Flower Flies. The larva of Eristalis tenax is the celebrated rat-tailed maggot; it feeds in shallow water breathing through a hollow, extensible tail which reaches to the surface. Bembex julii is one of the four new species of wasps that Fabre discovered. He named three of them for his son, Jules, who died at 15. At the end of the first volume of SOUVENIRS ENTOMOLOGIQUES, *in a note dated "Orange, 3 April 1879," Fabre wrote: "I was to write this book for you, to whom its stories gave such delight; and you were to continue it one day. Alas, you went to a happier home, knowing nothing of the book but its first lines! May your name at least figure in it, borne by some of those industrious and beautiful Wasps whom you loved so well!" This portion of this first volume of his ten-volume work later appeared in the English translation,* THE HUNTING WASPS *from which the following selection is taken.*

One of my favourite spots for the observations which I will now describe is not far from Avignon, on the right bank of the Rhone, opposite the mouth of the Durance. It is the Bois des Issarts. Let not the reader mistake the value of this word *bois,* which usually suggests a carpet of cool moss and the shade of tall trees, with a dim light filtering through the leaves. The scorched plains where the Cicada grates out his ditty on the pale olive-tree know none of these delicious retreats filled with cool shadow.

The Bois des Issarts is a coppice of holm-oaks, no higher than one's head and sparingly distributed in scanty clumps which, even at their feet, hardly temper the force of the sun's rays. When I used to settle myself in some part of the coppice suitable for my observations, on certain afternoons in the dog-days of July and August, I had the shelter of a large umbrella, which later, in the most unexpected fashion, lent me a very precious aid of a different kind, as my story will show in good time. If I neglected to furnish myself with this embarrassing adjunct to a long walk, my only resource against sunstroke was to lie down at full length behind some sandy knoll; and, when the veins in my temples were throbbing to bursting point, my last hope lay in putting my head down a Rabbit-burrow. Such are one's means of keeping cool in the Bois des Issarts.

The soil not occupied by those clumps of woody vegetation is almost bare and consists of fine, dry, very loose sand, which the wind heaps into little dunes wherever the stems and roots of the holm-oak interfere with its dissemination. The sides of these sand-dunes are generally very smooth, because of the extreme lightness of the materials, which slide down into the smallest depression and of their own accord restore the evenness of the surface. You need but push your finger into the sand and take it out again to bring about an immediate land-slip which fills up the hole and restores things to their original condition without leaving a visible trace. But, at a certain depth, which varies according to the more or less recent date of the last rains, the sand retains a lingering dampness which keeps it in its place and gives it a consistency that enables it to have small excavations made in it without a subsequent collapse of walls and roof. A blazing sun, a gloriously blue sky, sandy slopes that yield without the least difficulty to the strokes of the Wasp's rake, game galore for the grub's food, a peaceful site hardly ever disturbed by the foot of man: all the good things are combined in this Bembex paradise. Let us watch the industrious insect at work.

If the reader will sit with me under the umbrella or consent to share my Rabbit-burrow, this is the sight which he is invited to behold, at the end of July: a Bembex (*B. rostrata*) arrives suddenly, I know not whence, and alights, without preliminary investigations or the least hesitation, at a spot which to my eyes differs in no respect from the rest of the sandy surface. With her fore-tarsi, which are armed with rows of stiff hairs and suggest at the same time a broom, a brush and a rake, she works at clearing her subterranean dwelling. The insect stands on its four hind-legs, holding the two at the back a little wide apart, while the front ones alternately scratch and sweep the shifting sand. The precision and quickness of the performance could not be greater if the circular movement of the tarsi were worked by a spring. The sand, shot backwards under the abdomen, passes through the arch of the hind-legs, gushes like a fluid in a continuous stream, describes its parabola and falls to the ground some seven or eight inches away. This spray of dust, kept up evenly for five or ten minutes at a time, is enough to show the dazzling rapidity of the tools employed. I know no other example of this swiftness, which nevertheless in no way detracts from the easy grace and the free movement of the insect, as it advances and retires first on this side, then on that, without discontinuing its parabolic streams of sand.

The soil excavated is of the lightest kind. As the Wasp digs, the sand near by slips back and fills the cavity. Amongst the rubbish that falls are tiny bits of wood, decayed leaf-stalks and particles of grit larger than the rest. The Bembex takes them up in her mandibles and carries them away, moving backwards as she goes; then she returns to her sweeping, but never going to any depth and making no attempt to bury herself underground. What is her object in thus labouring entirely on the surface? It would be impossible to tell from this first glance; but, after spending many days with my beloved Wasps and grouping together the scattered facts resulting from my observations, I seem to catch a glimpse of the reason for the present proceedings.

The Wasp's nest is certainly there, a few inches below the ground; in a little cell dug in the cool, firm sand lies an egg, perhaps a grub for which the mother caters from day to day, bringing it Flies, the unvarying food of the Bembex in their first state. The mother has to be able at any moment to enter the nest, as she flies up carrying in her legs the nurseling's daily portion of game, even as the bird of prey enters its eyrie with the food for its young in its talons. But, while the bird returns to a home on some inaccessible ledge of rock, with no difficulty to overcome but that of the weight and encumbrance of the captured prey, the Bembex has each time to undertake rough miner's work and open up anew a gallery blocked and closed by the mere fact that the sand gives way as the insect proceeds. In that underground dwelling, the only room with steady walls is the spacious cell where the larva lives amid the remnants of its fortnight's feast; the narrow corridor which the mother enters to reach the flat at the back or to come out and go hunting collapses each time, at least in the front part dug out of very dry sand, which repeated exits and entrances make looser still. Each time therefore that the Wasp goes in or out, she has to clear herself a passage through the debris.

Going out presents no difficulty, even should the sand retain the consistency which it might have at the start, when first disturbed: the insect's movements are free, it is safe under cover, it can take its time and use its tarsi and mandibles without undue hurry. Going in is a very different matter. The Bembex is hampered by her prey, which her legs hold clasped to her body; and the miner is thus deprived of the free use of her tools. And a still graver circumstance is this: brazen parasites, veritable bandits in ambush, crouch here and there in the neighbourhood of the burrow, spying on the mother Wasp as she makes her laborious entrance, so that they may rush in and lay their egg on the piece of game at the very moment when it is about to disappear down the corridor. If they succeed, the Wasp's nurseling, the son of the house, will perish, starved by its gluttonous fellow-boarders.

The Bembex seems aware of these dangers and makes arrangements for her entrance to be effected swiftly, without serious obstacles, in short, for the sand blocking the door to yield to a mere push of her head, aided by a brisk sweep of her front tarsi. With this object, the materials at the approaches to the home are subjected to a sort of sifting. At leisure moments, under a kindly sun, when the larva has its food and does not need her attentions, the mother rakes the ground in front of her door; she removes little bits of wood, any extra-large particles of gravel, any leaves that might get in the way and bar her passage at the dangerous moment of her return. The Bembex whom we have just seen so zealously employed was busy at this work of sifting: to facilitate the access to her home, the materials of the corridor have to be dug up, carefully sorted and rid of anything likely to obstruct the road. Who indeed can tell whether, by that nimble eagerness, that joyous activity, the insect is not expressing in its own way its maternal satisfaction, its happiness in watching over the roof of the cell to which the precious egg has been entrusted?

As the Wasp is confining herself to her duties outside the house, without trying to penetrate into the sand, everything must be in order inside and there is no hurry about anything. We should only wait in vain: the insect would tell us nothing more for the time being. Let us therefore examine the underground dwelling. If we scrape the dune lightly with the blade of a knife at the point where the Bembex was busiest, we soon discover the entrance-corridor, which, though blocked for part of the way down, is nevertheless recognizable by the distinctive appearance of the materials moved. This passage, which is as wide as one's finger and straight or winding, longer or shorter according to the nature and the accidents of the ground, measures eight to twelve inches. It leads to a single chamber, hollowed in the damp sand, whose walls are not coated with any kind of mortar likely to prevent a subsidence or to lend a polish to the rough surface. The ceiling will do, if it can hold out while the larva is growing up; it does not matter what falls in afterwards, when the larva is enclosed in its stout cocoon, a sort of safe which we shall see it building. The workmanship of the cell, therefore, is very rustic: the whole thing is reduced to a rough excavation, of no definite shape, with a low roof and space enough to contain two or three walnuts.

In this retreat lies a piece of game, one only, quite small and quite insufficient for the greedy nurseling which it is meant to feed. It is a golden-green Fly, a Green-bottle (*Lucilia Cæsar*), who lives on putrid flesh. The Fly served up as food is absolutely motionless. Is she quite dead, or only paralysed? This question will be cleared up later. For the moment, we will note the presence, on the side of the game, of a cylindrical egg, white, very slightly curved and a couple of millimetres long. It is the egg of the Bembex. As we expected from the mother's behaviour, there is nothing urgent indoors: the egg is laid and provided with a first ration apportioned to the requirements of the feeble grub which will hatch twenty-four hours hence. The Bembex had no need to reenter the underground passage for some time and was confining herself to keeping a good look-out all round, or perhaps to digging fresh burrows and continuing to lay her eggs, one by one, each in a cell to itself.

This peculiarity of beginning the provisioning with a single head of small game is not confined to the Rostrate Bembex. All the other species do the same thing. If we open the cell of any Bembex shortly after the egg is laid, we shall always find the tiny cylinder glued to the side of a Fly, who constitutes the entire provision; moreover, this initial ration is invariably small, as though the mother went in search of the tenderest mouthfuls for the feeble nurseling. Besides, another reason, the abiding freshness of the food, might easily prompt her to make this choice. We will look into that later. This first portion, always a scanty one, varies greatly in nature, according to the frequency of this or that kind of game in the neighbourhood of the nest. It is sometimes a Green-bottle, sometimes a Stomoxys, or some small Eristalis, sometimes a dainty Bee-fly clad in black velvet; but the most usual dish is a slim-bellied Sphærophoria.

This general fact, to which there is no exception, of the victualling of the egg with a single Fly, a ration infinitely too small for a larva blessed with a voracious appetite, at once puts us on the track of the most remarkable habit of the Bembex. Wasps whose larvæ live on prey heap up in each cell the number of victims necessary for the rearing of the grub; they lay the egg on one of the bodies and close the dwelling, which they do not enter again. From that moment, the larva hatches and develops alone, having before it from the very beginning the whole stock of provisions which it is to consume. The Bembex form an exception to this rule. The cell is first stocked with a single head of game, always small in size, and the egg is laid on it. When that is done, the mother leaves the burrow, which closes of itself; besides, before going away, the insect is careful to rake over the outside, so as to smooth the surface and hide the entrance from any eye but her own.

Two or three days elapse; the egg hatches and the little larva eats up the choice ration served to it. Meanwhile the mother remains in the neighbourhood and you see her sometimes feeding herself by sipping the sugary exudations of the field eringo, sometimes settling happily on the burning sand, no doubt watching the outside of the house. Every now and again she sifts the sand at the entrance; then she flies away and disappears, perhaps to dig other cells elsewhere and to stock them in the same way. But, however long she may stay away, she never forgets the young larva so scantily provided for; the instinct of a mother tells her the hour when the grub has finished its food and is calling for fresh nourishment. She therefore returns to the nest, of which she is wonderfully capable of discovering the invisible entrance; she goes down into the earth, this time carrying a bulkier piece of game. After depositing her prey, she again leaves the house and waits outside till the moment arrives to serve a third course. This moment is not slow in coming, for the larva devours its food with a lusty appetite. Again the mother appears with fresh provisions.

During nearly a fortnight, while the larva is growing up, the meals thus follow in succession, one by one, as needed, and coming closer together as the nurseling waxes bigger. Towards the end of the fortnight, it takes all the mother's activity to satisfy the appetite of the glutton, who crawls heavily along, with his great lumbering belly, amid the scorned leavings: rejected wings and legs and horny abdominal segments. You see her at every moment returning with a recent capture, at every moment setting out again upon the chase. In short, the Bembex brings up her family from day to day, without storing up provisions in advance, just as the bird does, which feeds its nestlings from hand to mouth. Of the many proofs that are evidence of this method of upbringing, a very singular method for a Wasp who feeds her off-spring on prey, I have already mentioned the presence of the egg in a cell containing no provisions but one small Fly, never more. And here is another one, which can be verified at any time.

Let us look into the burrow of a Wasp who stocks her grubs' pro-visions in advance: if we select the moment when the insect is going in with its prey, we shall find in the cell a certain number of victims, the commencement of a larder, but never at that time a grub, nor even an egg, for this is not laid until the provisions are quite complete. When the egg is laid, the cell is closed and the mother does not return to it. It is therefore only in burrows where the mother's visits are no longer necessary that we can find larvæ side by side with larger or smaller stacks of food. On the other hand, let us inspect the home of a Bembex at the moment when she is entering with the fruits of her hunting. We are certain of finding in the cell a larva, big or little as the case may be, among remnants of provisions already consumed. The portion which the mother is now bringing is therefore intended to prolong a meal which has already lasted several days and which is to continue for some time further with the produce of future hunting-expeditions. Should we be fortunate enough to make this search towards the end of the larva's infancy—an advantage which I have enjoyed as often as I wished to—we shall find, on a copious heap of remnants, a large and portly grub, to which the mother is still bringing fresh victuals. The Bembex does not cease her catering and does not leave the cell for good until the larva, distended by a purply paste, refuses its food and lies down, stuffed to repletion, on the jumble of legs and wings of the game which it has devoured.

Each time that the mother enters the burrow on returning from the chase, she brings but a single Fly. If it were possible, by counting the remnants contained in a cell whose occupant is full-grown, to tell the number of victims supplied to the larva, we should know how often at the least the Wasp visited her burrow after laying the egg. Unfor-tunately, these broken victuals, chewed and chewed again at moments of scarcity, are for the most part unrecognizable. But, if we open a cell with a less forward nurseling, the provisions lend themselves to ex-amination, some of them being still whole or nearly whole, while others, more numerous, are represented by fragments in a state of preservation that enables them to be identified. Incomplete though it be, the list obtained under these conditions is surprising and shows what activity the Wasp must display to satisfy the needs of such a table. I will set forth one of the bills of fare which I have observed.

At the end of September, around the larva of a Jules' Bembex (*Bembex Julii*), which has reached almost a third of the size which it will finally attain, I find the following heads of game: six *Echinomyia rubescens* (two whole and four in pieces); four *Syrphus corollæ* (two complete, the other two broken up); three *Gonia atra* (all three untouched: one of them had that moment been brought along by the mother, which led to my discovering the burrow); two *Pollenia rufescens* (one untouched, the other partly eaten); one Bombylius (reduced to pulp); two *Echinomyia intermedia* (in bits); and two *Pollenia floralis* (likewise in bits): twenty pieces in all. This certainly makes a both plentiful and varied bill of fare; but, as the larva was only a third of its ultimate size, the complete menu might easily number as many as sixty items.

It is not at all difficult to verify this sumptuous figure: I will myself take the place of the Bembex in her maternal functions and supply the larva with food till it is ready to burst. I move the cell into a little cardboard box which I furnish with a layer of sand. I place the larva on this bed, with all due consideration for its delicate skin. Around it, without omitting a single fragment, I arrange the provisions with which it was supplied. Then I go home, still holding the box in my hand, to avoid any shaking which might turn the house upside down and endanger my charge during a walk of several miles. Any one who had met me on the dusty Nîmes Road, dropping with fatigue and religiously carrying in my hand, as the sole fruit of my laborious trip, an ugly grub battening on a heap of Flies, would certainly have smiled at my simplicity.

The journey was effected without damage: when I reached home, the larva was placidly eating its Flies as though nothing had happened. On the third day of captivity, the provisions taken from the burrow were finished; the grub was rummaging with its pointed mouth among the heap of remains without finding anything to suit it; the dry particles taken hold of, all horny, juiceless bits, were rejected with disgust. The moment has come for me to continue the food supply. The first Flies within reach shall form my prisoner's diet. I kill them by pressing them in my fingers, but without crushing them. The first ration consists of three *Eristalis tenax* and one *Sarcophaga*. This is all gobbled up in twenty-four hours. Next day, I provide two Eristales, or Drone-flies, and four House-flies. It was enough for the day, but left nothing over.

I went on like this for eight days, giving the grub a larger portion every morning. On the ninth day, the larva refused all food and began to spin its cocoon. The full record of this eight days' feast amounts to sixty-two pieces, composed mainly of Drone-flies and House-flies, which, added to the twenty items found whole or in pieces in the cell brings up the total to eighty-two.

It is possible that I did not rear my larva with the wholesome frugality and the wise economy which the mother would have shown; there was perhaps some waste in the daily provisions served all at one time and left entirely to the grub's discretion. In some respects I feel inclined to believe that things do not happen just like that in the maternal cell, for my notes contain such details as the following. In the alluvial sands of the Durance, I discover a burrow which the Wasp (*Bembex oculata*) has just entered with a *Sarcophaga agricola*. Inside, I find a larva, numerous fragments and a few whole Flies, namely, four *Sphærophoria scripta*, one *Onesia viarum* and two *Sarcophaga agricola,* including the one which the Bembex has just brought along before my eyes. Now it is worthy of remark that half of this game, namely, the Sphærophoriæ, is right at the end of the cell, under the larva's very teeth, whereas the other half is still in the passage, on the threshold of the cell, and therefore beyond the reach of the grub, which is unable to change its position. It seems to me then that, when game is plentiful, the mother lays her captures on the threshold of the cell for the time and forms a reserve on which she draws as and when necessary, especially on rainy days when all labour is at a standstill.

WASPS OF THE BOIS DES ISSARTS

« 14 »

During the long years Fabre taught at the Lycée of Avignon, his steps often took him downstream along the right bank of the Rhone to the Bois des Issarts. This sun-scorched area was a favorite of the wasps. The Scolia wasps, Fabre mentions, are hunters of beetle larvae; the Cetoniae are day-flying beetles usually found around flowers where they feed on pollen. Favier was the aged ex-soldier who, for a time, worked as Fabre's gardener and helper at Sérignan. MORE HUNTING WASPS *is the source of this selection.*

I open my old note book; and I see myself once more, on the 6th of August, 1857, in the Bois des Issarts, that famous copse near Avignon which I have celebrated in my essay on the Bembex-wasps. Once again, my head crammed with entomological projects, I am at the beginning of my holidays which, for two months, will allow me to indulge in the insect's company.

A fig for Mariotte's flask and Toricelli's tube! This is the thrice-blest period when I cease to be a schoolmaster and become a schoolboy, the schoolboy in love with animals. Like a madder-cutter off for his day's work, I set out carrying over my shoulder a solid digging-implement, the local *luchet,* and on my back my game-bag with boxes, bottles, trowel, glass tubes, tweezers, lenses and other *impedimenta.* A large umbrella saves me from sunstroke. It is the most scorching hour of the hottest day in the year. Exhausted by the heat, the Cicadæ are silent. The bronze-eyed Gad-flies seek a refuge from the pitiless sun under the roof of my silken shelter; other large Flies, the sobre-hued Pangoniæ, dash themselves recklessly against my face.

The spot at which I have installed myself is a sandy clearing which I had recognized the year before as a site beloved of the Scoliæ. Here and there are scattered thickets of holm-oak, whose dense undergrowth shelters a bed of dead leaves and a thin layer of mould. My memory has served me well. Here, sure enough, as the heat grows a little less, appear, coming I know not from whence, some Two-banded Scoliæ. The number increases; and it is not long before I see very nearly a dozen of them about me, close enough for observation. By their smaller size and more buoyant flight, they are easily known for males. Almost grazing the ground, they fly softly, going to and fro, passing and repassing in every direction. From time to time one of them alights on the ground, feels the sand with his antennæ and seems to be enquiring into what is happening in the depths of the soil; then he resumes his flight, alternately coming and going.

What are they waiting for? What are they seeking in these evolutions of theirs, which are repeated a hundred times over? Food? No, for close beside them stand several eryngo-stems, whose sturdy clusters are the Wasps' usual resource at this season of parched vegetation; and not one of them settles upon the flowers, not one of them seems to care about their sugary exudations. Their attention is engrossed elsewhere. It is the ground, it is the stretch of sand which they are so assiduously exploring; what they are waiting for is the arrival of some female, who, bursting the cocoon, may appear from one moment to the next, issuing all dusty from the ground. She will not be given time to brush herself or to wash her eyes: three or four or more of them will be there at once, eager to dispute her possession. I am too familiar with the amorous contests of the Hymenopteron clan to allow myself to be mistaken. It is the rule for the males, who are the earlier of the two, to keep a close guard around the natal spot and watch for the emergence of the females, whom they pester with their pursuit the moment they reach the light of day. This is the motive of the interminable ballet of my Scoliæ. Let us have patience: perhaps we shall witness the nuptials.

The hours go by; the Pangoniæ and the Gad-flies desert my umbrella; the Scoliæ grow weary and gradually disappear. It is finished. I shall see nothing more to-day. I repeat my laborious expedition to the Bois des Issarts over and over again; and each time I see the males as assiduous as ever in skimming over the ground. My perseverance deserved to succeed. It did, though the success was very incomplete. Let me describe it, such as it was; the future will fill up the gaps.

A female issues from the soil before my eyes. She flies away, followed by several males. With the *luchet* I dig at the point of emergence; and, as the excavation progresses, I sift between my fingers the rubbish of sand mixed with mould. In the sweat of my brow, as I may justly say, I must have removed nearly a cubic yard of material, when at last I make a find. This is a recently ruptured cocoon, to the side of which adheres an empty skin, the last remnant of the game on which the larva fed that wrought the said cocoon. Considering the good condition of its silken fabric, this cocoon may have belonged to the Scolia who has just quitted her underground dwelling before my eyes. As for the skin accompanying it, this has been so much spoilt by the moisture of the soil and by the grassy roots that I cannot determine its origin exactly. The cranium, however, which is better-preserved, the mandibles and certain details of the general configuration lead me to suspect the larva of a Lamellicorn.

It is getting late. This is enough for to-day. I am worn out, but amply repaid for my exertions by a broken cocoon and the puzzling skin of a wretched grub. Young people who make a hobby of natural history, would you like to discover whether the sacred fire flows in your veins? Imagine yourselves returning from such an expedition. You are carrying on your shoulder the peasant's heavy spade; your loins are stiff with the laborious digging which you have just finished in a crouching position; the heat of an August afternoon has set your brain simmering; your eyelids are tired by the itch of an inflammation resulting from the overpowering light in which you have been working; you have a devouring thirst; and before you lies the dusty prospect of the miles that divide you from your well-earned rest. Yet something stings within you; forgetful of your present woes you are absolutely glad of your excursion. Why? Because you have in your possession a shred of rotten skin. If this is so, my young friends, you may go ahead, for you will do something, though I warn you that this does not mean, by a long way, that you will get on in the world.

I examined this shred of skin with all the care that it deserved. My first suspicions were confirmed: a Lamellicorn, a Scarabæid in the larval state, is the first food of the Wasp whose cocoon I have just unearthed. But which of the Scarabæidæ? And does this cocoon, my precious booty, really belong to the Scolia? The problem is beginning to take shape. To attempt its solution we must go back to the Bois des Issarts.

I did go back and so often that my patience ended by being exhausted before the problem of the Scoliæ had received a satisfactory solution. The difficulties are great indeed, under the conditions. Where am I to dig in the indefinite stretch of sandy soil to light upon a spot frequented by the Scoliæ? The *luchet* is driven into the ground at random; and almost invariably I find none of what I am seeking. To be sure, the males, flying level with the ground, give me a hint, at the outset, with their certainty of instinct, as to the spots where the females ought to be; but their hints are very vague, because they go so far in every direction. If I wished to examine the soil which a single male explores in his flight, with its constantly changing course, I should have to turn over, to the depth of perhaps a yard, at least four poles of earth. This is too much for my strength and the time at my disposal. Then, as the season advances, the males disappear, whereupon I am suddenly deprived of their hints. To know more or less where I should thrust my *luchet,* I have only one resource left, which is to watch for the females emerging from the ground or else entering it. With a great expenditure of time and patience I have at last had this windfall, very rarely, I admit.

The Scoliæ do not dig a burrow which can be compared with that of the other Hunting Wasps; they have no fixed residence, with an unimpeded gallery opening on the outer world and giving access to the cells, the abodes of the larvæ. They have no entrance- and exit-doors, no corridor built in advance. If they have to make their way underground, any point not hitherto turned over serves their purpose, provided that it be not too hard for their digging-tools, which, for that matter, are very powerful; if they have to come out, the point of exit is no less indifferent. The Scolia does not bore the soil through which she passes: she excavates and ploughs it with her legs and forehead; and the stuff shifted remains where it lies, behind her, forthwith blocking the passage which she has followed. When she is about to emerge into the outer world, her advent is heralded by the fresh soil which heaps itself into a mound as though heaved up by the snout of some tiny Mole. The insect sallies forth; and the mound collapses, completely filling up the exit-hole. If the Wasp is entering the ground, the digging-operations, undertaken at an arbitrary point, quickly yield a cavity in which the Scolia disappears, separated from the surface by the whole track of shifted material.

I can easily trace her passage through the thickness of the soil by certain long, winding cylinders, formed of loose materials in the midst of compact and stable earth. These cylinders are numerous; they sometimes run to a depth of twenty inches; they extend in all directions, fairly often crossing one another. Not one of them ever exhibits so much as a suspicion of an open gallery. They are obviously not permanent ways of communication with the outer world, but hunting-trails which the insect has followed once, without going back to them. What was the Wasp seeking when she riddled the soil with these tunnels which are now full of running sands? No doubt the food for her family, the larva of which I possess the empty skin, now an unrecognizable shred.

I begin to see a little light: the Scoliæ are underground workers. I already expected as much, having before now captured Scoliæ soiled with little earthy encrustations on the joints of the legs. The Wasp, who is so careful to keep clean, taking advantage of the least leisure to brush and polish herself, could never display such blemishes unless she were a devoted earth-worker. I used to suspect their trade; now I know it. They live underground, where they burrow in search of Lamellicorn-grubs, just as the Mole burrows in search of the White Worm. It is even possible that, after receiving the embraces of the males, they but very rarely return to the surface, absorbed as they are by their maternal duties; and this, no doubt, is why my patience becomes exhausted in watching for their entrance and their emergence.

It is in the subsoil that they establish themselves and travel to and fro; with the help of their powerful mandibles, their hard cranium, their strong, prickly legs, they easily make themselves paths in the loose earth. They are living ploughshares. By the end of August, therefore, the female population is for the most part underground, busily occupied in egg-laying and provisioning. Everything seems to tell me that I should watch in vain for the appearance of a few females in the broad daylight; I must resign myself to excavating at random.

The result was hardly commensurate with the labour which I expended on digging. I found a few cocoons, nearly all broken, like the one which I already possessed, and, like it, bearing on their side the tattered skin of a larva of the same Scarabæid. Two of these cocoons which are still intact contained a dead adult Wasp. This was actually the Two-banded Scolia, a precious discovery which changed my suspicions into a certainty.

I also unearthed some cocoons, slightly different in appearance, containing an adult inmate, likewise dead, in whom I recognized the Interrupted Scolia. The remnants of the provisions again consisted of the empty skin of a larva, also a Lamellicorn, but not the same as the one hunted by the first Scolia. And this was all. Now here, now there, I shifted a few cubic yards of soil, without managing to find fresh provisions with the egg or the young larva. And yet it was the right season, the egg-laying season, for the males, numerous at the outset, had grown rarer day by day until they disappeared entirely. My lack of success was due to the uncertainty of my excavations, in which I had nothing to guide me over the indefinite area covered.

If I could at least identify the Scarabæidæ whose larvæ form the prey of the two Scoliæ, the problem would be half solved. Let us try. I collect all that the *luchet* has turned up: larvæ, nymphs and adult Beetles. My booty comprises two species of Lamellicorns: *Anoxia villosa* and *Euchlora Julii,* both of whom I find in the perfect state, usually dead, but sometimes alive. I obtain a few of their nymphs, a great piece of luck, for the larval skin which accompanies them will serve me as a standard of comparison. I come upon plenty of larvæ, of all ages. When I compare them with the cast garment abandoned by the nymphs, I recognize some as belonging to the Anoxia and the rest to the Euchlora.

With these data, I perceive with absolute certainty that the empty skin adhering to the cocoon of the Interrupted Scolia belongs to the Anoxia. As for the Euchlora, she is not involved in the problem: the larva hunted by the Two-banded Scolia does not belong to her any more than it belongs to the Anoxia. Then with which Scarabæid does the empty skin which is still unknown to me correspond? The Lamellicorn whom I am seeking must exist in the ground which I have been exploring, because the Two-banded Scolia has established herself there. Later—oh, very long afterwards!—I recognized where my search was at fault. In order not to find a network of roots beneath my *luchet* and to render the work of excavation lighter, I was digging the bare places, at some distance from the thickets of holm-oak; and it was just in those thickets, which are rich in vegetable mould, that I should have sought. There, near the old stumps, in the soil consisting of dead leaves and rotting wood, I should certainly have come upon the larva so greatly desired, as will be proved by what I have still to say.

Here ends what my earlier investigations taught me. There is reason to believe that the Bois des Issarts would never have furnished me with the precise data, in the form in which I wanted them. The remoteness of the spot, the fatigue of the expeditions, which the heat rendered intensely exhausting, the impossibility of knowing which points to attack would undoubtedly have discouraged me before the problem had advanced a step farther. Studies such as these call for home leisure and application, for residence in a country village. You are then familiar with every spot in your own grounds and the surrounding country and you can go to work with certainty.

Twenty-three years have passed; and here I am at Sérignan, where I have become a peasant, working by turns on my writing-pad and my cabbage-patch. On the 14th of August, 1880, Favier clears away a heap of mould consisting of vegetable refuse and of leaves stacked in a corner against the wall of the paddock. This clearance is considered necessary because Bull, when the lovers' moon arrives, uses this hillock to climb to the top of the wall and thence to repair to the canine wedding the news of which is brought to him by the effluvia borne upon the air. His pilgrimage fulfilled, he returns, with a discomfited look and a slit ear, but always ready, once he has had his feed, to repeat the escapade. To put an end to this licentious behaviour, which has cost him so many gaping wounds, we decided to remove the heap of soil which serves him as a ladder of escape.

Favier calls me while in the midst of his labours with the spade and barrow:

"Here's a find, sir, a great find! Come and look."

I hasten to the spot. The find is a magnificent one indeed and of a nature to fill me with delight, awakening all my old recollections of the Bois des Issarts. Any number of females of the Two-banded Scolia, disturbed at their work, are emerging here and there from the depth of the soil. The cocoons also are plentiful, each lying next to the skin of the victim on which the larva has fed. They are all open but still fresh: they date from the present generation; the Scoliæ whom I unearth have quitted them not long since. I learnt later, in fact, that the hatching took place in the course of July.

In the same heap of mould is a swarming colony of Scarabæidæ in the form of larvæ, nymphs and adult insects. It includes the largest of our Beetles, the common Rhinoceros Beetle, or *Oryctes nasicornis*. I find some who have been recently liberated, whose wing-cases, of a glossy brown, now see the sunlight for the first time; I find others enclosed in their earthen shell, almost as big as a Turkey's egg. More frequent is her powerful larva, with its heavy paunch, bent into a hook. I note the presence of a second bearer of the nasal horn, *Oryctes Silenus,* who is much smaller than her kinswoman, and of *Pentodon punctatus,* a Scarabæid who ravages my lettuces.

But the predominant population consists of Cetoniæ, or Rosechafers, most of them enclosed in their egg-shaped shells, with earthen walls encrusted with dung. There are three different species: *C. aurata, C. morio* and *C. floricola.* Most of them belong to the first species. Their larvæ, which are easily recognized by their singular talent for walking on their backs with their legs in the air, are numbered by the hundred. Every age is represented, from the new-born grub to the podgy larva on the point of building its shell.

This time the problem of the victuals is solved. When I compare the larval slough sticking to the Scolia's cocoons with the Cetonia-larvæ or, better, with the skin cast by these larvæ, under cover of the cocoon, at the moment of the nymphal transformation, I establish an absolute identity. The Two-banded Scolia rations each of her eggs with a Cetonia-grub. Behold the riddle which my irksome searches in the Bois des Issarts had not enabled me to solve. To-day, at my threshold, the difficult problem becomes child's play. I can investigate the question easily to the fullest possible extent; I need not put myself out at all; at any hour of the day, at any period that seems favourable, I have the requisite elements before my eyes. Ah, dear village, so poor, so countrified, how happily inspired was I when I came to ask of you a hermit's retreat, where I could live in the company of my beloved insects and, in so doing, set down not too unworthily a few chapters of their wonderful history!

THE LAST MOULT OF THE LOCUST

« 15 »

The secret of the grasshopper's moult, so vividly described here by Fabre, lies in the fact that its limbs and antennae and body are drawn from the outer skeleton which has encased them while in a soft and pliable condition. The saw-like tibia, while soft, slips from its case without difficulty. The chitin of the new skeleton-and-skin-in-one hardens after the insect is completely moulted. This selection is from THE LIFE OF THE GRASSHOPPER.

I have just beheld a stirring sight: the last moult of a Locust, the extraction of the adult from his larval wrapper. It is magnificent. The object of my enthusiasm is the Grey Locust, the giant among our Acridians, who is common on the vines at vintage-time, in September. On account of his size—he is as long as my finger—he is a better subject for observation than any other of his tribe.

The fat, ungraceful larva, a rough draft of the perfect insect, is usually pale-green; but some also are bluish-green, dirty-yellow, red-brown or even ashen-grey, like the grey of the adult. The corselet is strongly keeled and notched, with a sprinkling of fine white worm-holes. The hind-legs, powerful as those of mature age, have a great haunch striped with red and a long shank shaped like a two-edged saw.

The wing-cases, which in a few days will project well beyond the tip of the abdomen, are in their present state two skimpy, triangular pinions, touching back to back along their upper edges and continuing the keel of the corselet. Their free ends stand up like a pointed gable. These two coat-tails, of which the material seems to have been clipped short with ridiculous meanness, just cover the creature's nakedness at the small of the back. They shelter two lean strips, the germs of the wings, which are even more exiguous. In brief, the sumptuous, slender sails of the near future are at present sheer rags, of such meagre dimensions as to be grotesque. What will come out of these miserable envelopes? A marvel of stately elegance.

Let us observe the proceedings in detail. Feeling itself ripe for transformation, the creature clutches the trelliswork of the cage with its hinder and intermediary legs. The fore-legs are folded and crossed over the breast and are not employed in supporting the insect, which hangs in a reversed position, back downwards. The triangular pinions, the sheaths of the wing-cases, open their peaked roof and separate sideways; the two narrow strips, the germs of the wings, stand in the centre of the uncovered space and diverge slightly. The position for the moult has now been taken with the necessary stability.

The first thing to be done is to burst the old tunic. Behind the corselet, under the pointed roof of the prothorax, pulsations are produced by alternate inflation and deflation. A similar operation is performed in front of the neck and probably also under the entire covering of the shell that is to be split. The delicacy of the membranes at the joints enables us to perceive what is going on at these bare points, but the harness of the corselet hides it from us in the central portion.

It is there that the insect's reserves of blood flow in waves. The rising tide expresses itself in blows of an hydraulic battering-ram. Distended by this rush of humours, by this injection wherein the organism concentrates its energies, the skin at last splits along a line of least resistance prepared by life's subtle previsions. The fissure yawns all along the corselet, opening precisely over the keel, as though the two symmetrical halves had been soldered. Unbreakable any elsewhere, the wrapper yields at this median point which is kept weaker than the rest. The split is continued some little way back and runs between the fastenings of the wings; it goes up the head as far as the base of the antennæ, where it sends a short ramification to the right and left.

Through this break the back is seen, quite soft, pale, hardly tinged with grey. Slowly it swells into a larger and larger hunch. At last it is wholly released. The head follows, extracted from its mask, which remains in its place, intact in the smallest particular, but looking strange with its great glassy eyes that do not see. The sheaths of the antennæ, with not a wrinkle, with nothing out of order and with their normal position unchanged, hang over this dead face, which is now translucent.

Therefore, in emerging from their narrow sheaths, which enclosed them with such absolute precision, the antennary threads encountered no resistance capable of turning their scabbards inside out, or disturbing their shape, or even wrinkling them. Without injuring the twisted containers, the contents, equal in size and themselves twisted, have managed to slip out as easily as a smooth, straight object would do, if sliding in a loose sheath. The extraction-mechanism will be still more remarkable in the case of the hind-legs.

Meanwhile it is the turn of the fore-legs and then of the intermediary legs to shed armlets and gauntlets, always without the least rent, however small, without a crease of rumpled material, without a trace of any change in the natural position. The insect is now fixed to the top of the cage only by the claws of the long hind-legs. It hangs perpendicularly, head downwards, swinging like a pendulum, if I touch the wire-gauze. Four tiny hooks are what it hangs by. If they gave way, if they became unfastened, the insect would be lost, for it is incapable of unfurling its enormous wings anywhere except in space. But they will hold: life, before withdrawing from them, left them stiff and solid, so as to be able firmly to support the struggles that are to follow.

The wing-cases and wings now emerge. These are four narrow strips, faintly grooved and looking like bits of paper ribbon. At this stage, they are scarcely a quarter of their final length. So limp are they that they bend under their own weight and sprawl along the insect's sides in the opposite direction to the normal. Their free end, which should be turned backwards, now points towards the head of the Locust, who is hanging upside down. Imagine four blades of thick grass, bent and battered by a rainstorm, and you will have a fair picture of the pitiable bunch formed by the future organs of flight.

It must be no light task to bring things to the requisite stage of perfection. The deeper-seated changes are already well-started, solidifying liquid mucilages, bringing order out of chaos; but so far nothing outside betrays what is happening in that mysterious laboratory where everything seems lifeless.

Meanwhile, the hind-legs become released. The great thighs appear in view, tinted on their inner surface with a pale pink, which will soon turn into a streak of bright crimson. The emergence is easy, the bulky haunch clearing the way for the tapering knuckle.

It is different with the shank. This, in the adult insect, bristles throughout its length with a double row of hard, pointed spikes. Moreover, the lower extremity ends in four large spurs. It is a genuine saw, but with two parallel sets of teeth and so powerful that, if we dismiss the size from our minds, it might be compared with the rough saw wielded by a quarryman.

The larva's shin is similarly constructed, so that the object to be extracted is contained in a sheath as awkwardly shaped as itself. Each spur is enclosed in a similar spur, each tooth fits into the hollow of a similar tooth; and the moulding is so exact that we should obtain no more intimate contact if, instead of the envelope waiting to be shed, we coated the limb with a layer of varnish distributed uniformly with a fine brush.

Nevertheless the sawlike tibia slips out of its long, narrow case without catching in it at any point whatever. If I had not seen this happen over and over again, I could never have believed it: the discarded legging is quite intact all the way down. Neither the terminal spurs nor the two rows of spikes have caught in the delicate mould. The saw has respected the dainty scabbard which a puff of my breath is enough to tear; the formidable rake has slipped through without leaving the least scratch behind it.

I was far from expecting such a result as this. Because of the spiked armour, I imagined that the leg would strip in scales which came loose of themselves or yielded to rubbing, like dead cuticle. How greatly did the reality exceed my expectations!

From the spurs and spikes of the infinitely thin matrix there emerge spurs and spikes that make the leg capable of cutting soft wood. This is done without violence or the least inconvenience; and the discarded garment remains where it is, hanging by the claws to the top of the cage, uncreased and untorn. The magnifying-glass shows not a trace of rough usage. As the thing was before the excoriation, so it remains afterwards. The legging of dead skin continues, down to the pettiest details, an exact replica of the live leg.

If any one suggested that we should extract a saw from some sort of goldbeater's-skin sheath which had been exactly moulded on the steel and that we should perform the operation without producing the least tear, we should burst out laughing: the thing is so flagrantly impossible. Life makes light of these impossibilities; it has methods of realizing the absurd, in case of need. And the Locust's leg tells us so.

A RECOLLECTION OF CHILDHOOD

« 16 »

Another engaging picture of Fabre as a small boy is presented by this recollection of childhood taken from THE LIFE OF THE FLY. *At the time of which he writes, he had returned from his grandparents' farm and was just beginning his schooling at Saint-Leons. The bird that he encountered has several common names in English in addition to the Saxicola. They include the Wheatear, the Stone-Chat, the Whin-Chat and the Fallow-Finch. The Greenland Wheatear, Oenathe oenathe leucorhoa, breeding in Greenland and Arctic North America and migrating to Europe, sometimes strays southward into the United States. It is an active, ground-feeding bird, a little smaller than a bluebird with a white rump-patch as its outstanding field mark.*

Almost as much as insects and birds—the former so dear to the child, who loves to rear his Cockchafers and Rose-beetles on a bed of hawthorn in a box pierced with holes; the latter an irresistible temptation, with their nests and their eggs and their little ones opening tiny yellow beaks—the mushroom early won my heart with its varied shapes and colours. I can still see myself as an innocent small boy sporting my first braces and beginning to know my way through the cabalistic mazes of my reading-book, I see myself in ecstasy before the first bird's-nest found and the first mushroom gathered. Let us relate these grave events. Old age loves to meditate the past.

O happy days when curiosity awakens and frees us from the limbo of unconsciousness, your distant memory makes me live my best years over again. Disturbed at its siesta by some wayfarer, the Partridge's young brood hastily disperses. Each pretty little ball of down scurries off and disappears in the brushwood; but, when quiet is restored, at the first summoning note they all return under the mother's wing. Even so, recalled by memory, do my recollections of childhood return, those other fledglings which have lost so many of their feathers on the brambles of life. Some, which have hardly come out of the bushes, have aching heads and tottering steps; some are missing, stifled in some dark corner of the thicket; some remain in their full freshness. Now of those which have escaped the clutches of time the liveliest are the first-born. For them the soft wax of childish memory has been converted into enduring bronze.

On that day, wealthy and leisured, with an apple for my lunch and all my time to myself, I decided to visit the brow of the neighbouring hill, hitherto looked upon as the boundary of the world. Right at the top is a row of trees which, turning their backs to the wind, bend and toss about as though to uproot themselves and take to flight. How often, from the little window in my home, have I not seen them bowing their heads in stormy weather; how often have I not watched them writhing like madmen amid the snow-dust which the north wind's besom raises and smooths along the hill-side! What are they doing up there, those desolate trees? I am interested in their supple backs, to-day still and upright against the blue of the sky, to-morrow shaken when the clouds pass overhead. I am gladdened by their calmness; I am distressed by their terrified gestures. They are my friends. I have them before my eyes at every hour of the day. In the morning, the sun rises behind their transparent screen and ascends in its glory. Where does it come from? I am going to climb up there and perhaps I shall find out.

I mount the slope. It is a lean grass-sward close-cropped by the sheep. It has no bushes, fertile in rents and tears, for which I should have to answer on returning home, nor any rocks, the scaling of which involves like dangers; nothing but large, flat stones, scattered here and there. I have only to go straight on, over smooth ground. But the sward is as steep as a sloping roof. It is long, ever so long; and my legs are very short. From time to time, I look up. My friends, the trees on the hill-top, seem to be no nearer. Cheerly, sonnie! Scramble away!

What is this at my feet? A lovely bird has flown from its hiding-place under the eaves of a big stone. Bless us, here's a nest made of hair and fine straw! It's the first I have ever found, the first of the joys which the birds are to bring me. And in this nest are six eggs, laid prettily side by side; and those eggs are a magnificent blue, as though steeped in a dye of celestial azure. Overpowered with happiness, I lie down on the grass and stare.

Meanwhile, the mother, with a little clap of her gullet—"Tack! Tack!"

—flies anxiously from stone to stone, not far from the intruder. My age knows no pity, is still too barbarous to understand maternal anguish. A plan is running in my head, a plan worthy of a little beast of prey. I will come back in a fortnight and collect the nestlings before they can fly away. In the meantime, I will just take one of those pretty blue eggs, only one, as a trophy. Lest it should be crushed, I place the fragile thing on a little moss in the scoop of my hand. Let him cast a stone at me that has not, in his childhood, known the rapture of finding his first nest.

My delicate burden, which would be ruined by a false step, makes me give up the remainder of the climb. Some other day I shall see the trees on the hill-top over which the sun rises. I go down the slope again. At the bottom, I meet the parish-priest's curate reading his breviary as he takes his walk. He sees me coming solemnly along, like a relic-bearer; he catches sight of my hand hiding something behind my back:

"What have you there, my boy?" he asks.

All abashed, I open my hand and show my blue egg on its bed of moss.

"Ah!" says his reverence. "A Saxicola's egg! Where did you get it?"

"Up there, father, under a stone."

Question follows question; and my peccadillo stands confessed. By chance I found a nest which I was not looking for. There were six eggs in it. I took one of them—here it is—and I am waiting for the rest to hatch. I shall go back for the others when the young birds have their quill-feathers.

"You mustn't do that, my little friend," replies the priest. "You mustn't rob the mother of her brood; you must respect the innocent little ones; you must let God's birds grow up and fly from the nest. They are the joy of the fields and they clear the earth of its vermin. Be a good boy, now, and don't touch the nest."

I promise and the curate continues his walk. I come home with two good seeds cast on the fallows of my childish brain. An authoritative word has taught me that spoiling birds'-nests is a bad action. I did not quite understand how the bird comes to our aid by destroying vermin, the scourge of the crops; but I felt, at the bottom of my heart, that it is wrong to afflict the mothers.

"Saxicola," the priest had said, on seeing my find.

"Hullo!" said I to myself. "Animals have names, just like ourselves. Who named them? What are all my different acquaintances in the woods and meadows called? What does Saxicola mean?"

Years passed and Latin taught me that Saxicola means an inhabitant of the rocks. My bird, in fact, was flying from one rocky point to the other while I lay in ecstasy before its eggs; its house, its nest, had the rim of a large stone for a roof. Further knowledge gleaned from books taught me that the lover of stony hill-sides is also called the *Motteux,* or Clodhopper, because, in the ploughing-season, she flies from clod to clod, inspecting the furrows rich in unearthed grubworms. Lastly, I came upon the Provençal expression *Cul-blanc,* which is also a picturesque term, suggesting the patch on the bird's rump which spreads out like a white Butterfly flitting over the fields.

Thus did the vocabulary come into being that would one day allow me to greet by their real names the thousand actors on the stage of the fields, the thousand little flowers that smile at us from the wayside. The word which the curate had spoken without attaching the least importance to it revealed a world to me, the world of plants and animals designated by their real names. To the future must belong the task of deciphering some pages of the immense lexicon; for to-day I will content myself with remembering the Saxicola, or Stone-chat.

THE SONG OF THE CICADA

« 17 »

As anyone knows who has listened to the shrill din of the summer cicadas, these insects are lovers of heat. The hotter the day, the more vehement their music becomes. Thus, Fabre's sun-baked Provence village was to the liking of the insect musicians. His amusing efforts to outdo the din of the sap-drinking cicadas, related here, is taken from THE LIFE OF THE GRASSHOPPER.

My neighbours the peasants say that, at harvest-time, the Cicada sings, "*Sego, sego, sego!* Reap, reap, reap!" to encourage them to work. Whether harvesters of wheat or harvesters of thought, we follow the same occupation, one for the bread of the stomach, the other for the bread of the mind. I can understand their explanation, therefore; and I accept it as an instance of charming simplicity.

Science asks for something better; but she finds in the insect a world that is closed to us. There is no possibility of divining or even suspecting the impression produced by the clash of the cymbals upon those who inspire it. All that I can say is that their impassive exterior seems to denote complete indifference. Let us not insist too much: the private feelings of animals are an unfathomable mystery.

Another reason for doubt is this: those who are sensitive to music always have delicate hearing; and this hearing, a watchful sentinel, should give warning of any danger at the least sound. The birds, those skilled songsters, have an exquisitely fine sense of hearing. Should a leaf stir in the branches, should two wayfarers exchange a word, they will be suddenly silent, anxious, on their guard. How far the Cicada is from such sensibility!

He has very clear sight. His large faceted eyes inform him of what happens on the right and what happens on the left; his three stemmata, like little ruby telescopes, explore the expanse above his head. The moment he sees us coming, he is silent and flies away. But place yourself behind the branch on which he is singing, arrange so that you are not within reach of the five visual organs; and then talk, whistle, clap your hands, knock two stones together. For much less than this, a bird, though it would not see you, would interrupt its singing and fly away terrified. The imperturbable Cicada goes on rattling as though nothing were afoot.

Of my experiments in this matter, I will mention only one, the most memorable. I borrow the municipal artillery, that is to say, the mortars which are made to thunder forth on the feast of the patron-saint. The gunner is delighted to load them for the benefit of the Cicadæ and to come and fire them off at my place. There are two of them, crammed as though for the most solemn rejoicings. No politician making the circuit of his constituency in search of reelection was ever honoured with so much powder. We are careful to leave the windows open, to save the panes from breaking. The two thundering engines are set at the foot of the plane-trees in front of my door. No precautions are taken to mask them: the Cicadæ singing in the branches overhead cannot see what is happening below.

We are an audience of six. We wait for a moment of comparative quiet. The number of singers is checked by each of us, as are the depth and rhythm of the song. We are now ready, with ears pricked up to hear what will happen in the aerial orchestra. The mortar is let off, with a noise like a genuine thunder-clap.

There is no excitement whatever up above. The number of executants is the same, the rhythm is the same, the volume of sound the same. The six witnesses are unanimous: the mighty explosion has in no way affected the song of the Cicadæ. And the second mortar gives an exactly similar result.

What conclusion are we to draw from this persistence of the orchestra, which is not at all surprised or put out by the firing of a gun? Am I to infer from it that the Cicada is deaf? I will certainly not venture so far as that; but, if any one else, more daring than I, were to make the assertion, I should really not know what arguments to employ in contradicting him. I should be obliged at least to concede that the Cicada is extremely hard of hearing and that we may apply to him the familiar saying, to bawl like a deaf man.

When the Blue-winged Locust takes his luxurious fill of sunshine on a gravelly path and with his great hind-shanks rubs the rough edge of his wing-cases; when the Green Tree-frog, suffering from as chronic a cold as the *Cacan,* swells his throat among the leaves and distends it into a resounding bladder at the approach of a storm, are they both calling to their absent mates? By no means. The bow-strokes of the first produce hardly a perceptible stridulation; the throaty exuberance of the second is no more effective: the object of their desire does not come.

Does the insect need these sonorous outbursts, these loquacious avowals, to declare its flame? Consult the vast majority, whom the meeting of the two sexes leaves silent. I see in the Grasshopper's fiddle, the Tree-frog's bagpipes and the cymbals of the *Cacan* but so many methods of expressing the joy of living, the universal joy which every animal species celebrates after its kind.

If any one were to tell me that the Cicadæ strum on their noisy instruments without giving a thought to the sound produced and for the sheer pleasure of feeling themselves alive, just as we rub our hands in a moment of satisfaction, I should not be greatly shocked. That there may be also a secondary object in their concert, an object in which the dumb sex is interested, is quite possible, quite natural, though this has not yet been proved.

THE PRAYING MANTIS

« 18 »

The European mantis of which Fabre writes, Mantis religiosa, *is now a naturalized insect citizen of the United States. About the time of the Spanish-American War, a nurseryman at Rochester, N.Y., noticed these striking creatures among his trees. They had come from the south of France in the form of egg-cases attached to packing material around nursery stock. Since then, the Rochester colony has spread northward until the insects are now found across the Canadian line. At almost the same time the European mantis was found at Rochester, an oriental mantis,* Tenodera sinensis, *was discovered by another nurseryman outside of Philadelphia, Pa. It, too, had crossed the ocean in the form of an egg-case attached to packing material. This mantis has now spread into New England and west along the Great Lakes. A native species,* Stagmomantis carolina, *is indigenous to the South. It is found as far north as southern New Jersey. Mention is made in this selection, from* THE LIFE OF THE GRASSHOPPER, *of the big Grey Locust,* Pachytylus cinerescens *(Fab). It should be noted that the "Fab"—the abbreviation of the name of the namer of the insect—does not stand for Fabre but for Johann Christian Fabricius (1745–1808) the Danish entomologist and friend of Linnaeus who contributed so many scientific names to the lists of entomology.*

Another creature of the south is at least as interesting as the Cicada, but much less famous, because it makes no noise. Had Heaven granted it a pair of cymbals, the one thing needed, its renown would eclipse the great musician's, for it is most unusual in both shape and habits. Folk hereabouts call it *lou Prègo-Diéu,* the animal that prays to God. Its official name is the Praying Mantis (*M. religiosa,* LIN.).

The language of science and the peasant's artless vocabulary agree in this case and represent the queer creature as a pythoness delivering her oracles or an ascetic rapt in pious ecstasy. The comparison dates a long way back. Even in the time of the Greeks the insect was called Μάντις, the divine, the prophet. The tiller of the soil is not particular about analogies: where points of resemblance are not too clear, he will make up for their deficiencies. He saw on the sun-scorched herbage an insect of imposing appearance, drawn up majestically in a half-erect posture. He noticed its gossamer wings, broad and green, trailing like long veils of finest lawn; he saw its fore-legs, its arms so to speak, raised to the sky in a gesture of invocation. That was enough; popular imagination did the rest; and behold the bushes from ancient times stocked with Delphic priestesses, with nuns in orison.

Good people, with your childish simplicity, how great was your mistake! Those sanctimonious airs are a mask for Satanic habits; those arms folded in prayer are cut-throat weapons: they tell no beads, they slay whatever passes within range. Forming an exception which one would never have suspected in the herbivorous order of the Orthoptera, the Mantis feeds exclusively on living prey. She is the tigress of the peaceable entomological tribes, the ogress in ambush who levies a tribute of fresh meat. Picture her with sufficient strength; and her carnivorous appetites, combined with her traps of horrible perfection, would make her the terror of the country-side. The *Prègo-Diéu* would become a devilish vampire.

Apart from her lethal implement, the Mantis has nothing to inspire dread. She is not without a certain beauty, in fact, with her slender figure, her elegant bust, her pale-green colouring and her long gauze wings. No ferocious mandibles, opening like shears; on the contrary, a dainty pointed muzzle that seems made for billing and cooing. Thanks to a flexible neck, quite independent of the thorax, the head is able to move freely, to turn to right or left, to bend, to lift itself. Alone among insects, the Mantis directs her gaze; she inspects and examines; she almost has a physiognomy.

Great indeed is the contrast between the body as a whole, with its very pacific aspect, and the murderous mechanism of the forelegs, which are correctly described as raptorial. The haunch is uncommonly long and powerful. Its function is to throw forward the rat-trap, which does not await its victim but goes in search of it. The snare is decked out with some show of finery. The base of the haunch is adorned on the inner surface with a pretty, black mark, having a white spot in the middle; and a few rows of bead-like dots complete the ornamentation.

The thigh, longer still, a sort of flattened spindle, carries on the front half of its lower surface two rows of sharp spikes. In the inner row there are a dozen, alternately black and green, the green being shorter than the black. This alternation of unequal lengths increases the number of cogs and improves the effectiveness of the weapon. The outer row is simpler and has only four teeth. Lastly, three spurs, the longest of all, stand out behind the two rows. In short, the thigh is a saw with two parallel blades, separated by a groove in which the leg lies when folded back.

The leg, which moves very easily on its joint with the thigh, is likewise a double-edged saw. The teeth are smaller, more numerous and closer together than those on the thigh. It ends in a strong hook whose point vies with the finest needle for sharpness, a hook fluted underneath and having a double blade like a curved pruning-knife.

This hook, a most perfect instrument for piercing and tearing, has left me many a painful memory. How often, when Mantis-hunting, clawed by the insect which I had just caught and not having both hands at liberty, have I been obliged to ask somebody else to release me from my tenacious captive! To try to free yourself by force, without first disengaging the claws implanted in your flesh, would expose you to scratches similar to those produced by the thorns of a rose-tree. None of our insects is so troublesome to handle. The Mantis claws you with her pruning-hooks, pricks you with her spikes, seizes you in her vice and makes self-defence almost impossible if, wishing to keep your prize alive, you refrain from giving the pinch of the thumb that would put an end to the struggle by crushing the creature.

When at rest, the trap is folded and pressed back against the chest and looks quite harmless. There you have the insect praying. But, should a victim pass, the attitude of prayer is dropped abruptly. Suddenly unfolded, the three long sections of the machine throw to a distance their terminal grapnel, which harpoons the prey and, in returning, draws it back between the two saws. The vice closes with a movement like that of the fore-arm and the upper arm; and all is over: Locusts, Grasshoppers and others even more powerful, once caught in the mechanism with its four rows of teeth, are irretrievably lost. Neither their desperate fluttering nor their kicking will make the terrible engine release its hold.

An uninterrupted study of the Mantis' habits is not practicable in the open fields; we must rear her at home. There is no difficulty about this: she does not mind being interned under glass, on condition that she be well fed. Offer her choice viands, served up fresh daily, and she will hardly feel her absence from the bushes.

As cages for my captives I have some ten large wire-gauze dish-covers, the same that are used to protect meat from the Flies. Each stands in a pan filled with sand. A dry tuft of thyme and a flat stone on which the laying may be done later constitute all the furniture. These huts are placed in a row on the large table in my insect laboratory, where the sun shines on them for the best part of the day. I instal my captives in them, some singly, some in groups.

It is in the second fortnight of August that I begin to come upon the adult Mantis in the withered grass and on the brambles by the roadside. The females, already notably corpulent, are more frequent from day to day. Their slender companions, on the other hand, are rather scarce; and I sometimes have a good deal of difficulty in making up my couples, for there is an appalling consumption of these dwarfs in the cages. Let us keep these atrocities for later and speak first of the females.

They are great eaters, whose maintenance, when it has to last for some months, is none too easy. The provisions, which are nibbled at disdainfully and nearly all wasted, have to be renewed almost every day. I trust that the Mantis is more economical on her native bushes. When game is not plentiful, no doubt she devours every atom of her catch; in my cages she is extravagant, often dropping and abandoning the rich morsel after a few mouthfuls, without deriving any further benefit from it. This appears to be her particular method of beguiling the tedium of captivity.

To cope with these extravagant ways I have to employ assistants. Two or three small local idlers, bribed by the promise of a slice of melon or bread-and-butter, go morning and evening to the grass-plots in the neighbourhood and fill their game-bags—cases made of reed-stumps— with live Locusts and Grasshoppers. I on my side, net in hand, make a daily circuit of my enclosure, in the hope of obtaining some choice morsel for my boarders.

These tit-bits are intended to show me to what lengths the Mantis' strength and daring can go. They include the big Grey Locust (*Pachytylus cinerescens,* FAB.), who is larger than the insect that will consume him; the White-faced Decticus, armed with a vigorous pair of mandibles whereof our fingers would do well to fight shy; the quaint Tryxalis, who wears a pyramid-shaped mitre on her head; the Vine Ephippiger, who clashes cymbals and sports a sword at the bottom of her pot-belly. To this assortment of game that is not any too easy to tackle, let us add two monsters, two of the largest Spiders of the district: the Silky Epeira, whose flat, festooned abdomen is the size of a franc piece; and the Cross Spider, or Diadem Epeira, who is hideously hairy and obese.

I cannot doubt that the Mantis attacks such adversaries in the open, when I see her, under my covers, boldly giving battle to whatever comes in sight. Lying in wait among the bushes, she must profit by the fat prizes offered by chance even as, in the wire cage, she profits by the treasures due to my generosity. Those big hunts, full of danger, are no new thing; they form part of her normal existence. Nevertheless they appear to be rare, for want of opportunity, perhaps to the Mantis' deep regret.

Locusts of all kinds, Butterflies, Dragon-flies, large Flies, Bees and other moderate-sized captures are what we usually find in the lethal limbs. Still the fact remains that, in my cages, the daring huntress recoils before nothing. Sooner or later, Grey Locust and Decticus, Epeira and Tryxalis are harpooned, held tight between the saws and crunched with gusto. The facts are worth describing.

At the sight of the Grey Locust who has heedlessly approached along the trelliswork of the cover, the Mantis gives a convulsive shiver and suddenly adopts a terrifying posture. An electric shock would not produce a more rapid effect. The transition is so abrupt, the attitude so threatening that the observer beholding it for the first time at once hesitates and draws back his fingers, apprehensive of some unknown danger. Old hand as I am, I cannot even now help being startled, should I happen to be thinking of something else.

You see before you, most unexpectedly, a sort of bogey-man or Jack-in-the-box. The wing-covers open and are turned back on either side, slantingly; the wings spread to their full extent and stand erect like parallel sails or like a huge heraldic crest towering over the back; the tip of the abdomen curls upwards like a crosier, rises and falls, relaxing with short jerks and a sort of sough, a "Whoof! Whoof!" like that of a Turkey-cock spreading his tail. It reminds one of the puffing of a startled Adder.

Planted defiantly on its four hind-legs, the insect holds its long bust almost upright. The murderous legs, originally folded and pressed together upon the chest, open wide, forming a cross with the body and revealing the arm-pits decorated with rows of beads and a black spot with a white dot in the centre. These two faint imitations of the eyes in a Peacock's tail, together with the dainty ivory beads, are warlike ornaments kept hidden at ordinary times. They are taken from the jewel-case only at the moment when we have to make ourselves brave and terrible for battle.

Motionless in her strange posture, the Mantis watches the Locust, with her eyes fixed in his direction and her head turning as on a pivot whenever the other changes his place. The object of this attitudinizing is evident: the Mantis wants to strike terror into her dangerous quarry, to paralyze it with fright, for, unless demoralized by fear, it would prove too formidable.

Does she succeed in this? Under the shiny head of the Decticus, behind the long face of the Locust, who can tell what passes? No sign of excitement betrays itself to our eyes on those impassive masks. Nevertheless it is certain that the threatened one is aware of the danger. He sees standing before him a spectre, with uplifted claws, ready to fall upon him; he feels that he is face to face with death; and he fails to escape while there is yet time. He who excels in leaping and could so easily hop out of reach of those talons, he, the big-thighed jumper, remains stupidly where he is, or even draws nearer with a leisurely step.

They say that little birds, paralysed with terror before the open jaws of the Snake, spell-bound by the reptile's gaze, lose their power of flight and allow themselves to be snapped up. The Locust often behaves in much the same way. See him within reach of the enchantress. The two grapnels fall, the claws strike, the double saws close and clutch. In vain the poor wretch protests: he chews space with his mandibles and, kicking desperately, strikes nothing but the air. His fate is sealed. The Mantis furls her wings, her battle-standard; she resumes her normal posture; and the meal begins.

In attacking the Tryxalis and the Ephippiger, less dangerous game than the Grey Locust and the Decticus, the spectral attitude is less imposing and of shorter duration. Often the throw of the grapnels is sufficient. This is likewise so in the case of the Epeira, who is grasped round the body with not a thought of her poison-fangs. With the smaller Locusts, the usual fare in my cages as in the open fields, the Mantis seldom employs her intimidation-methods and contents herself with seizing the reckless one that passes within her reach.

When the prey to be captured is able to offer serious resistance, the Mantis has at her service a pose that terrorizes and fascinates her quarry and gives her claws a means of hitting with certainty. Her rat-traps close on a demoralized victim incapable of defence. She frightens her victim into immobility by suddenly striking a spectral attitude.

The wings play a great part in this fantastic pose. They are very wide, green on the outer edge, colourless and transparent every elsewhere. They are crossed lengthwise by numerous veins, which spread in the shape of a fan. Other veins, transversal and finer, intersect the first at right angles and with them form a multitude of meshes. In the spectral attitude, the wings are displayed and stand upright in two parallel planes that almost touch each other, like the wings of a Butterfly at rest. Between them the curled tip of the abdomen moves with sudden starts. The sort of breath which I have compared with the puffing of an Adder in a posture of defence comes from this rubbing of the abdomen against the nerves of the wings. To imitate the strange sound, all that you need do is to pass your nail quickly over the upper surface of an unfurled wing.

Wings are essential to the male, a slender pigmy who has to wander from thicket to thicket at mating-time. He has a well-developed pair, more than sufficient for his flight, the greatest range of which hardly amounts to four or five of our paces. The little fellow is exceedingly sober in his appetites. On rare occasions, in my cages, I catch him eating a lean Locust, an insignificant, perfectly harmless creature. This means that he knows nothing of the spectral attitude, which is of no use to an unambitious hunter of his kind.

On the other hand, the advantage of the wings to the female is not very obvious, for she is inordinately stout at the time when her eggs ripen. She climbs, she runs; but, weighed down by her corpulence, she never flies. Then what is the object of wings, of wings, too, which are seldom matched for breadth?

The question becomes more significant if we consider the Grey Mantis (*Ameles decolor*), who is closely akin to the Praying Mantis. The male is winged and is even pretty quick at flying. The female, who drags a great belly full of eggs, reduces her wings to stumps and, like the cheese-makers of Auvergne and Savoy, wears a short-tailed jacket. For one who is not meant to leave the dry grass and the stones, this abbreviated costume is more suitable than superfluous gauze furbelows. The Grey Mantis is right to retain but a mere vestige of the cumbrous sails.

Is the other wrong to keep her wings, to exaggerate them, even though she never flies? Not at all. The Praying Mantis hunts big game. Sometimes a formidable prey appears in her hiding-place. A direct attack might be fatal. The thing to do is first to intimidate the new-comer, to conquer his resistance by terror. With this object she suddenly unfurls her wings into a ghost's winding-sheet. The huge sails incapable of flight are hunting-implements. This stratagem is not needed by the little Grey Mantis, who captures feeble prey, such as Gnats and new-born Locusts. The two huntresses, who have similar habits and, because of their stoutness, are neither of them able to fly, are dressed to suit the difficulties of the ambuscade. The first, an impetuous amazon, puffs her wings into a threatening standard; the second, a modest fowler, reduces them to a pair of scanty coat-tails.

In a fit of hunger, after a fast of some days' duration, the Praying Mantis will gobble up a Grey Locust whole, except for the wings, which are too dry; and yet the victim of her voracity is as big as herself, or even bigger. Two hours are enough for consuming this monstrous head of game. An orgy of the sort is rare. I have witnessed it once or twice and have always wondered how the gluttonous creature found room for so much food and how it reversed in its favour the axiom that the cask must be greater than its contents. I can but admire the lofty privileges of a stomach through which matter merely passes, being at once digested, dissolved and done away with.

The usual bill of fare in my cages consists of Locusts of greatly varied species and sizes. It is interesting to watch the Mantis nibbling her Acridian, firmly held in the grip of her two murderous fore-legs. Notwithstanding the fine, pointed muzzle, which seems scarcely made for this gorging, the whole dish disappears, with the exception of the wings, of which only the slightly fleshy base is consumed. The legs, the tough skin, everything goes down. Sometimes the Mantis seizes one of the big hinder thighs by the knuckle-end, lifts it to her mouth, tastes it and crunches it with a little air of satisfaction. The Locust's fat and juicy thigh may well be a choice morsel for her, even as a leg of mutton is for us.

The prey is first attacked in the neck. While one of the two lethal legs holds the victim transfixed through the middle of the body, the other presses the head and makes the neck open upwards. The Mantis' muzzle roots and nibbles at this weak point in the armour with some persistency. A large wound appears in the head. The Locust gradually ceases kicking and becomes a lifeless corpse; and, from this moment, freer in its movements, the carnivorous insect picks and chooses its morsel.

The Mantis naturally wants to devour the victuals in peace, without being troubled by the plunges of a victim who absolutely refuses to be devoured. A meal liable to interruptions lacks savour. Now the principal means of defence in this case are the hind-legs, those vigorous levers which can kick out so brutally and which moreover are armed with toothed saws that would rip open the Mantis' bulky paunch if by ill-luck they happen to graze it. What shall we do to reduce them to help-lessness, together with the others, which are not dangerous but trouble-some all the same, with their desperate gesticulations?

Strictly speaking, it would be practicable to cut them off one by one. But that is a long process and attended with a certain risk. The Mantis has hit upon something better. She has an intimate knowledge of the anatomy of the spine. By first attacking her prize at the back of the half-opened neck and munching the cervical ganglia, she destroys the muscular energy at its main seat; and inertia supervenes, not suddenly and completely, for the clumsily-constructed Locust has not the Bee's exquisite and frail vitality, but still sufficiently, after the first mouth-fuls. Soon the kicking and the gesticulating die down, all movement ceases and the game, however big it be, is consumed in perfect quiet.

MATING OF THE MANTIS

« *19* »

*In regions where the mantis is found, museums and zoos are flooded
with telephone calls each autumn when the insects reach full size. Few
other species arouse more general interest than these large, striking appearing members of the Orthoptera order. I have given several chapters
relating to the Praying Mantis, partly because of the unusual interest
in the insect and partly because Fabre is at his best in describing the
strange habits of the creature. His outlook on the insects was sympa-
thetic but not sentimental. He recorded facts. The following selection
is from the seventh chapter of* THE LIFE OF THE GRASSHOPPER.

The little that we have seen of the Mantis' habits hardly tallies with
what we might have expected from her popular name. To judge by the
term *Prègo-Diéu,* we should look to see a placid insect, deep in pious
contemplation; and we find ourselves in the presence of a cannibal, of
a ferocious spectre munching the brain of a panic-stricken victim. Nor
is even this the most tragic part. The Mantis has in store for us, in her
relations with her own kith and kin, manners even more atrocious than
those prevailing among the Spiders, who have an evil reputation in this
respect.

To reduce the number of cages on my big table and give myself a
little more space while still retaining a fair-sized menagerie, I install
several females, sometimes as many as a dozen, under one cover. So
far as accommodation is concerned, no fault can be found with the
common lodging. There is room and to spare for the evolutions of my
captives, who naturally do not want to move about much with their
unwieldy bellies. Hanging to the trelliswork of the dome, motionless
they digest their food or else await an unwary passer-by. Even so do
they act when at liberty in the thickets.

Cohabitation has its dangers. I know that even Donkeys, those peace-loving animals, quarrel when hay is scarce in the manger. My boarders, who are less complaisant, might well, in a moment of dearth, become sour-tempered and fight among themselves. I guard against this by keeping the cages well supplied with Locusts, renewed twice a day. Should civil war break out, famine cannot be pleaded as the excuse.

At first, things go pretty well. The community lives in peace, each Mantis grabbing and eating whatever comes near her, without seeking strife with her neighbours. But this harmonious period does not last long. The bellies swell, the eggs are ripening in the ovaries, marriage and laying-time are at hand. Then a sort of jealous fury bursts out, though there is an entire absence of males who might be held responsible for feminine rivalry. The working of the ovaries seems to pervert the flock, inspiring its members with a mania for devouring one another. There are threats, personal encounters, cannibal feasts. Once more the spectral pose appears, the hissing of the wings, the fearsome gesture of the grapnels outstretched and uplifted in the air. No hostile demonstration in front of a Grey Locust or White-faced Decticus could be more menacing.

For no reason that I can gather, two neighbours suddenly assume their attitude of war. They turn their heads to right and left, provoking each other, exchanging insulting glances. The "Puff! Puff!" of the wings rubbed by the abdomen sounds the charge. When the duel is to be limited to the first scratch received, without more serious consequences, the lethal fore-arms, which are usually kept folded, open like the leaves of a book and fall back sideways, encircling the long bust. It is a superb pose, but less terrible than that adopted in a fight to the death.

Then one of the grapnels, with a sudden spring, shoots out to its full length and strikes the rival; it is no less abruptly withdrawn and resumes the defensive. The adversary hits back. The fencing is rather like that of two Cats boxing each other's ears. At the first blood drawn from her flabby paunch, or even before receiving the least wound, one of the duellists confesses herself beaten and retires. The other furls her battle-standard and goes off elsewhither to meditate the capture of a Locust, keeping apparently calm, but ever ready to repeat the quarrel.

Very often, events take a more tragic turn. At such times, the full posture of the duels to the death is assumed. The murderous fore-arms are unfolded and raised in the air. Woe to the vanquished! The other seizes her in her vice and then and there proceeds to eat her, beginning at the neck, of course. The loathsome feast takes place as calmly as though it were a matter of crunching up a Grasshopper. The diner enjoys her sister as she would a lawful dish; and those around do not protest, being quite willing to do as much on the first occasion.

Oh, what savagery! Why, even Wolves are said not to eat one another. The Mantis has no such scruples; she banquets off her fellows when there is plenty of her favourite game, the Locust, around her. She practises the equivalent of cannibalism, that hideous peculiarity of man.

These aberrations, these child-bed cravings can reach an even more revolting stage. Let us watch the pairing and, to avoid the disorder of a crowd, let us isolate the couples under different covers. Each pair shall have its own home, where none will come to disturb the wedding. And let us not forget the provisions, with which we will keep them well supplied, so that there may be no excuse of hunger.

It is near the end of August. The male, that slender swain, thinks the moment propitious. He makes eyes at his strapping companion; he turns his head in her direction; he bends his neck and throws out his chest. His little pointed face wears an almost impassioned expression. Motionless, in this posture, for a long time he contemplates the object of his desire. She does not stir, is as though indifferent. The lover, however, has caught a sign of acquiescence, a sign of which I do not know the secret. He goes nearer; suddenly he spreads his wings, which quiver with a convulsive tremor. That is his declaration. He rushes, small as he is, upon the back of his corpulent companion, clings on as best he can, steadies his hold. As a rule, the preliminaries last a long time. At last, coupling takes place and is also long drawn out, lasting sometimes for five or six hours.

Nothing worthy of attention happens between the two motionless partners. They end by separating, but only to unite again in a more intimate fashion. If the poor fellow is loved by his lady as the vivifier of her ovaries, he is also loved as a piece of highly-flavoured game. And, that same day, or at latest on the morrow, he is seized by his spouse, who first gnaws his neck, in accordance with precedent, and then eats him deliberately, by little mouthfuls, leaving only the wings. Here we have no longer a case of jealousy in the harem, but simply a depraved appetite.

I was curious to know what sort of reception a second male might expect from a recently fertilized female. The result of my enquiry was shocking. The Mantis, in many cases, is never sated with conjugal raptures and banquets. After a rest that varies in length, whether the eggs be laid or not, a second male is accepted and then devoured like the first. A third succeeds him, performs his function in life, is eaten and disappears. A fourth undergoes a like fate. In the course of two weeks I thus see one and the same Mantis use up seven males. She takes them all to her bosom and makes them all pay for the nuptial ecstasy with their lives.

Orgies such as this are frequent, in varying degrees, though there are exceptions. On very hot days, highly charged with electricity, they are almost the general rule. At such times the Mantes are in a very irritable mood. In the cages containing a large colony, the females devour one another more than ever; in the cages containing separate pairs, the males, after coupling, are more than ever treated as an ordinary prey.

I should like to be able to say, in mitigation of these conjugal atrocities, that the Mantis does not behave like this in a state of liberty; that the male, after doing his duty, has time to get out of the way, to make off, to escape from his terrible mistress, for in my cages he is given a respite, lasting sometimes until next day. What really occurs in the thickets I do not know, chance, a poor resource, having never instructed me concerning the love-affairs of the Mantis when at large. I can only go by what happens in the cages, where the captives, enjoying plenty of sunshine and food and spacious quarters, do not seem to suffer from homesickness in any way. What they do here they must also do under normal conditions.

Well, what happens there utterly refutes the idea that the males are given time to escape. I find, by themselves, a horrible couple engaged as follows. The male, absorbed in the performance of his vital functions, holds the female in a tight embrace. But the wretch has no head; he has no neck; he has hardly a body. The other, with her muzzle turned over her shoulder continues very placidly to gnaw what remains of the gentle swain. And, all the time, that masculine stump, holding on firmly, goes on with the business!

Love is stronger than death, men say. Taken literally, the aphorism has never received a more brilliant confirmation. A headless creature, an insect amputated down to the middle of the chest, a very corpse persists in endeavouring to give life. It will not let go until the abdomen, the seat of the procreative organs, is attacked.

Eating the lover after consummation of marriage, making a meal of the exhausted dwarf, henceforth good for nothing, can be understood, to some extent, in the insect world, which has no great scruples in matters of sentiment; but gobbling him up during the act goes beyond the wildest dreams of the most horrible imagination. I have seen it done with my own eyes and have not yet recovered from my astonishment.

THE MANTIS EGG-CASE

« 20 »

Like most members of the Orthoptera order, the mantis produces a large number of eggs. I once counted the number of balls of froth these insects had left attached to twigs and weed-stems in an area only seventy paces wide and one hundred paces long. There were more than 200 cases, each holding from 125 to 350 eggs. Thus in this area there were well over 50,000 mantis eggs ready to hatch in the spring. The mortality is very great during the early lives of the insects. Even during the winter a number of the egg-masses are destroyed by mice and woodpeckers. The egg-cases are attached to everything from grass-stems to the twigs of trees twenty feet in the air. Around Rochester, the European mantis seems to prefer birch twigs as supports for its pecan-shaped egg-masses. The Rumford Fabre refers to is Benjamin Thompson (1753–1814) an American Loyalist who became Minister for War in Bavaria where he was created Count Rumford. He was celebrated as a scientist and is credited with the discovery of the con-vertability of mechanical energy into heat. This account of the making of the egg-case by the mantis is also from THE LIFE OF THE GRASS-HOPPER.

Let us show the insect of the tragic amours under a more attractive aspect. Its nest is a marvel. In scientific language it is called *ootheca*, the egg-case. I shall not overwork this outlandish term. We do not say, "the Chaffinch's egg-case," when we mean, "the Chaffinch's nest": why should I be obliged to talk about a case when I speak of the Mantis? It may sound more learned; but that is not my business.

The nest of the Praying Mantis is found more or less everywhere in sunny places, on stones, wood, vine-stocks, twigs, dry grass and even on products of human industry, such as bits of brick, strips of coarse linen or the hard, shrivelled leather of an old boot. Any support serves, without distinction, so long as there is an uneven surface to which the bottom of the nest can be fixed, thus securing a solid foundation.

The usual dimensions are four centimetres in length and two in width. The colour is as golden as a grain of wheat. When set alight, the material burns readily and exhales a faint smell of singed silk. The substance is in fact akin to silk; only, instead of being drawn into thread, it has curdled into a frothy mass. When the nest is fixed to a branch, the base goes round the nearest twigs, envelops them and assumes a shape which varies in accordance with the support encountered; when it is fixed to a flat surface, the under side, which is always moulded on the support, is itself flat. The nest thereupon takes the form of a semi-ellipsoid, more or less blunt at one end, tapering at the other and often ending in a short, curved tail.

Whatever the support, the upper surface of the nest is systematically convex. We can distinguish in it three well-marked longitudinal zones. The middle one, which is narrower than the others, is composed of little plates or scales arranged in pairs and overlapping like the tiles of a roof. The edges of these plates are free, leaving two parallel rows of slits or fissures through which the young emerge at hatching-time. In a recently-abandoned nest, this middle zone is furry with gossamer skins, discarded by the larvæ. These cast skins flutter at the least breath and soon vanish when exposed to rough weather. I will call it the exit-.zone, because it is only along this median belt that the liberation of the young takes place, thanks to the outlets contrived before-hand.

In every other part the cradle of the numerous family presents an impenetrable wall. The two side zones, in fact, which occupy the greater part of the semiellipsoid, have perfect continuity of surface. The little Mantes, so feeble at the start, could never make their way out through so tough a substance. All that we see on it is a number of fine, transversal furrows, marking the various layers of which the mass of eggs consists.

Cut the nest across. It will now be perceived that the eggs, taken together, form an elongated kernel, very hard and firm and coated on the sides with a thick, porous rind, like solidified foam. Above are curved plates, set very closely and almost independent of one another; their edges end in the exit-zone, where they form a double row of small, imbricated scales.

The eggs are buried in a yellow matrix of horny appearance. They are placed in layers, shaped like segments of a circle, with the ends containing the heads converging towards the exit-zone. This arrangement tells us how the deliverance is accomplished. The new-born larvæ will slip into the space left between two adjoining plates, a prolongation of the kernel, where they will find a narrow passage, difficult to go through, but just sufficient when we bear in mind the curious provision of which we shall speak presently; and by so doing they will reach the middle belt. Here, under the imbricated scales, two outlets open for each layer of eggs. Half of the larvæ undergoing their liberation will emerge through the right door, half through the left. And this is repeated for each layer from end to end of the nest.

To sum up these structural details, which are rather difficult to grasp for any one who has not the thing in front of him: lying along the axis of the nest and shaped like a date-stone is the cluster of eggs, grouped in layers. A protecting rind, a sort of solidified foam, surrounds this cluster, except at the top along the median line, where the frothy rind is replaced by thin plates set side by side. The free ends of these plates form the exit-zone outside; they are imbricated in two series of scales and leave a couple of outlets, narrow clefts, for each layer of eggs.

The most striking part of my researches was being present at the construction of the nest and seeing how the Mantis goes to work to produce so complex a building. I managed it with some difficulty, for the laying takes place without warning and nearly always at night. After much useless waiting, chance at last favoured me. On the 5th of September, one of my boarders, who had been fertilized on the 29th of August, decided to lay her eggs before my eyes at about four o'clock in the afternoon.

Before watching her labour, let us note one thing: all the nests that I have obtained in the cages—and there are a good many of them—have as their support, with not a single exception, the wire gauze of the covers. I had taken care to place at the Mantes' disposal a few rough bits of stone, a few tufts of thyme, foundations very often used in the open fields. My captives preferred the wire network, whose meshes furnish a perfectly safe support as the soft material of the building becomes encrusted in them.

The nests, under natural conditions, enjoy no shelter; they have to endure the inclemencies of winter, to withstand rain, wind, frost and snow without coming loose. Therefore the mother always chooses an uneven support for the nest, so that the foundations can be wedged into it and a firm hold obtained. But, when circumstances permit, the better is preferred to the middling and the best to the better; and this must be the reason why the trelliswork of the cages is invariably adopted.

The only Mantis that I have been allowed to observe while engaged in laying does her work upside down, hanging from the top of the cage. My presence, my magnifying-glass, my investigations do not disturb her at all, so great is her absorption in her labour. I can raise the trellised dome, tilt it, turn it over, spin it this way and that, without the insect's suspending its task for a moment. I can take my forceps and lift the long wings to see what is happening underneath. The Mantis takes no notice. Up to this point, all is well: the mother does not move and impassively endures all the indiscretions of which I am guilty as an observer. And yet things do not go quite as I could wish, for the operation is too rapid and is too difficult to follow.

The end of the abdomen is immersed the whole time in a sea of foam, which prevents us from grasping the details of the process with any clearness. This foam is greyish-white, a little sticky and almost like soapsuds. When it first appears, it adheres slightly to a straw which I dip into it, but, two minutes afterwards, it is solidified and no longer sticks to the straw. In a very short time, its consistency is that which we find in an old nest.

The frothy mass consists mainly of air imprisoned in little bubbles. This air, which gives the nest a volume much greater than that of the Mantis' belly, obviously does not come from the insect, though the foam appears at the entrance of the genital organs; it is taken from the atmosphere. The Mantis, therefore, builds above all with air, which is eminently suited to protect the nest against the weather. She discharges a sticky substance, similar to the caterpillars' silk-fluid; and with this composition, which amalgamates instantly with the outer air, she produces foam.

She whips her product just as we whip white of egg to make it rise and froth. The tip of the abdomen, opening with a long cleft, forms two lateral ladles which meet and separate with a constant, rapid movement, beating the sticky fluid and turning it into foam as it is discharged outside. In addition, between the two flapping ladles, we see the internal organs rising and falling, appearing and disappearing, after the manner of a piston-rod, without being able to distinguish their precise action, drowned as they are in the opaque stream of foam.

The end of the abdomen, ever throbbing, quickly opening and closing its valves, swings from right to left and left to right like a pendulum. The result of each swing is a layer of eggs inside and a transversal furrow outside. As the abdomen advances in the arc described, suddenly and at very close intervals it dips deeper into the foam, as though it were pushing something to the bottom of the frothy mass. Each time, no doubt, an egg is laid; but things happen so fast and under conditions so unfavourable to observation that I never once succeed in seeing the ovipositor at work. I can judge of the arrival of the eggs only by the movements of the tip of the abdomen, which suddenly drives down and immerses itself more deeply.

At the same time, the viscous stuff is poured forth in intermittent waves and whipped and turned into foam by the two terminal valves. The froth obtained spreads over the sides of the layer of eggs and at the base, where I see it, pressed back by the abdomen, projecting through the meshes of the gauze. Thus the spongy covering is gradually brought into being as the ovaries are emptied.

I imagine, without being able to rely on direct observation, that for the central kernel, where the eggs are contained in a more homogeneous material than the rind, the Mantis employs her product as it is, without beating it up and making it foam. When the eggs are deposited, the two valves would produce foam to cover them. Once again, however, all this is very difficult to follow under the veil of the bubbling mass.

In a new nest, the exit-zone is coated with a layer of fine porous matter, of a pure, dull, almost chalky white, which contrasts with the dirty white of the remainder of the nest. It is like the composition which confectioners make out of whipped white of egg, sugar and starch, with which to ornament their cakes. This snowy covering is very easily crumbled and removed. When it is gone, the exit-zone is clearly defined, with its two rows of plates with free edges. The weather, the wind and the rain sooner or later remove it in strips and flakes; and therefore the old nests retain no traces of it.

At the first inspection, one might be tempted to look upon this snowy matter as a different substance from the remainder of the nest. But can it be that the Mantis really employs two different products? By no means. Anatomy, to begin with, assures us of the unity of the materials. The organ that secretes the substance of the nest consists of twisted cylindrical tubes, divided into two sections of twenty each. All are filled with a colourless, viscous fluid, exactly similar in appearance wherever we look. There is nowhere any sign of a product with a chalky colouring.

The manner in which the snowy ribbon is formed also makes us reject the theory of different materials. We see the Mantis' two caudal threads sweeping the surface of the foamy mass, skimming, so to speak, the top of the froth, collecting it and retaining it along the back of the nest to form a band that looks like a ribbon of icing. What remains after this sweeping, or what trickles from the band before it sets, spreads over the sides in a thin wash of bubbles so fine that they cannot be seen without the magnifying-glass.

The surface of a muddy stream containing clay will be covered with coarse and dirty foam, churned up by the rushing torrent. On this foam, soiled with earthy materials, we see here and there masses of beautiful white froth, with smaller bubbles. Selection is due to the difference in density; and so the snow-white foam in places lies on top of the dirty foam whence it proceeds. Something similar happens when the Mantis builds her nest. The twin ladles reduce to foam the sticky spray from the glands. The thinnest and lightest portion, made whiter by its more delicate porousness, rises to the surface, where the caudal threads sweep it up and gather it into a snowy ribbon along the back of the nest.

Until now, with a little patience, observation has been practicable and has given satisfactory results. It becomes impossible when we come to the very complex structure of that middle zone where exits are contrived for the emergence of the larvæ under the shelter of a double row of imbricated plates. The little that I am able to make out amounts to this: the tip of the abdomen, split wide from top to bottom, forms a sort of buttonhole whose upper end remains almost fixed while the lower end, in swinging, produces foam and immerses eggs in it. It is that upper end which is undoubtedly responsible for the work of the middle zone. I always see it in the extension of that zone, in the midst of the fine white foam collected by the caudal filaments. These, one on the right, the other on the left, mark the boundaries of the band. They feel its edges; they seem to be testing the work. I can easily imagine them two long and exquisitely delicate fingers controlling the difficult business of construction.

But how are the two rows of scales obtained and the fissures, the exit-doors, which they shelter? I do not know. I cannot even guess. I leave the rest of the problem to others.

What a wonderful mechanism is this which emits so methodically and swiftly the horny matrix of the central kernel, the protecting froth, the white foam of the median ribbon, the eggs and the fertilizing fluid and which at the same time is able to build overlapping plates, imbricated scales and alternating open fissures! We are lost in admiration. And yet how easily the work is done! The Mantis hangs motionless on the wire gauze which is the foundation of her nest. She gives not a glance at the edifice that is rising behind her; her legs are not called upon for assistance of any kind. The thing works of itself. We have here not an industrial task requiring the cunning of instinct; it is a purely automatic process, regulated by the insect's tools and organization. The nest, with its highly complicated structure, proceeds solely from the play of the organs, even as in our own industries we manufacture by machinery a host of objects whose perfection would outwit our manual dexterity.

From another point of view, the Mantis' nest is more remarkable still. We see in it a superb application of one of the most beautiful principles of physics, that of the conservation of heat. The Mantis anticipated us in a knowledge of non-conducting bodies.

We owe to Rumford, the natural philosopher, the following curious experiment, which fittingly demonstrates the low conductivity of the air. The illustrious scientist dropped a frozen cheese into a mass of foam supplied by well-beaten eggs. The whole was subjected to the heat of an oven. The result in a short time was an *omelette soufflée* hot enough to burn the tongue, with the cheese in the middle as cold as at the beginning. The air contained in the bubbles of the surrounding froth explains the strange phenomenon. As an exceedingly poor thermal conductor, it had arrested the heat of the oven and prevented it from reaching the frozen substance in the centre.

Now what does the Mantis do? Precisely the same as Rumford: she whips her white of egg into an *omelette soufflée,* to protect the eggs collected into a central kernel. Her aim, it is true, is reversed: her coagulated foam is intended to ward off the cold, not the heat. But a protection against one is a protection against the other; and the ingenious physicist, had he wished, could easily with the same frothy wrapper have maintained the heat of a body in cold surroundings.

Rumford knew the secrets of the stratum of air thanks to the accumulated knowledge of his ancestors, his own researches and his own studies. How is it that for no one knows how many centuries the Mantis has beaten our natural philosophers in the matter of this delicate problem of heat? How did she come to think of wrapping a blanket of foam around her mass of eggs, which, fixed without any shelter to a twig or stone, has to endure the rigours of winter with impunity?

The other Mantidæ of my neighbourhood, the only ones of whom I can speak with full knowledge, use the non-conducting wrapper of solidified foam or do without it, according as the eggs are destined to live through the winter or not. The little Grey Mantis, who differs so greatly from the other owing to the almost entire absence of wings in the female, builds a nest not quite so big as a cherry-stone and covers it very cleverly with a rind of froth. Why this beaten-up envelope? Because the nest of the Grey Mantis, like that of the Praying Mantis, has to last through the winter, exposed on its bough or stone to all the dangers of the bad weather.

On the other hand, in spite of her size, which is equal to that of the Praying Mantis, *Empusa pauperata,* who is the most curious of our insects, builds a nest as small as that of the Grey Mantis. It is a very modest edifice, consisting of a small number of cells set side by side in three or four rows joined together. Here there is no frothy envelope at all, though the nest, like those mentioned above, is fixed in an exposed situation on some twig or broken stone. This absence of a nonconducting mattress points to a difference in climatic conditions. The Empusa's eggs, in fact, hatch soon after they are laid, during the fine weather. Not having to undergo the inclemencies of winter, they have no protection but the slender sheath of their cases.

Are these scrupulous and rational precautions, which rival Rumford's *omelette soufflée,* a casual result, one of those numberless combinations turned out by the wheel of fortune? If so, let us not shrink from any absurdity, but recognize straightway that the blindness of chance is endowed with marvellous foresight.

The blunt end of the nest is the first part built by the Praying Mantis and the tapering end the last. The latter is often prolonged into a sort of spur made by drawing out the final drop of albuminous fluid used. To complete the whole thing demands about two hours of concentrated work, free from interruption.

As soon as the laying is finished, the mother withdraws, callously. I expected to see her return and display some tender feeling for the cradle of her family. But there is not the least sign of maternal joy. The work is done and possesses no further interest for her. Some Locusts have come up. One even perches on the nest. The Mantis pays no attention to the intruders. They are peaceful, it is true. Would she drive them away if they were dangerous and if they looked like ripping open the egg-casket? Her impassive behaviour answers no. What is the nest to her henceforth? She knows it no more.

I have spoken of the repeated coupling of the Praying Mantis and of the tragic end of the male, who is nearly always devoured like an ordinary piece of game. In the space of a fortnight I have seen the same female marry again as many as seven times over. Each time the easily-consoled widow ate up her mate. Such habits make one assume repeated layings; and these do, in fact, take place, though they are not the general rule. Among my mothers, some gave me only one nest; others supplied me with two, both equally large. The most fertile produced three, of which the first two were of normal size, while the third was reduced to half the usual dimensions.

The last-mentioned insect shall tell us the population which the Mantis' ovaries are capable of producing. Reckoning by the transversal furrows of the nest, we can easily count the layers of eggs. These are more or less rich according to their position at the middle of the ellipsoid or at the ends. The numbers of the eggs in the biggest and in the smallest layer furnish an average from which we can approximately deduce the total. In this way I find that a good-sized nest contains about four hundred eggs. The mother with the three nests, the last of which was only half the size of the others, therefore left as her offspring no fewer than a thousand germs; those who laid twice left eight hundred; and the less fertile mothers three to four hundred.

THE HATCHING OF THE MANTIS

« 21 »

Hardly larger than mosquitoes when they hatch from the hardened froth-mass that has been their insulated winter home, the young insects disappear among the grass and weeds. They are rarely noticed until late in summer when their large size makes them conspicuous and their newly-acquired wings enable them to travel about over greater distances. The chalcis parasites that prey upon the mantis are species of a widely distributed group. Some have the ability of laying eggs that multiply themselves, 2,000 or more individuals resulting from a single egg. This selection from THE LIFE OF THE GRASSHOPPER *originally appeared in the ninth chapter of that book.*

The eggs of the Praying Mantis usually hatch in bright sunshine, at about ten o'clock on a mid-June morning. The median band or exit-zone is the only portion of the nest that affords an outlet to the youngsters.

From under each scale of that zone we see slowly appearing a blunt, transparent protuberance, followed by two large black specks, which are the eyes. Softly the new-born grub slips under the thin plate and half-releases itself. Is it the little Mantis in his larval form, so nearly allied to that of the adult? Not yet. It is a transition organism. The head is opalescent, blunt, swollen, with palpitations caused by the flow of the blood. The rest is tinted reddish-yellow. It is quite easy to distinguish, under a general overall, the large black eyes clouded by the veil that covers them, the mouth-parts flattened against the chest, the legs plastered to the body from front to back. Altogether, with the exception of the very obvious legs, the whole thing, with its big blunt head, its eyes, its delicate abdominal segmentation and its boatlike shape, reminds us somewhat of the first state of the Cicadæ on leaving the egg, a state which is pictured exactly by a tiny, finless fish.

Here then is a second instance of an organization of very brief dura-tion having as its function to bring into the light of day, through narrow and difficult passes, a microscopic creature whose limbs, if free, would, because of their length, be an insurmountable impediment. To enable him to emerge from the exiguous tunnel of his twig, a tunnel bristling with woody fibres and blocked with shells already empty, the Cicada is born swathed in bands and endowed with a boat shape, which is eminently suited to slipping easily through an awkward passage. The young Mantis is exposed to similar difficulties. He has to emerge from the depths of the nest through narrow, winding ways, in which full-spread, slender limbs would not be able to find room. The high stilts, the murderous harpoons, the delicate antennæ, organs which will be most useful presently, in the brushwood, would now hinder the emer-gence, would make it very laborious, impossible. The creature therefore comes into existence swaddled and furthermore takes the shape of a boat.

The case of the Cicada and the Mantis opens up a new vein to us in the inexhaustible entomological mine. I extract from it a law which other and similar facts, picked up more or less everywhere, will cer-tainly not fail to confirm. The true larva is not always the direct product of the egg. When the new-born grub is likely to experience special dif-ficulties in effecting its deliverance, an accessory organism, which I shall continue to call the primary larva, precedes the genuine larval state and has as its function to bring to the light of day the tiny creature which is incapable of releasing itself.

To go on with our story, the primary larvæ show themselves under the thin plates of the exit-zone. A vigorous flow of humours occurs in the head, swelling it out and converting it into a diaphanous and ever-throbbing blister. In this way the splitting-apparatus is prepared. At the same time, the little creature, half-caught under its scale, sways, pushes forward, draws back. Each swaying is accompanied by an in-crease of the swelling in the head. At last the prothorax arches and the head is bent low towards the chest. The tunic bursts across the prothorax. The little animal tugs, wriggles, sways, bends and straightens itself again. The legs are drawn from their sheaths; the antennæ, two long parallel threads, are likewise released. The creature is now fastened to the nest only by a worn-out cord. A few shakes complete the deliverance.

We here have the insect in its genuine larval form. All that remains behind is a sort of irregular cord, a shapeless clout which the least breath blows about like a flimsy bit of fluff. It is the exit-tunic violently shed and reduced to a mere rag.

The hatching does not take place all over the nest at one time, but rather in sections, in successive swarms which may be separated by intervals of two days or more. The pointed end, containing the last eggs, usually begins. This inversion of chronological order, calling the last to the light of day before the first, may well be due to the shape of the nest. The thin end, which is more accessible to the stimulus of a fine day, wakes up before the blunt end, which is larger and does not so soon acquire the necessary amount of heat.

Sometimes, however, although still broken up in swarms, the hatching embraces the whole length of the exit-zone. A striking sight indeed is the sudden exodus of a hundred young Mantes. Hardly does the tiny creature show its black eyes under a scale before others appear instantly, in their numbers. It is as though a certain shock were being communicated from one to another, as though an awakening signal were transmitted, so swiftly does the hatching spread all round. Almost in a moment the median band is covered with young Mantes who run about feverishly, stripping themselves of their rent garments.

The nimble little creatures do not stay long on the nest. They let themselves drop off or else clamber into the nearest foliage. All is over in less than twenty minutes. The common cradle resumes its peaceful condition, prior to furnishing a new legion a few days later; and so on until all the eggs are finished.

I have witnessed this exodus as often as I wished to, either out of doors, in my enclosure, where I had deposited in sunny places the nests gathered more or less everywhere during my winter leisure, or else in the seclusion of a greenhouse, where I thought, in my simplicity, that I should be better able to protect the budding family. I have witnessed the hatching twenty times if I have once; and I have always beheld a scene of unforgettable carnage. The round-bellied Mantis may procreate germs by the thousands: she will never have enough to cope with the devourers who are destined to decimate the breed from the moment that it leaves the egg.

The Ants above all are zealous exterminators. Daily I surprise their ill-omened visits on my rows of nests. It is vain for me to intervene, however seriously; their assiduity never slackens. They seldom succeed in making a breach in the fortress: that is too difficult; but, greedy of the dainty flesh in course of formation inside, they await a favourable opportunity, they lie in wait for the exit.

Despite my daily watchfulness, they are there the moment that the young Mantes appear. They grab them by the abdomen, pull them out of their sheaths, cut them up. You see a piteous fray between tender babes gesticulating as their only means of defence and ferocious brigands carrying their *spolia opima* at the end of their mandibles. In less than no time the massacre of the innocents is consummated; and all that remains of the flourishing family is a few scattered survivors who have escaped by accident.

The future assassin, the scourge of the insect race, the terror of the Locust on the brushwood, the dread devourer of fresh meat, is herself devoured, from her birth, by one of the least of that race, the Ant. The ogress, prolific to excess, sees her family thinned by the dwarf. But the slaughter is not long continued. So soon as she has acquired a little firmness from the air and strengthened her legs, the Mantis ceases to be attacked. She trots about briskly among the Ants, who fall back as she passes, no longer daring to tackle her. With her grappling-legs brought close to her chest, like arms ready for self-defence, already she strikes awe into them by her proud bearing.

A second connoisseur in tender meats pays no heed to these threats. This is the little Grey Lizard, the lover of sunny walls. Apprised I know not how of the quarry, here he comes, picking up one by one, with the tip of his slender tongue, the stray insects that have escaped the Ants. They make a small mouthful but an exquisite one, so it seems, to judge by the blinking of the reptile's eye. For each little wretch gulped down, its lid half-closes, a sign of profound satisfaction. I drive away the bold Lizard who ventures to perpetrate his raid before my eyes. He comes back again and, this time, pays dearly for his rashness. If I let him have his way, I should have nothing left.

Is this all? Not yet. Another ravager, the smallest of all but not the least formidable, has anticipated the Lizard and the Ant. This is a very tiny Hymenopteron armed with a probe, a Chalcis, who establishes her eggs in the newly-built nest. The Mantis' brood shares the fate of the Cicada's: parasitic vermin attack the eggs and empty the shells. Out of all that I have collected I often obtain nothing or hardly anything. The Chalcis has been that way.

Let us gather up what the various exterminators, known or unknown, have left me. When newly hatched, the larva is of a pale hue, white faintly tinged with yellow. The swelling of its head soon diminishes and disappears. Its colour is not long in darkening and turns light-brown within twenty-four hours. The little Mantis very nimbly lifts up her grappling-legs, opens and closes them; she turns her head to right and left; she curls her abdomen. The fully-developed larva has no greater litheness and agility. For a few minutes the family stops where it is, swarming over the nest; then it scatters at random on the ground and the plants hard by.

THE BOOK THAT CHANGED
MY LIFE

« 22 »

Leon Dufour was born at Saint-Sever, in the Landes, in 1870. He served
with distinction as an army surgeon during several campaigns. In 1823,
the year J. Henri Fabre was born, he took part in the Spanish campaign
after which he retired to the practice of medicine in his native town.
Through his study of the habits of the insects, Dufour attained great
eminence as a naturalist during the mid-decades of the Nineteenth
Century. He died in 1865. This account of the influence on Fabre of
one of his books is taken from the first chapter of THE HUNTING WASPS.

There are for each one of us, according to his turn of mind, certain books
that open up horizons hitherto undreamed of and mark an epoch in our
mental life. They fling wide the gates of a new world wherein our intellectual powers are henceforth to be employed; they are the spark which
lights the fuel on a hearth doomed, without its aid, to remain indefinitely
bleak and cold. And it is often chance that places in our hands those
books which mark the beginning of a new era in the evolution of our
ideas. The most casual circumstances, a few lines that happen somehow to come before our eyes decide our future and plant us in the ap-
pointed groove.

One winter evening, when the rest of the household was asleep, as I
sat reading beside a stove whose ashes were still warm, my book made
me forget for a while the cares of the morrow: those heavy cares of a
poor professor of physics who, after piling up diplomas and for a quarter of a century performing services of uncontested merit, was receiving
for himself and his family a stipend of sixteen hundred francs, or less
than the wages of a groom in a decent establishment. Such was the

disgraceful parsimony of the day where education was concerned; such was the edict of our government red-tape: I was an irregular, the offspring of my solitary studies. And so I was forgetting the poverty and anxieties of a professor's life, amid my books, when I chanced to turn over the pages of an entomological essay that had fallen into my hands I forget how.

It was a monograph by the then father of entomology, the venerable scientist Léon Dufour, on the habits of a Wasp that hunted Buprestis-beetles. Certainly, I had not waited till then to interest myself in insects; from my early childhood, I had delighted in Beetles, Bees and Butterflies; as far back as I can remember, I see myself in ecstasy before the splendour of a Ground-beetle's wing-cases or the wings of *Papilio machaon,* the Swallowtail. The fire was laid; the spark to kindle it was absent. Léon Dufour's essay provided that spark.

New lights burst forth: I received a sort of mental revelation. So there was more in science than the arranging of pretty Beetles in a cork box and giving them names and classifying them; there was something much finer: a close and loving study of insect life, the examination of the structure and especially the faculties of each species. I read of a magnificent instance of this, glowing with excitement as I did so. Some time after, aided by those lucky circumstances which he who seeks them eagerly is always able to find, I myself published an entomological article, a supplement to Léon Dufour's. This first work of mine won honourable mention from the Institute of France and was awarded a prize for experimental physiology. But soon I received a far more welcome recompense, in the shape of a most eulogistic and encouraging letter from the very man who had inspired me. From his home in the Landes, the revered master sent me a warm expression of his enthusiasm and urged me to go on with my studies. Even now, at that sacred recollection, my old eyes fill with happy tears.

THE RED ANTS

« 23 »

This, Fabre's only study of the ants, is taken from the sixth chapter of THE MASON-BEES. *While the last word remains to be spoken on the complex subject of ant-orientation, it is now known that the sense of smell, located in the antennae of the insects, enables most species to find their way about by following chemical trails. Such trails, laid down by the insects, sometimes last for a surprisingly long time. Dr. T. C. Schnierla, leading authority on the life of the army ant, found trails these insects laid down in the jungle had a strong smell like rancid butter and could be followed by the ants after a lapse of weeks and even months.* Polyergus rufescens, *the red slave-maker Fabre observed, has its counterpart in the United States,* Polyergus lucidus. *It is the same ant with a different name.*

Among the treasures of my harmas-laboratory, I place in the first rank an Ant-hill of *Polyergus rufescens,* the celebrated Red Ant, the slave-hunting Amazon. Unable to rear her family, incapable of seeking her food, of taking it even when it is within her reach, she needs servants who feed her and undertake the duties of housekeeping. The Red Ants make a practice of stealing children to wait on the community. They ransack the neighbouring Ant-hills, the home of a different species; they carry away nymphs, which soon attain maturity in the strange house and become willing and industrious servants.

When the hot weather of June and July sets in, I often see the Amazons leave their barracks of an afternoon and start on an expedition. The column measures five or six yards in length. If nothing worthy of attention be met upon the road, the ranks are fairly well maintained; but, at the first suspicion of an Ant-hill, the vanguard halts and deploys in a swarming throng, which is increased by the others as they come up hurriedly. Scouts are sent out; the Amazons recognize that they are on a wrong track; and the column forms again. It resumes its march, crosses the garden-paths, disappears from sight in the grass, reappears farther on, threads its way through the heaps of dead leaves, comes out again and continues its search. At last, a nest of Black Ants is discovered. The Red Ants hasten down to the dormitories where the nymphs lie and soon emerge with their booty. Then we have, at the gates of the underground city, a bewildering scrimmage between the defending blacks and the attacking reds. The struggle is too unequal to remain indecisive. Victory falls to the reds, who race back to their abode, each with her prize, a swaddled nymph, dangling from her mandibles. The reader who is not acquainted with these slave-raiding habits would be greatly interested in the story of the Amazons. I relinquish it, with much regret: it would take us too far from our subject, namely, the return to the nest.

The distance covered by the nymph-stealing column varies: it all depends on whether Black Ants are plentiful in the neighbourhood. At times, ten or twenty yards suffice; at others, it requires fifty, a hundred or more. I once saw the expedition go beyond the garden. The Amazons scaled the surrounding wall, which was thirteen feet high at that point, climbed over it and went on a little farther, into a corn-field. As for the route taken, this is a matter of indifference to the marching column. Bare ground, thick grass, a heap of dead leaves or stones, brickwork, a clump of shrubs: all are crossed without any marked preference for one sort of road rather than another.

What is rigidly fixed is the path home, which follows the outward track in all its windings and all its crossings, however difficult. Laden with their plunder, the Red Ants return to the nest by the same road, often an exceedingly complicated one, which the exigencies of the chase compelled them to take originally. They repass each spot which they passed at first; and this is to them a matter of such imperative necessity that no additional fatigue nor even the gravest danger can make them alter the track.

Let us suppose that they have crossed a thick heap of dead leaves, representing to them a path beset with yawning gulfs, where every moment some one falls, where many are exhausted as they struggle out of the hollows and reach the heights by means of swaying bridges, emerging at last from the labyrinth of lanes. No matter: on their return, they will not fail, though weighed down with their burden, once more to struggle through that weary maze. To avoid all this fatigue, they would have but to swerve slightly from the original path, for the

good, smooth road is there, hardly a step away. This little deviation
never occurs to them.

I came upon them one day when they were on one of their raids.
They were marching along the inner edge of the stone-work of the
garden-pond, where I have replaced the old batrachians by a colony of
Gold-fish. The wind was blowing very hard from the north and, taking
the column in flank, sent whole rows of the Ants flying into the water.
The fish hurried up; they watched the performance and gobbled up the
drowning insects. It was a difficult bit; and the column was decimated
before it had passed. I expected to see the return journey made by another road, which would wind round and avoid the fatal cliff. Not at
all. The nymph-laden band resumed the parlous path and the Gold-
fish received a double windfall: the Ants and their prizes. Rather than
alter its track, the column was decimated a second time.

It is not easy to find the way home again after a distant expedition,
during which there have been various sorties, nearly always by different
paths; and this difficulty makes it absolutely necessary for the Amazons
to return by the same road by which they went. The insect has no
choice of route, if it would not be lost on the way: it must come back by
the track which it knows and which it has lately travelled. The Processionary Caterpillars, when they leave their nest and go to another
branch, on another tree, in search of a type of leaf more to their taste,
carpet the course with silk and are able to return home by following
the threads stretched along their road. This is the most elementary
method open to the insect liable to stray on its excursions: a silken path
brings it home again. The Processionaries, with their unsophisticated
traffic-laws, are very different from the Mason-bees and others, who
have a special sense to guide them.

The Amazon, though belonging to the Hymenopteron clan, herself
possesses rather limited homing-faculties, as witness her compulsory
return by her former trail. Can she imitate, to a certain extent, the
Processionaries' method, that is to say, does she leave, along the road
traversed, not a series of conducting threads, for she is not equipped for
that work, but some odorous emanation, for instance, some formic scent,
which would allow her to guide herself by means of the olfactory sense?
This view is pretty generally accepted. The Ants, people say, are guided
by the sense of smell; and this sense of smell appears to have its seat
in the antennæ, which we see in continual palpitation. It is doubtless
very reprehensible, but I must admit that the theory does not inspire
me with overwhelming enthusiasm. In the first place, I have my sus-

picions about a sense of smell seated in the antennæ: I have given my reasons before; and, next, I hope to prove by experiment that the Red Ants are not guided by a scent of any kind.

To lie in wait for my Amazons, for whole afternoons on end, often unsuccessfully, meant taking up too much of my time. I engaged an assistant whose hours were not so much occupied as mine. It was my granddaughter Lucie, a little rogue who liked to hear my stories of the Ants. She had been present at the great battle between the reds and blacks and was much impressed by the rape of the long-clothes babies. Well-coached in her exalted functions, very proud of already serving that august lady, Science, my little Lucie would wander about the garden, when the weather seemed propitious, and keep an eye on the Red Ants, having been commissioned to reconnoitre carefully the road to the pillaged Ant-hill. She had given proof of her zeal; I could rely upon it.

One day, while I was spinning out my daily quota of prose, there came a banging at my study-door:

"It's I, Lucie! Come quick: the reds have gone into the blacks' house. Come quick!"

"And do you know the road they took?"

"Yes, I marked it."

"What! Marked it? And how?"

"I did what Hop-o'-My-Thumb did: I scattered little white stones along the road."

I hurried out. Things had happened as my six-year-old colleague said. Lucie had secured her provision of pebbles in advance and, on seeing the Amazon regiment leave barracks, had followed them step by step and placed her stones at intervals along the road covered. The Ants had made their raid and were beginning to return along the track of tell-tale pebbles. The distance to the nest was about a hundred paces, which gave me time to make preparations for an experiment previously contemplated.

I take a big broom and sweep the track for about a yard across. The dusty particles on the surface are thus removed and replaced by others. If they were tainted with any odorous effluvia, their absence will throw the Ants off the track. I divide the road, in this way, at four different points, a few feet apart.

The column arrives at the first section. The hesitation of the Ants is evident. Some recede and then return, only to recede once more; others wander along the edge of the cutting; others disperse sideways and seem to be trying to skirt the unknown country. The head of the column, at first closed up to a width of a foot or so, now scatters to three or four yards. But fresh arrivals gather in their numbers before the obstacle; they form a mighty array, an undecided horde. At last, a few Ants venture into the swept zone and others follow, while a few have meantime gone ahead and recovered the track by a circuitous route. At the other cuttings, there are the same halts, the same hesitations; nevertheless, they are crossed, either in a straight line or by going round. In spite of my snares, the Ants manage to return to the nest; and that by way of the little stones.

The result of the experiment seems to argue in favour of the sense of smell. Four times over, there are manifest hesitations wherever the road is swept. Though the return takes place, nevertheless, along the original track, this may be due to the uneven work of the broom, which has left certain particles of the scented dust in position. The Ants who went round the cleared portion may have been guided by the sweepings removed to either side. Before, therefore, pronouncing judgment for or against the sense of smell, it were well to renew the experiment under better conditions and to remove everything containing a vestige of scent.

A few days later, when I have definitely decided on my plan, Lucie resumes her watch and soon comes to tell me of a sortie. I was counting on it, for the Amazons rarely miss an expedition during the hot and sultry afternoons of June and July, especially when the weather threatens storm. Hop-o'-My-Thumb's pebbles once more mark out the road, on which I choose the point best-suited to my schemes.

A garden-hose is fixed to one of the feeders of the pond; the sluice is opened; and the Ants' path is cut by a continuous torrent, two or three feet wide and of unlimited length. The sheet of water flows swiftly and plentifully at first, so as to wash the ground well and remove anything that may possess a scent. This thorough washing lasts for nearly a quarter of an hour. Then, when the Ants draw near, returning from the plunder, I let the water flow more slowly and reduce its depth, so as not to overtax the strength of the insects. Now we have an obstacle which the Amazons must surmount, if it is absolutely necessary for them to follow the first trail.

This time, the hesitation lasts long and the stragglers have time to come up with the head of the column. Nevertheless, an attempt is made to cross the torrent by means of a few bits of gravel projecting above the water; then, failing to find bottom, the more reckless of the

Ants are swept off their feet and, without loosing hold of their prizes, drift away, land on some shoal, regain the bank and renew their search for a ford. A few straws borne on the waters stop and become so many shaky bridges, on which the Ants climb. Dry olive-leaves are converted into rafts, each with its load of passengers. The more venturesome, partly by their own efforts, partly by good luck, reach the opposite bank without adventitious aid. I see some who, dragged by the current to one or the other bank, two or three yards off, seem very much concerned as to what they shall do next. Amid this disorder, amid the dangers of drowning, not one lets go her booty. She would not dream of doing so: death sooner than that! In a word, the torrent is crossed somehow or other along the regular track.

The scent of the road cannot be the cause of this, it seems to me, for the torrent not only washed the ground some time beforehand, but also pours fresh water on it all the time that the crossing is taking place. Let us now see what will happen when the formic scent, if there really be one on the trail, is replaced by another, much stronger odour, one perceptible to our own sense of smell, which the first is not, at least not under present conditions.

I wait for a third sortie and, at one point in the road taken by the Ants, rub the ground with some handfuls of freshly-gathered mint. I cover the track, a little farther on, with the leaves of the same plant. The Ants, on their return, cross the section over which the mint was rubbed without apparently giving it a thought; they hesitate in front of the section heaped up with leaves and then go straight on.

After these two experiments, first with the torrent of water which washes away all trace of smell from the ground and then with the mint which changes the smell, I think that we are no longer at liberty to quote scent as the guide of the Ants that return to the nest by the road which they took at starting. Further tests will tell us more about it.

Without interfering with the soil, I now lay across the track some large sheets of paper, newspapers, keeping them in position with a few small stones. In front of this carpet, which completely alters the appearance of the road, without removing any sort of scent that it may possess, the Ants hesitate even longer than before any of my other snares, including the torrent. They are compelled to make manifold attempts, reconnaissances to right and left, forward movements and repeated retreats, before venturing altogether into the unknown zone. The paper straits are crossed at last and the march resumed as usual.

Another ambush awaits the Amazons some distance farther on. I have divided the track by a thin layer of yellow sand, the ground itself being grey. This change of colour alone is enough for a moment to disconcert the Ants, who again hesitate in the same way, though not for so long, as they did before the paper. Eventually, this obstacle is overcome like the others.

As neither the stretch of sand nor the stretch of paper got rid of any scented effluvia with which the trail may have been impregnated, it is patent that, as the Ants hesitated and stopped in the same way as before, they find their way not by sense of smell, but really and truly by sense of sight; for, every time that I alter the appearance of the track in any way whatever—whether by my destructive broom, my streaming water, my green mint, my paper carpet or my golden sand—the returning column calls a halt, hesitates and attempts to account for the changes that have taken place. Yes, it is sight, but a very dull sight, whose horizon is altered by the shifting of a few bits of gravel. To this short sight, a strip of paper, a bed of mint-leaves, a layer of yellow sand, a stream of water, a furrow made by the broom, or even lesser modifications are enough to transform the landscape; and the regiment, eager to reach home as fast as it can with its loot, halts uneasily on beholding this unfamiliar scenery. If the doubtful zones are at length passed, it is due to the fact that fresh attempts are constantly being made to cross the doctored strips and that at last a few Ants recognize well-known spots beyond them. The others, relying on their clearer-sighted sisters, follow.

Sight would not be enough, if the Amazon had not also at her service a correct memory for places. The memory of an Ant! What can that be? In what does it resemble ours? I have no answers to these questions; but a few words will enable me to prove that the insect has a very exact and persistent recollection of places which it has once visited. Here is something which I have often witnessed. It sometimes happens that the plundered Ant-hill offers the Amazons a richer spoil than the invading column is able to carry away. Or, again, the region visited is rich in Ant-hills. Another raid is necessary, to exploit the site thoroughly. In such cases, a second expedition takes place, sometimes on the next day, sometimes two or three days later. This time, the column does no reconnoitring on the way: it goes straight to the spot known to abound in nymphs and travels by the identical path which it followed before. It has sometimes happened that I have marked with small stones, for a distance of twenty yards, the road pursued a couple of days earlier and have then found the Amazons proceeding by the same route, stone by stone:

"They will go first here and then there," I said, according to the position of the guide-stones.

And they would, in fact, go first here and then there, skirting my line of pebbles, without any noticeable deviation.

Can one believe that odoriferous emanations diffused along the route are going to last for several days? No one would dare to suggest it. It must, therefore, be sight that directs the Amazons, sight assisted by its memory for places. And this memory is tenacious enough to retain the impression until the next day and later; it is scrupulously faithful, for it guides the column by the same path as on the day before, across the thousand irregularities of the ground.

How will the Amazon behave when the locality is unknown to her? Apart from topographical memory, which cannot serve her here, the region in which I imagine her being still unexplored, does the Ant possess the Mason-bee's sense of direction, at least within modest limits, and is she able thus to regain her Ant-hill or her marching column?

The different parts of the garden are not all visited by the marauding legions to the same extent: the north side is exploited by preference, doubtless because the forays in that direction are more productive. The Amazons, therefore, generally direct their troops north of their barracks; I seldom see them in the south. This part of the garden is, if not wholly unknown, at least much less familiar to them than the other. Having said that, let us observe the conduct of the strayed Ant.

I take up my position near the Ant-hill; and, when the column returns from the slave-raid, I force an Ant to step on a leaf which I hold out to her. Without touching her, I carry her two or three paces away from her regiment: no more than that, but in a southerly direction. It is enough to put her astray, to make her lose her bearings entirely. I see the Amazon, now replaced on the ground, wander about at random, still, I need hardly say, with her booty in her mandibles; I see her hurry away from her comrades, thinking that she is rejoining them; I see her retrace her steps, turn aside again, try to the right, try to the left and grope in a host of directions, without succeeding in finding her whereabouts. The pugnacious, strong-jawed slave-hunter is utterly lost two steps away from her party. I have in mind certain strays who, after half an hour's searching, had not succeeded in recovering the route and were going farther and farther from it, still carrying the nymph in their teeth. What became of them? What did they do with their spoil? I had not the patience to follow those dull-witted marauders to the end.

Let us repeat the experiment, but place the Amazon to the north. After more or less prolonged hesitations, after a search now in this direction, now in that, the Ant succeeds in finding her column. She knows the locality.

Here, of a surety, is a Hymenopteron deprived of that sense of direction which other Hymenoptera enjoy. She has in her favour a memory for places and nothing more. A deviation amounting to two or three of our strides is enough to make her lose her way and to keep her from returning to her people, whereas miles across unknown country will not foil the Mason-bee.

THE CABBAGE-CATERPILLAR

« 24 »

The European cabbage butterfly that Linnaeus named Pieris brassicae *is almost twice as large as the familiar cabbage butterfly of America,* Pieris rapae. *The latter, incidentally, is an insect immigrant from Europe. Accidentally introduced, it appeared around Quebec in 1860. Eight years later it had crossed the Canadian border into the United States and now it is familiar wherever cabbage is grown from coast to coast. It is the only butterfly doing serious damage to our crops. The Larini, mentioned by Fabre in this chapter, are beetles living in the heads of thistles. The parasitic Microgaster is related to the insects that prey upon tomato worms, emerging to spin tiny white, oval cocoons clustered together on the backs of the doomed larvae. Fabre's study of the cabbage-caterpillar forms the fourteenth chapter of* THE LIFE OF THE CATERPILLAR.

The cabbage of our modern kitchen-gardens is a semi-artificial plant, the produce of our agricultural ingenuity quite as much as of the niggardly gifts of nature. Spontaneous vegetation supplied us with the long-stalked, scanty-leaved, ill-smelling wilding, as found, according to the botanists, on the ocean cliffs. He had need of a rare inspiration who first showed faith in this rustic clown and proposed to improve it in his garden-patch.

Progressing by infinitesimal degrees, culture wrought miracles. It began by persuading the wild cabbage to discard its wretched leaves, beaten by the sea-winds, and to replace them by others, ample and fleshy and close-fitting. The gentle cabbage submitted without protest. It deprived itself of the joys of light by arranging its leaves in a large, compact head, white and tender. In our day, among the successors of those first tiny hearts, are some that, by virtue of their massive bulk, have earned the glorious name of *chou quintal,* as who should say, a hun-dredweight of cabbage. They are real monuments of green stuff.

Later, man thought of obtaining a generous dish with the thousand little sprays of the inflorescence. The cabbage consented. Under the cover of the central leaves, it gorged with food its sheaves of blossom, its flower-stalks, its branches and worked the lot into a fleshy conglomeration. This is the cauliflower, the broccoli.

Differently entreated, the plant, economizing in the centre of its shoot, set a whole family of close-wrapped cabbages ladder-wise on a tall stem. A multitude of dwarf leaf-buds took the place of the colossal head. This is the Brussels sprout.

Next comes the turn of the stump, an unprofitable, almost wooden thing, which seemed never to have any other purpose than to act as a support for the plant. But the tricks of gardeners are capable of every-thing, so much so that the stalk yields to the grower's suggestions and becomes fleshy and swells into an ellipse similar to the turnip, of which it possesses all the merits of corpulence, flavour and delicacy; only the strange product serves as a base for a few sparse leaves, the last protests of a real stem that refuses to lose its attributes entirely. This is the cole-rape.

If the stem allows itself to be allured, why not the root? It does in fact, yield to the blandishments of agriculture: it dilates its pivot into a flat turnip, which half emerges from the ground. This is the rutabaga, or swede, the turnip-cabbage of our northern districts.

Incomparably docile under our nursing, the cabbage has given its all for our nourishment and that of our cattle: its leaves, its flowers, its buds, its stalk, its root; all that it now wants is to combine the ornamental with the useful, to smarten itself, to adorn our flower-beds and cut a good figure on a drawing-room table. It has done this to perfection, not with its flowers, which, in their modesty, continue intractible, but with its curly and variegated leaves, which have the undulating grace of Ostrich-feathers and the rich colouring of a mixed bouquet. None who beholds it in this magnificence will recognize the near relation of the vulgar "greens" that form the basis of our cabbage-soup.

The cabbage, first in order of date in our kitchen-gardens, was held in high esteem by classic antiquity, next after the bean and, later, the pea; but it goes much farther back, so far indeed that no memories of its acquisition remain. History pays but little attention to these details: it celebrates the battle-fields whereon we meet our death, it scorns to speak of the ploughed fields whereby we thrive; it knows the names of the kings' bastards, it cannot tell us the origin of wheat. That is the way of human folly.

This silence respecting the precious plants that serve as food is most regrettable. The cabbage in particular, the venerable cabbage, that denizen of the most ancient garden-plots, would have had extremely interesting things to teach us. It is a treasure in itself, but a treasure twice exploited, first by man and next by the caterpillar of the Pieris, the common Large White Butterfly whom we all know (*Pieris brassicæ*, LIN.). This caterpillar feeds indiscriminately on the leaves of all varieties of cabbage, however dissimilar in appearance: he nibbles with the same appetite red cabbage and broccoli, curly greens and savoy, swedes and turnip-tops, in short, all that our ingenuity, lavish of time and patience, has been able to obtain from the original plant since the most distant ages.

But what did the caterpillar eat before our cabbages supplied him with copious provender? Obviously the Pieris did not wait for the advent of man and his horticultural works in order to take part in the joys of life. She lived without us and would have continued to live without us. A Butterfly's existence is not subject to ours, but rightfully independent of our aid.

Before the white-heart, the cauliflower, the savoy and the others were invented, the Pieris' caterpillar certainly did not lack food: he browsed the wild cabbage of the cliffs, the parent of all the latter-day wealth; but, as this plant is not widely distributed and is, in any case, limited to certain maritime regions, the welfare of the Butterfly, whether on plain or hill, demanded a more luxuriant and more common plant for pasturage. This plant was apparently one of the Cruciferæ, more or less seasoned with sulphuretted essence, like the cabbages. Let us experiment on these lines.

I rear the Pieris' caterpillars from the egg upwards on the wall-rocket (*Diplotaxis tenuifolia*, DEC.), which imbibes strong spices along the edge of the paths and at the foot of the walls. Penned in a large, wire-gauze bell-cage, they accept this provender without demur; they nibble it with the same appetite as if it were cabbage; and they end by producing chrysalids and Butterflies. The change of fare causes not the least trouble.

I am equally successful with other crucifers of a less marked flavour: white mustard (*Sinapis incana*, LIN.), dyer's woad (*Isatis tinctoria*, LIN.), wild radish (*Raphanus raphanistrum*, LIN.), whitlow pepper-wort (*Lepidium draba*, LIN.), hedge-mustard (*Sisymbrium officinale*,

Scop.). On the other hand, the leaves of the lettuce, the bean, the pea, the corn-salad are obstinately refused. Let us be content with what we have seen: the fare has been sufficiently varied to show us that the Cabbage-caterpillar feeds exclusively on a large number of crucifers, perhaps even on all.

As these experiments are made in the enclosure of a bell-cage, one might imagine that captivity impels the flock to feed, in the absence of better things, on what it would refuse were it free to hunt for itself. Having naught else within their reach, the starvelings consume any and all Cruciferæ, without distinction of species. Can things sometimes be the same in the open fields, where I play none of my tricks? Can the family of the White Butterfly be settled on other crucifers than the cabbage? I start a quest along the paths near the gardens and end by finding on wild radish and white mustard colonies as crowded and prosperous as those established on cabbage.

Now, except when the metamorphosis is at hand, the caterpillar of the White Butterfly never travels: he does all his growing on the identical plant whereon he saw the light. The caterpillars observed on the wild radish, as well as other households, are not, therefore, emigrants who have come as a matter of fancy from some cabbage-patch in the neighbourhood: they have hatched on the very leaves where I find them. Hence I arrive at this conclusion: the White Butterfly, who is fitful in her flight, chooses cabbage first, to dab her eggs upon, and different Cruciferæ next, varying greatly in appearance.

How does the Pieris manage to know her way about her botanical domain? We have seen the Larini, those explorers of fleshy receptacles with an artichoke flavour, astonish us with their knowledge of the flora of the thistle tribe; but their lore might, at a pinch, be explained by the method followed at the moment of housing the egg. With their rostrum, they prepare niches and dig out basins in the receptacle exploited and consequently they taste the thing a little before entrusting their eggs to it. On the other hand, the Butterfly, a nectar-drinker, makes not the least enquiry into the savoury qualities of the leafage; at most, dipping her proboscis into the flowers, she abstracts a mouthful of syrup. This means of investigation, moreover, would be of no use to her, for the plant selected for the establishing of her family is, for the most part, not yet in flower. The mother flits for a moment around the plant; and that swift examination is enough: the emission of eggs takes place if the provender be found suitable.

The botanist, to recognize a crucifer, requires the indications provided by the flower. Here the Pieris surpasses us. She does not consult the seed-vessel, to see if it be long or short, nor yet the petals, four in number and arranged in a cross, because the plant, as a rule, is not in flower; and still she recognizes off-hand what suits her caterpillars, in spite of profound differences that would embarrass any but a botanical expert.

Unless the Pieris has an innate power of discrimination to guide her, it is impossible to understand the great extent of her vegetable realm. She needs for her family Cruciferæ, nothing but Cruciferæ; and she knows this group of plants to perfection. I have been an enthusiastic botanist for half a century and more. Nevertheless, to discover if this or that plant, new to me, is or is not one of the Cruciferæ, in the absence of flowers and fruits I should have more faith in the Butterfly's statements than in all the learned records of the books. Where science is apt to make mistakes, instinct is infallible.

The Pieris has two families a year: one in April and May, the other in September. The cabbage-patches are renewed in those same months. The Butterfly's calendar tallies with the gardener's: the moment that provisions are in sight, consumers are forthcoming for the feast.

The eggs are a bright orange-yellow and do not lack prettiness when examined under the lens. They are blunted cones, ranged side by side on their round base and adorned with finely-scored longitudinal ridges. They are collected in slabs, sometimes on the upper surface, when the leaf that serves as a support is spread wide, sometimes on the lower surface when the leaf is pressed to the next ones. Their number varies considerably. Slabs of a couple of hundred are pretty frequent; isolated eggs, or eggs collected in small groups, are, on the contrary, rare. The mother's output is affected by the degree of quietness at the moment of laying.

The outer circumference of the group is irregularly formed, but the inside presents a certain order. The eggs are here arranged in straight rows backing against one another in such a way that each egg finds a double support in the preceding row. This alternation, without being of an irreproachable precision, gives a fairly stable equilibrium to the whole.

To see the mother at her laying is no easy matter: when examined too closely, the Pieris decamps at once. The structure of the work, however, reveals the order of the operations pretty clearly. The ovipositor swings slowly first in this direction, then in that, by turns; and a new egg is lodged in each space between two adjoining eggs in the previous row. The extent of the oscillation determines the length of the row, which is longer or shorter according to the layer's fancy.

The hatching takes place in about a week. It is almost simultaneous for the whole mass: as soon as one caterpillar comes out of its egg, the others come out also, as though the natal impulse were communicated from one to the other. In the same way, in the nest of the Praying Mantis, a warning seems to be spread abroad, arousing every one of the population. It is a wave propagated in all directions from the point first struck.

The egg does not open by means of a dehiscence similar to that of the vegetable-pods whose seeds have attained maturity; it is the new-born grub itself that contrives an exit-way by gnawing a hole in its enclosure. In this manner, it obtains near the top of the cone a symmetrical dormer-window, clean-edged, with no joins nor unevenness of any kind, showing that this part of the wall has been nibbled away and swallowed. But for this breach, which is just wide enough for the deliverance, the egg remains intact, standing firmly on its base. It is now that the lens is best able to take in its elegant structure. What it sees is a bag made of ultra-fine goldbeater's-skin, translucent, stiff and white, retaining the complete form of the original egg. A score of streaked and knotted lines run from the top to the base. It is the wizard's pointed cap, the mitre with the grooves carved into jewelled chaplets. All said, the Cabbage-caterpillar's birth-casket is an exquisite work of art.

The hatching of the lot is finished in a couple of hours and the swarming family musters on the layer of swaddling-clothes, still in the same position. For a long time, before descending to the fostering leaf, it lingers on this kind of hot-bed, is even very busy there. Busy with what? It is browsing a strange kind of grass, the handsome mitres that remain standing on end. Slowly and methodically, from top to base, the new-born grubs nibble the wallets whence they have just emerged. By to-morrow, nothing is left of these but a pattern of round dots, the bases of the vanished sacks.

As his first mouthfuls, therefore, the Cabbage-caterpillar eats the membranous wrapper of his egg. This is a regulation diet, for I have never seen one of the little grubs allow itself to be tempted by the adjacent green stuff before finishing the ritual repast whereat skin bottles furnish forth the feast. It is the first time that I have seen a larva make a meal of the sack in which it was born. Of what use can this singular fare be to the budding caterpillar? I suspect as follows: the leaves of the cabbage are waxed and slippery surfaces and nearly always slant considerably. To graze on them without risking a fall, which would be fatal in earliest childhood, is hardly possible unless with moorings that afford a steady support. What is needed is bits of silk stretched along the road as fast as progress is made, something for the legs to grip, something to provide a good anchorage even when the grub is upside down. The silk-tubes, where those moorings are manufactured, must be very scantily supplied in a tiny, new-born animal; and it is expedient that they be filled without delay with the aid of a special form of nourishment. Then what shall the nature of the first food be? Vegetable matter, slow to elaborate and niggardly in its yield, does not fulfil the desired conditions at all well, for time presses and we must trust ourselves safely to the slippery leaf. An animal diet would be preferable: it is easier to digest and undergoes chemical changes in a shorter time. The wrapper of the egg is of a horny nature, as silk itself is. It will not take long to transform the one into the other. The grub therefore tackles the remains of its egg and turns it into silk to carry with it on its first journeys.

If my surmise is well-founded, there is reason to believe that, with a view to speedily filling the silk-glands to which they look to supply them with ropes, other caterpillars beginning their existence on smooth and steeply-slanting leaves also take as their first mouthful the membranous sack which is all that remains of the egg.

The whole of the platform of birth-sacks which was the first camping-ground of the White Butterfly's family is razed to the ground; naught remains but the round marks of the individual pieces that composed it. The structure of piles has disappeared; the prints left by the piles remain. The little caterpillars are now on the level of the leaf which shall henceforth feed them. They are a pale orange-yellow, with a sprinkling of white bristles. The head is a shiny black and remarkably powerful; it already gives signs of the coming gluttony. The little animal measures scarcely two millimetres in length.

The troop begins its steadying-work as soon as it comes into contact with its pasturage, the green cabbage-leaf. Here, there, in its immediate neighbourhood, each grub emits from its spinning-glands short cables so slender that it takes an attentive lens to catch a glimpse of them. This is enough to ensure the equilibrium of the almost imponderable atom.

The vegetarian meal now begins. The grub's length promptly increases from two millimetres to four. Soon, a moult takes place which alters its costume: its skin becomes speckled, on a pale-yellow ground, with a number of black dots intermingled with white bristles. Three or four days of rest are necessary after the fatigue of breaking cover. When this is over, the hunger-fit starts that will make a ruin of the cabbage within a few weeks.

What an appetite! What a stomach, working continuously day and night! It is a devouring laboratory, through which the foodstuffs merely pass, transformed at once. I serve up to my caged herd a bunch of leaves picked from among the biggest: two hours later, nothing remains but the thick midribs; and even these are attacked when there is any delay in renewing the victuals. At this rate, a "hundredweight-cabbage," doled out leaf by leaf, would not last my menagerie a week.

The gluttonous animal, therefore, when it swarms and multiplies, is a scourge. How are we to protect our gardens against it? In the days of Pliny, the great Latin naturalist, a stake was set up in the middle of the cabbage-bed to be preserved; and on this stake was fixed a Horse's skull bleached in the sun: a Mare's skull was considered even better. This sort of bogey was supposed to ward off the devouring brood.

My confidence in this preservative is but an indifferent one; my reason for mentioning it is that it reminds me of a custom still observed in our own days, at least in my part of the country. Nothing is so long-lived as absurdity. Tradition has retained, in a simplified form, the ancient defensive apparatus of which Pliny speaks. For the Horse's skull our people have substituted an eggshell on the top of a switch stuck among the cabbages. It is easier to arrange; also, it is quite as useful, that is to say, it has no effect whatever.

Everything, even the nonsensical, is capable of explanation with a little credulity. When I question the peasants, our neighbours, they tell me that the effect of the eggshell is as simple as can be: the Butterflies, attracted by the whiteness, come and lay their eggs on it. Broiled by the sun and lacking all nourishment on that thankless support, the little caterpillars die; and that makes so many fewer.

I insist; I ask them if they have ever seen slabs of eggs or masses of young caterpillars on those white shells.

"Never," they reply, with one voice.

"Well, then?"

"It was done in the old days and so we go on doing it: that's all we know; and that's enough for us."

I leave it at that, persuaded that the memory of the Horse's skull used once upon a time is ineradicable, like all the rustic absurdities implanted by the ages.

We have, when all is said, but one means of protection, which is to watch and inspect the cabbage-leaves assiduously and crush the slabs of eggs between our finger and thumb and the caterpillars with our feet. Nothing is so effective as this method, which makes great demands on one's time and vigilance. What pains to obtain an unspoilt cabbage!

And what a debt do we not owe to those humble scrapers of the soil, those ragged heroes who provide us with the wherewithal to live!

To eat and digest, to accumulate reserves whence the Butterfly will issue: that is the caterpillar's one and only business. The Cabbage-caterpillar performs it with insatiable gluttony. Incessantly it browses, incessantly digests: the supreme felicity of an animal which is little more than an intestine. There is never a distraction, unless it be certain see-saw movements which are particularly curious when several cater-pillars are grazing side by side, abreast. Then, at intervals, all the heads in the row are briskly lifted and as briskly lowered, time after time, with an automatic precision worthy of a Prussian drill-ground. Can it be their method of intimidating an always possible aggressor? Can it be a manifestation of gaiety, when the wanton sun warms their full paunches? Whether sign of fear or sign of bliss, this is the only exercise that the gluttons allow themselves until the proper degree of plumpness is at-tained.

After a month's grazing, the voracious appetite of my caged herd is assuaged. The caterpillars climb the trelliswork in every direction, walk about anyhow, with their forepart raised and searching space. Here and there, as they pass, the swaying herd put forth a thread. They wander restlessly, anxious to travel afar. The exodus now prevented by the trellised enclosure I once saw under excellent conditions. At the advent of the cold weather, I had placed a few cabbage-stalks, covered with caterpillars, in a small greenhouse. Those who saw the common kitchen vegetable sumptuously lodged under glass, in the company of the pelargonium and the Chinese primrose, were astonished at my curi-ous fancy. I let them smile. I had my plans: I wanted to find out how the family of the Large White Butterfly behaves when the cold weather sets in. Things happened just as I wished. At the end of November, the caterpillars, having grown to the desired extent, left the cabbages, one by one, and began to roam about the walls. None of them fixed himself there or made preparations for the transformation. I suspected that they wanted the choice of a spot in the open air, exposed to all the rigours of winter. I therefore left the door of the hothouse open. Soon, the whole crowd had disappeared.

I found them dispersed all over the neighbouring walls, some thirty yards off. The thrust of a ledge, the eaves formed by a projecting bit of mortar served them as a shelter where the chrysalid moult took place and where the winter was passed. The Cabbage-caterpillar possesses a robust constitution, unsusceptible to torrid heat or icy cold. All that he needs for his metamorphosis is an airy lodging, free from permanent damp.

The inmates of my fold, therefore, move about for a few days on the trelliswork, anxious to travel afar in search of a wall. Finding none and realizing that time presses, they resign themselves. Each one, supporting himself on the trellis, first weaves around himself a thin carpet of white silk, which will form the sustaining layer at the time of the laborious and delicate work of the nymphosis. He fixes his rear-end to this base by a silk pad and his fore-part by a strap that passes under his shoulders and is fixed on either side to the carpet. Thus slung from his three fastenings, he strips himself of his larval apparel and turns into a chrysalis in the open air, with no protection save that of the wall, which the caterpillar would certainly have found had I not interfered.

Of a surety, he would be short-sighted indeed that pictured a world of good things prepared exclusively for our advantage. The earth, the great foster-mother, has a generous breast. At the very moment when nourishing matter is created, even though it be with our own zealous aid, she summons to the feast host upon host of consumers, who are all the more numerous and enterprising in proportion as the table is more amply spread. The cherry of our orchards is excellent eating: a maggot contends with us for its possession. In vain do we weigh suns and planets: our supremacy, which fathoms the universe, cannot prevent a wretched worm from levying its toll on the delicious fruit. We make ourselves at home in a cabbage-bed: the sons of the Pieris make themselves at home there too. Preferring broccoli to wild radish, they profit where we have profited; and we have no remedy against their competition save caterpillar-raids and egg-crushing, a thankless, tedious and none too efficacious work.

Every creature has its claims on life. The Cabbage-caterpillar eagerly puts forth his own, so much so that the cultivation of the precious plant would be endangered if others concerned did not take part in its defence. These others are the auxiliaries, our helpers from necessity and not from sympathy. The words friend and foe, auxiliaries and ravagers are here the mere conventions of a language not always adapted to render the exact truth. He is our foe who eats or attacks our crops; our friend is he who feeds upon our foes. Everything is reduced to a frenzied contest of appetites.

In the name of the might that is mine, of trickery, of highway robbery, clear out of that, you, and make room for me: give me your seat at the banquet! That is the inexorable law in the world of animals and more or less, alas, in our own world as well!

Now, among our entomological auxiliaries, the smallest in size are the best at their work. One of them is charged with watching over the cabbages. She is so small, she works so discreetly that the gardener does not know her, has not even heard of her. Were he to see her by accident, flitting around the plant which she protects, he would take no notice of her, would not suspect the service rendered. I propose to set forth the tiny midget's deserts.

Scientists call her *Microgaster glomeratus*. What exactly was in the mind of the author of the name Microgaster, which means little belly? Did he intend to allude to the insignificance of the abdomen? Not so. However slight the belly may be, the insect nevertheless possesses one, correctly proportioned to the rest of the body, so that the classic denomination, far from giving us any information, might mislead us, were we to trust it wholly. Nomenclature, which changes from day to day and becomes more and more cacophonous, is an unsafe guide. Instead of asking the animal what its name is, let us begin by asking:

"What can you do? What is your business?"

Well, the Microgaster's business is to exploit the Cabbage-caterpillar, a clearly-defined business, admitting of no possible confusion. Would we behold her works? In the spring, let us inspect the neighbourhood of the kitchen-garden. Be our eye never so unobservant, we shall notice against the walls or on the withered grasses at the foot of the hedges some very small yellow cocoons, heaped into masses the size of a hazel-nut. Beside each group lies a Cabbage-caterpillar, sometimes dying, sometimes dead and always presenting a most tattered appearance. These cocoons are the work of the Microgaster's family, hatched or on the point of hatching into the perfect stage; the caterpillar is the dish whereon that family has fed during its larval state. The epithet *glomeratus,* which accompanies the name of Microgaster, suggests this conglomeration of cocoons. Let us collect the clusters as they are, without seeking to separate them, an operation which would demand both patience and dexterity, for the cocoons are closely united by the inextricable tangle of their surface-threads. In May, a swarm of pigmies will sally forth, ready to get to business in the cabbages.

Colloquial language uses the terms Midge and Gnat to describe the tiny insects which we often see dancing in a ray of sunlight. There is something of everything in those aerial ballets. It is possible that the persecutrix of the Cabbage-caterpillar is there, along with many another; but the name of Midge cannot properly be applied to her. He who says

Midge says Fly, Dipteron, two-winged insect; and our friend has four wings, one and all adapted for flying. By virtue of this characteristic and others no less important, she belongs to the order of Hymenoptera. No matter: as our language possesses no more precise term outside the scientific vocabulary, let us use the expression Midge, which pretty well conveys the general idea. Our Midge, the Microgaster, is the size of an average Gnat. She measures 3 or 4 millimetres. The two sexes are equally numerous and wear the same costume, a black uniform, all but the legs, which are pale red. In spite of this likeness, they are easily distinguished. The male has an abdomen which is slightly flattened and moreover curved at the tip; the female, before the laying, has hers full and perceptibly distended by its ovular contents. This rapid sketch of the insect should be enough for our purpose.

If we wish to know the grub and especially to inform ourselves of its manner of living, it is advisable to rear in a cage a numerous herd of Cabbage-caterpillars. Whereas a direct search on the cabbages in our garden would give us but a difficult and uncertain harvest, by this means we shall daily have as many as we wish before our eyes.

In the course of June, which is the time when the caterpillars quit their pastures and go far afield to settle on some wall or other, those in my fold, finding nothing better, climb to the dome of the cage to make their preparations and to spin a supporting network for the chrysalid's needs. Among these spinners we see some weaklings working listlessly at their carpet. Their appearance makes us deem them in the grip of a mortal disease. I take a few of them and open their bellies, using a needle by way of a scalpel. What comes out is a bunch of green entrails, soaked in a bright yellow fluid, which is really the creature's blood. These tangled intestines swarm with little, lazy grubs, varying greatly in number, from ten or twenty at least to sometimes half-a-hundred. They are the offspring of the Microgaster.

What do they feed on? The lens makes conscientious enquiries; nowhere does it manage to show me the vermin attacking solid nourishment, fatty tissues, muscles or other parts; nowhere do I see them bite, gnaw or dissect. The following experiment will tell us more fully: I pour into a watch-glass the crowds extracted from the hospitable paunches. I flood them with caterpillar's blood obtained by simple pricks; I place the preparation under a glass bell-jar, in a moist atmosphere, to prevent evaporation; I repeat the nourishing bath by means of fresh bleedings and give them the stimulant which they would have gained from the living caterpillar. Thanks to these precautions, my charges have all the appearance of excellent health; they drink and thrive. But this state of things cannot last long. Soon ripe for the transformation, my grubs leave the dining-room of the watch-glass as they would have left the caterpillar's belly; they come to the ground to try and weave their tiny cocoons. They fail in the attempt and perish. They have missed a suitable support, that is to say, the silky carpet provided by the dying caterpillar. No matter: I have seen enough to convince me. The larvæ of the Microgaster do not eat in the strict sense of the word: they live on soup; and that soup is the caterpillar's blood.

Examine the parasites closely and you shall see that their diet is bound to be a liquid one. They are little white grubs, neatly segmented, with a pointed fore-part splashed with tiny black marks, as though the atom had been slaking its thirst in a drop of ink. It moves its hindquarters slowly, without shifting its position. I place it under the microscope. The mouth is a pore, devoid of any apparatus for disintegration-work: it has no fangs, no horny nippers, no mandibles; its attack is just a kiss. It does not chew, it sucks, it takes discreet sips at the moisture all around it.

The fact that it refrains entirely from biting is confirmed by my autopsy of the stricken caterpillars. In the patient's belly, notwithstanding the number of nurselings who hardly leave room for the nurse's entrails, everything is in perfect order; nowhere do we see a trace of mutilation. Nor does aught on the outside betray any havoc within. The exploited caterpillars graze and move about peacefully, giving no sign of pain. It is impossible for me to distinguish them from the unscathed ones in respect of appetite and untroubled digestion.

When the time approaches to weave the carpet for the support of the chrysalis, an appearance of emaciation at last points to the evil that is at their vitals. They spin nevertheless. They are stoics who do not forget their duty in the hour of death. At last, they expire, quite softly, not of any wounds, but of anæmia, even as a lamp goes out when the oil comes to an end. And it has to be. The living caterpillar, capable of feeding itself and forming blood, is a necessity for the welfare of the grubs; it has to last about a month, until the Microgaster's offspring have achieved their full growth. The two calendars synchronize in a remarkable way. When the caterpillar leaves off eating and makes its preparations for the metamorphosis, the parasites are ripe for the exodus. The bottle dries up when the drinkers cease to need it; but until that moment it must remain more or less well-filled, although becoming limper daily. It is important, therefore, that the caterpillar's existence be not endangered

by wounds which, even though very tiny, would stop the working of
the blood-fountains. With this intent, the drainers of the bottle are, in
a manner of speaking, muzzled; they have by way of a mouth a pore
that sucks without bruising.

The dying caterpillar continues to lay the silk of his carpet with a
slow oscillation of the head. The moment now comes for the parasites
to emerge. This happens in June and generally at nightfall. A breach
is made on the ventral surface or else in the sides, never on the back:
one breach only, contrived at a point of minor resistance, at the junction
of two segments; for it is bound to be a toilsome business, in the absence of a set of filing-tools. Perhaps the worms take one another's places
at the point attacked and come by turns to work at it with a kiss.

In one short spell, the whole tribe issues through this single opening
and is soon wriggling about, perched on the surface of the caterpillar.
The lens cannot perceive the hole, which closes on the instant. There is
not even a hæmorrhage: the bottle has been drained too thoroughly.
You must press it between your fingers to squeeze out a few drops of
moisture and thus discover the spot of exit.

Around the caterpillar, who is not always quite dead and who sometimes even goes on weaving his carpet a moment longer, the vermin at
once begin to work at their cocoons. The straw-coloured thread, drawn
from the silk-glands by a backward jerk of the head, is first fixed to the
white network of the caterpillar and then produces adjacent warp-
beams, so that, by mutual entanglements, the individual works are
welded together and form an agglomeration in which each of the worms
has its own cabin. For the moment, what is woven is not the real
cocoon, but a general scaffolding which will facilitate the construction
of the separate shells. All these frames rest upon those adjoining and,
mixing up their threads, become a common edifice wherein each grub
contrives a shelter for itself. Here at last the real cocoon is spun, a pretty
little piece of closely-woven work.

In my rearing-jars, I obtain as many groups of those tiny shells as my
future experiments can wish for. Three-fourths of the caterpillars have
supplied me with them, so ruthless has been the toll of the spring births.
I lodge these groups, one by one, in separate glass tubes, thus forming
a collection on which I can draw at will, while, in view of my experiments, I keep under observation the whole swarm produced by one
caterpillar.

The adult Microgaster appears a fortnight later, in the middle of
June. There are fifty in the first tube examined. The riotous multitude

is in the full enjoyment of the pairing-season, for the two sexes always figure among the guests of any one caterpillar. What animation! What an orgy of love! The carnival of those pigmies bewilders the observer and makes his head swim.

Most of the females, wishful of liberty, plunge down to the waist between the glass of the tube and the plug of cotton-wool that closes the end turned to the light; but the lower halves remain free and form a circular gallery in front of which the males hustle one another, take one another's places and hastily operate. Each bides his turn, each attends to his little matters for a few moments and then makes way for his rivals and goes off to start again elsewhere. The turbulent wedding lasts all the morning and begins afresh next day, a mighty throng of couples embracing, separating and embracing once more.

There is every reason to believe that, in gardens, the mated ones, finding themselves in isolated couples, would keep quieter. Here, in the tube, things degenerate into a riot because the assembly is too numerous for the narrow space.

What is lacking to complete its happiness? Apparently, a little food, a few sugary mouthfuls extracted from the flowers. I serve up some provisions in the tubes: not drops of honey, in which the puny creatures would get stuck, but little strips of paper spread with that dainty. They come to them, take their stand on them and refresh themselves. The fare appears to agree with them. With this diet, renewed as the strips dry up, I can keep them in very good condition until the end of my inquisition.

There is another arrangement to be made. The colonists in my spare tubes are restless and quick of flight; they will have to be transferred presently to sundry vessels without my risking the loss of a good number, or even the whole lot, a loss which my hands, my forceps and other means of coercion would be unable to prevent by checking the nimble movements of the tiny prisoners. The irresistible attraction of the sunlight comes to my aid. If I lay one of my tubes horizontally on the table, turning one end towards the full light of a sunny window, the captives at once make for this brighter end and play about there for a long while, without seeking to retreat. If I turn the tube in the opposite direction, the crowd immediately shifts its quarters and collects at the other end. The brilliant sunlight is its great joy. With this bait, I can send it whithersoever I please.

We will therefore place the new receptacle, jar or test-tube, on the table, pointing the closed end towards the window. At its mouth, we open one of the full tubes. No other precaution is needed: even though the mouth leaves a large interval free, the swarm hastens into the lighted chamber. All that remains to be done is to close the apparatus before moving it. The observer is now in control of the multitude, without appreciable losses, and is able to question it at will.

We will begin by asking:

"How do you manage to lodge your germs inside the caterpillar?"

This question and others of the same category, which ought to take precedence of everything else, are generally neglected by the impaler of insects, who cares more for the niceties of nomenclature than for glorious realities. He classifies his subjects, dividing them into regiments with barbarous labels, a work which seems to him the highest expression of entomological science. Names, nothing but names: the rest hardly counts. The persecutor of the Pieris used to be called Microgaster, that is to say, little belly: to-day she is called Apantales, that is to say, the incomplete. What a fine step forward! We now know all about it!

Can our friend at least tell us how "the little belly" or "the incomplete" gets into the caterpillar? Not a bit of it! A book which, judging by its recent date, should be the faithful echo of our actual knowledge, informs us that the Microgaster inserts her eggs direct into the caterpillar's body. It goes on to say that the parasitic vermin inhabit the chrysalis, whence they make their way out by perforating the stout horny wrapper. Hundreds of times have I witnessed the exodus of the grubs ripe for weaving their cocoons; and the exit has always been made through the skin of the caterpillar and never through the armour of the chrysalis. The fact that its mouth is a mere clinging pore, deprived of any offensive weapon, would even lead me to believe the grub is incapable of perforating the chrysalid's covering.

This proved error makes me doubt the other proposition, though logical, after all, and agreeing with the methods followed by a host of parasites. No matter: my faith in what I read in print is of the slightest; I prefer to go straight to facts. Before making a statement of any kind, I want to see, what I call seeing. It is a slower and more laborious process; but it is certainly much safer.

I will not undertake to lie in wait for what takes place on the cabbages in the garden: that method is too uncertain and besides does not lend itself to precise observation. As I have in hand the necessary materials, to wit, my collection of tubes swarming with the parasites newly hatched into the adult form, I will operate on the little table in my animals' laboratory. A jar with a capacity of about a litre is placed on the table, with the bottom turned towards the window in the sun. I put into it a cabbage-leaf covered with caterpillars, sometimes fully developed, sometimes half-way, sometimes just out of the egg. A strip of honeyed paper will serve the Microgaster as a dining-room, if the experiment is destined to take some time. Lastly, by the method of transfer which I described above, I send the inmates of one of my tubes into the apparatus. Once the jar is closed, there is nothing left to do but to let things take their course and to keep an assiduous watch, for days and weeks, if need be. Nothing worth remarking can escape me.

The caterpillars graze placidly, heedless of their terrible attendants. If some giddy-pates in the turbulent swarm pass over the caterpillars' spines, these draw up their fore-part with a jerk and as suddenly lower it again; and that is all: the intruders forthwith decamp. Nor do the latter seem to contemplate any harm: they refresh themselves on the honey-smeared strip, they come and go tumultuously. Their short flights may land them, now in one place, now in another, on the browsing herd, but they pay no attention to it. What we see is casual meetings, not deliberate encounters.

In vain I change the flock of caterpillars and vary their age; in vain I change the squad of parasites: in vain I follow events in the jar for long hours, morning and evening, both in a dim light and in the full glare of the sun: I succeed in seeing nothing, absolutely nothing, on the parasite's side, that resembles an attack. No matter what the ill-informed authors say—ill-informed because they had not the patience to see for themselves —the conclusion at which I arrive is positive: to inject the germs, the Microgaster never attacks the caterpillars.

The invasion, therefore, is necessarily effected through the Butterfly's eggs themselves, as experiment will prove. My broad jar would tell against the inspection of the troop, kept at too great a distance by the glass enclosure; and I therefore select a tube an inch wide. I place in this a shred of cabbage-leaf, bearing a slab of eggs, as laid by the Butterfly. I next introduce the inmates of one of my spare vessels. A strip of paper smeared with honey accompanies the new arrivals.

This happens early in July. Soon, the females are there, fussing about, sometimes to the extent of blackening the whole slab of yellow eggs. They inspect the treasure, flutter their wings and brush their hind-legs against each other, a sign of keen satisfaction. They sound the heap, probe the interstices with their antennæ and tap the individual eggs with their palpi; then, this one here, that one there, they quickly apply the tip of their abdomen to the egg selected. Each time, we see a slender, horny prickle darting from the ventral surface, close to the end. This is the instrument that deposits the germ under the film of the egg; it is the inoculation-needle. The operation is performed calmly and methodically, even when several mothers are working at one and the same time. Where one has been, a second goes, followed by a third, a fourth and others yet, nor am I able definitely to see the end of the visits paid to the same egg. Each time, the needle enters and inserts a germ.

It is impossible, in such a crowd, for the eye to follow the successive mothers who hasten to lay in each; but there is one quite practicable method by which we can estimate the number of germs introduced into a single egg, which is, later, to open the ravaged caterpillars and count the worms which they contain. A less repugnant means is to number the little cocoons heaped up around each dead caterpillar. The total will tell us how many germs were injected, some by the same mother returning several times to the egg already treated, others by different mothers. Well, the number of these cocoons varies greatly. Generally, it fluctuates in the neighbourhood of twenty, but I have come across as many as sixty-five; and nothing tells me that this is the extreme limit. What hideous industry for the extermination of a Butterfly's progeny!

I am fortunate at this moment in having a highly-cultured visitor, versed in the profundities of philosophic thought. I make way for him before the apparatus wherein the Microgaster is at work. For an hour and more, standing lens in hand, he, in his turn, looks and sees what I have just seen; he watches the layers who go from one egg to the other, make their choice, draw their slender lancet and prick what the stream of passers-by, one after the other, have already pricked. Thoughtful and a little uneasy, he puts down his lens at last. Never had he been vouchsafed so clear a glimpse as here, in my finger-wide tube, of the masterly brigandage that runs through all life down to that of the very smallest.

HOW I MET THE MASON-BEE

« 25 »

At the time of the events recorded here, Fabre had just obtained a diploma from the normal school at Avignon and begun his teaching career as a schoolmaster at Carpentras. He was then nineteen years old. His salary, of which he spent his whole first month's income to purchase an illustrated book on insects, was $140 a year. THE MASON-BEES *is the source of the following selection.*

It was when I first began to teach, about 1843, that I made the Mason-bee's acquaintance. I had left the normal school at Vaucluse, some months before, with my diploma and all the simple enthusiasm of my eighteen years, and had been sent to Carpentras, there to manage the primary school attached to the college.

Among the subjects taught, one in particular appealed to both master and pupils. This was open-air geometry, practical surveying. The college had none of the necessary outfit; but, with my fat pay—seven hundred francs a year, if you please!—I could not hesitate over the expense. A surveyor's chain and stakes, arrows, level, square and compass were bought with my money. A microscopic graphometer, not much larger than the palm of one's hand and costing perhaps five francs, was provided by the establishment. There was no tripod to it; and I had one made. In short, my equipment was complete.

And so, when May came, once every week we left the gloomy school-room for the fields. It was a regular holiday. The boys disputed for the honour of carrying the stakes, divided into bundles of three; and more than one shoulder, as we walked through the town, felt the reflected glory of those erudite rods. I myself—why conceal the fact?—was not without a certain satisfaction as I piously carried that most delicate and precious apparatus, the historic five-franc graphometer. The scene of operations was an untilled, flinty plain, a *harmas,* as we call it in the district. Here, no curtain of green hedges or shrubs prevented me from keeping an eye upon my staff; here—an indispensable condition—I had not the irresistible temptation of the unripe apricots to fear for my scholars. The plain stretched far and wide, covered with nothing but flowering thyme and rounded pebbles. There was ample scope for every imaginable polygon; trapezes and triangles could be combined in all sorts of ways. The inaccessible distances had ample elbow-room; and there was even an old ruin, once a pigeon-house, that lent its perpendicular to the graphometer's performances.

Well, from the very first day, my attention was attracted by something suspicious. If I sent one of the boys to plant a stake, I would see him stop frequently on his way, bend down, stand up again, look about and stoop once more, neglecting his straight line and his signals. Another, who was told to pick up the arrows, would forget the iron pin and take up a pebble instead; and a third, deaf to the measurements of angles, would crumble a clod of earth between his fingers. Most of them were caught licking a bit of straw. The polygon came to a full stop, the diagonals suffered. What could the mystery be?

I enquired; and everything was explained. A born searcher and observer, the scholar had long known what the master had not yet heard of, namely, that there was a big black Bee who made clay nests on the pebbles in the harmas. These nests contained honey; and my surveyors used to open them and empty the cells with a straw. The honey, although rather strong-flavoured, was most acceptable. I acquired a taste for it myself and joined the nest-hunters, putting off the polygon till later. It was thus that I first saw Réaumur's Mason-bee, knowing nothing of her history and, for that matter, knowing nothing of her historian.

The magnificent Bee herself, with her dark-violet wings and black-velvet raiment, her rustic edifices on the sun-blistered pebbles amid the thyme, her honey, providing a diversion from the severities of the compass and the square, all made a great impression on my mind; and I wanted to know more than I had learned from the schoolboys, which was just how to rob the cells of their honey with a straw. As it happened, my bookseller had a gorgeous work on insects for sale. It was called *Histoire naturelle des animaux articulés,* by de Castelnau, E. Blanchard and Lucas, and boasted a multitude of most attractive illustrations; but the price of it, the price of it! No matter; was not my splendid income supposed to cover everything, food for the mind as well as food for the body? Anything extra that I gave to the one I could save upon the other: a method of balancing painfully familiar to those who look to science for their livelihood. The purchase was effected. That day my professional emoluments were severely strained: I devoted a month's salary to the acquisition of the book. I had to resort to miracles of economy for some time to come before making up the enormous deficit.

The book was devoured; there is no other word for it. In it, I learned the name of my black Bee; I read for the first time various details of the habits of insects; I found, surrounded in my eyes with a sort of halo, the revered names of Réaumur, Huber and Léon Dufour; and, while I turned over the pages for the hundredth time, a voice within me seemed to whisper:

"You also shall be of their company!"

EXPERIMENTS WITH MASON-BEES

« 26 »

"The best of witnesses," Fabre used to say, "are experiments." In this chapter, taken from THE MASON-BEES, *he is drawing from such witnesses one of the most amazing of all examples of the limitations of instinct, the story of an insect that bites its way through masonry and then dies a prisoner within a thin paper shell. One of Fabre's revered predecessors in the study of the insects was Réne Antoine Ferchault de Réaumur, French inventor and scientist. During his lifetime, from 1683 to 1757, Réaumur conducted many experiments, particularly with bees and ants. In several of his books, as in this chapter, Fabre pays homage to the pioneer work he accomplished.*

As the nests of the Mason-bee of the Walls are erected on small-sized pebbles, which can be easily carried wherever you like and moved about from one place to another, without disturbing either the work of the builder or the repose of the occupants of the cells, they lend themselves readily to practical experiment, the only method that can throw a little light on the nature of instinct. To study the insect's mental faculties to any purpose, it is not enough for the observer to be able to profit by some happy combination of circumstances: he must know how to produce other combinations, vary them as much as possible and to test them by substitution and interchange. Lastly, to provide science with a solid basis of facts, he must experiment. In this way, the evidence of formal records will one day dispel the fantastic legends with which our books are crowded: the Sacred Beetle calling on his comrades to lend a helping hand in dragging his pellet out of a rut; the Sphex cutting up her fly so as to be able to carry him despite the obstacle of the wind; and all the other fallacies which are the stock-in-trade of those who wish to see in the animal world what is not really there. In this way, again, materials will be prepared which will one day be worked up by the hand of a master and consign hasty and unfounded theories to oblivion.

Réaumur, as a rule, confines himself to stating facts as he sees them in the normal course of events and does not try to probe deeper into the insect's ingenuity by means of artificially produced conditions. In his time, everything had yet to be done; and the harvest was so great that the illustrious harvester went straight to what was most urgent, the gathering of the crop, and left his successors to examine the grain and the ear in detail. Nevertheless, in connection with the Chalicodoma of the Walls, he mentions an experiment made by his friend, Duhamel. He tells us how a Mason-bee's nest was enclosed in a glass funnel, the mouth of which was covered merely with a bit of gauze. From it there issued three males, who, after vanquishing mortar as hard as stone, either never thought of piercing the flimsy gauze or else deemed the work beyond their strength. The three bees died under the funnel. Réaumur adds that insects generally know only how to do what they have to do in the ordinary course of nature.

The experiment does not satisfy me, for two reasons: first, to ask workers equipped with tools for cutting clay as hard as granite to cut a piece of gauze does not strike me as a happy inspiration; you cannot expect a navvy's pickaxe to do the same work as a dressmaker's scissors. Secondly, the transparent glass prison seems to me ill-chosen. As soon as the insect has made a passage through the thickness of its earthen dome, it finds itself in broad daylight; and to it daylight means the final deliverance, means liberty. It strikes against an invisible obstacle, the glass; and to it glass is nothing at all and yet an obstruction. On the far side, it sees free space, bathed in sunshine. It wears itself out in efforts to fly there, unable to understand the futile nature of its attempts against that strange barrier which it cannot see. It perishes, at last, of exhaustion, without, in its obstinacy, giving a glance at the gauze closing the conical chimney. I must devise a means of renewing the experiment under better conditions.

The obstacle which I select is ordinary brown paper, stout enough to keep the insect in the dark and thin enough not to offer serious resistance to the prisoner's efforts. As there is a great difference, in so far as the actual nature of the barrier is concerned, between a paper partition and a clay ceiling, let us begin by enquiring if the Mason-bee of the Walls knows how or rather is able to make her way through one of these partitions. The mandibles are pickaxes suitable for breaking through hard mortar: are they also scissors capable of cutting a thin membrane?

This is the point to look into first of all.

In February, by which time the insect is in its perfect state, I take a certain number of cocoons, without damaging them, from their cells and insert them each in a separate stump of reed, closed at one end by the natural wall of the node and open at the other. These pieces of reed represent the cells of the nest. The cocoons are introduced with the insect's head turned toward the opening. Lastly, my artificial cells are closed in different ways. Some receive a stopper of kneaded clay, which, when dry, will correspond in thickness and consistency with the mortar ceiling of the natural nest. Others are plugged with a cylinder of sorghum, at least a centimetre thick; and the remainder with a disk of brown paper solidly fastened by the edge. All these bits of reed are placed side by side in a box, standing upright, with the roof of my making at the top. The insects, therefore, are in the exact position which they occupied in the nest. To open a passage, they must do what they would have done without my interference, they must break through the wall situated above their heads. I shelter the whole under a wide bell-glass and wait for the month of May, the period of the deliverance.

The results far exceed my anticipations. The clay stopper, the work of my fingers, is perforated with a round hole, differing in no wise from that which the Mason-bee contrives through her native mortar dome. The vegetable barrier, new to my prisoners, namely, the sorghum cylinder, also opens with a neat orifice, which might have been the work of a punch. Lastly, the brown-paper cover allows the Bee to make her exit not by bursting through, by making a violent rent, but once more by a clearly-defined round hole. My Bees therefore are capable of a task for which they were not born; to come out of their reed cells they do what probably none of their race did before them; they perforate the wall of sorghum-pith, they make a hole in the paper barrier, just as they would have pierced their natural clay ceiling. When the moment comes to free themselves, the nature of the impediment does not stop them, provided that it be not beyond their strength; and henceforth the argument of incapacity cannot be raised when a mere paper barrier is in question.

In addition to the cells made out of bits of reed, I put under the bell-glass, at the same time, two nests which are intact and still resting on their pebbles. To one of them I have attached a sheet of brown paper pressed close against the mortar dome. In order to come out, the insect will have to pierce first the dome and then the paper, which follows without any intervening space. Over the other, I have placed a little brown-paper cone, gummed to the pebble. There is here, therefore, as in the first case, a double wall—a clay partition and a paper partition—with this difference, that the two walls do not come immediately after each other, but are separated by an empty space of about a centimetre at the bottom, increasing as the cone rises.

The results of these two experiments are quite different. The Bees in the nest to which a sheet of paper was tightly stuck come out by piercing the two enclosures, of which the outer wall, the paper wrapper, is perforated with a very clean round hole, as we have already seen in the reed cells closed with a lid of the same material. We thus become aware, for the second time, that, when the Mason-bee is stopped by a paper barrier, the reason is not her incapacity to overcome the obstacle. On the other hand, the occupants of the nest covered with the cone, after making their way through the earthen dome, finding the sheet of paper at some distance, do not even try to perforate this obstacle, which they would have conquered so easily had it been fastened to the nest. They die under the cover without making any attempt to escape. Even so did Réaumur's Bees perish in the glass funnel, where their liberty depended only upon their cutting through a bit of gauze.

This fact strikes me as rich in inferences. What? Here are sturdy insects, to whom boring through granite is mere play, to whom a stopper of soft wood and a paper partition are walls quite easy to perforate despite the novelty of the material; and yet these vigorous housebreakers allow themselves to perish stupidly in the prison of a paper bag, which they could have torn open with one stroke of their mandibles? They are capable of tearing it, but they do not dream of doing so! There can be only one explanation of this suicidal inaction. The insect is well-endowed with tools and instinctive faculties for accomplishing the final act of its metamorphosis, namely, the act of emerging from the cocoon and from the cell. Its mandibles provide it with scissors, file, pickaxe and lever wherewith to cut, gnaw through and demolish either its cocoon and its mortar enclosure or any other not too obstinate barrier substituted for the natural covering of the nest. Moreover—and this is an important proviso, but for which the outfit would be useless—it has, I will not say the will to use those tools, but a secret stimulus inviting it to employ them. When the hour for the emergence arrives, this stimulus is aroused and the insect sets to work to bore a passage. It little cares in this case whether the material to be pierced be the natural mortar, sorghum-pith, or paper; the lid that holds it imprisoned does not resist for long. Nor even does it care if the obstacle be increased in thickness and a paper wall be added outside the wall of clay: the two barriers, with no interval between them, form but one to the Bee, who passes through them because the act of getting out is still one act and one only. With the paper cone, whose wall is a little way off, the conditions are changed, though the total thickness of wall is really the same. Once outside its earthen abode, the insect has done all that it was destined to do in order to release itself; to move freely on the mortar dome represents to it the end of the release, the end of the act of boring. Around the nest a new barrier appears, the wall made by the paper bag; but, in order to pierce this, the insect would have to repeat the act which it has just accomplished, the act which it is not intended to perform more than once in its life; it would, in short, have to make into a double act that which by nature is a single one; and the insect cannot do this, for the sole reason that it has not the wish to. The Mason-bee perishes for lack of the smallest gleam of intelligence.

THE FOAMY CICADELLA

« 27 »

*Ever since the days of Aristotle, tiny snow-white masses of foam on
the stems of grasses and weeds have attracted wide attention. They
are the homes of soft-bodied little insects, the immature froghoppers.
Within the protection of the foam, they grow and molt until they
emerge as winged adults. The insects belong to the tree-hopper group.
Up until recent times, many scientists have maintained that the little
froghopper produces the foam by lashing about with its tail, turning
liquid into a froth in the manner of an egg-beater. Fabre was nearer
the truth. The froghopper is a bubble-blower. However, researches
since Fabre's day have revealed that a series of overlapping plates beneath the abdomen of the insect form a chamber which acts as a pump
in blowing the bubbles. A series of magnified closeup photographs of
the froghopper in action can be found in* NEAR HORIZONS, THE STORY
OF AN INSECT GARDEN. *Secretions given off by tiny glands in the abdomen of the froghopper apparently account for the stiffness and persistency of the foam. Fabre's story of the Foamy Cicadella is taken
from* THE LIFE OF THE GRASSHOPPER.

In April, let us consider the fields for a while, keeping our eyes on
the ground, as befits the eager observer of insect-life. We shall not fail
to see, here and there, on the grass, little masses of white foam. It might
easily be taken for a spray of frothy spittle from the lips of a passer-by;
but there is so much of it that we soon abandon this first idea. Never
would human saliva suffice for so lavish an expenditure of foam, even
if some one with nothing better to do were to devote all his disgusting
and misdirected zeal to the effort.

While recognizing that man is blameless in the matter, the northern
peasant has not relinquished the name suggested by the appearance:

he calls those strange flakes "Cuckoo-spit," after the bird whose note is then proclaiming the awakening of spring. The vagrant creature, unequal to the toils and delights of housekeeping, ejects it at random, so they say, as it pays its flying visits to the homes of others, in search of a resting-place for its egg.

The interpretation does credit to the Cuckoo's salivary powers, but not to the interpreter's intelligence. The other popular denomination is worse still: "Frog-spit!" My dear good people, what on earth has the Frog or his slaver to do with it?

The shrewder Provençal peasant also knows that vernal foam; but he is too cautious to give it any wild names. My rustic neighbours, when I ask them about Cuckoo-spit and Frog-spit, begin to smile and see nothing in those words but a poor joke. To my questions on the nature of the thing they reply:

"I don't know."

Exactly! That's the sort of answer I like, an answer not complicated with grotesque explanations.

Would you know the real perpetrator of this spittle? Rummage about the frothy mass with a straw. You will extract a little yellow, pot-bellied, dumpy creature, shaped like a Cicada without wings. That's the foam-producer.

When laid naked on another leaf, she brandishes the pointed tip of her little round paunch. This at once betrays the curious machine which we shall see at work presently. When older and still operating under the cover of its foam, the little thing becomes a nymph, turns green in colour and gives itself stumps of wings fixed scarfwise on its sides. From underneath its blunted head there projects, when it is working, a little gimlet, a beak similar to that of the Cicadæ.

In its adult form the insect is, in fact, a sort of very small-sized Cicada, for which reason the entomologist capable of shaking off the trammels of nonsensical nomenclature calls it simply the Foamy Cicadella. For this euphonic name, the diminutive of Cicada, the others have substituted that horrible word Aphrophora. Orthodox science says, *Aphrophora spumaria,* meaning Foamy Foambearer. The ear is none the better for this improvement. Let us content ourselves with Cicadella, which respects the tympanum and does not reduplicate the foam.

I have consulted my few books as to the habits of the Cicadella. They tell me that she punctures plants and makes the sap exude in foamy flakes. Under this cover, the insect lives sheltered from the heat. A work recently compiled has one curious piece of information: it tells me that

I must get up early in the morning, inspect my crops, pick any twig with foam on it and at once plunge it into a cauldron of boiling water.

Oh, my poor Cicadella, this is a bad lookout! The author does not do things by halves. I see him rising before the dawn, lighting a stove on wheels and pushing his infernal contrivance through the midst of his lucern, his clover and his peas, to boil you on the spot. He will have his work cut out for him. I remember a certain patch of sainfoin of which almost every stalk had its foam-flakes. Had the stewing-process been necessary, one might just as well have reaped the field and turned the whole crop into herb-tea.

Why these violent measures? Are you so very dangerous to the harvest, my pretty little Cicada? They accuse you of draining the plant which you attack. Upon my word, they are right: you drain it almost as dry as the Flea does the Dog. But to touch another's grass—you know it: doesn't the fable say so?—is a heinous crime, an offence which can be punished by nothing less drastic than boiling water.

Let us waste no more time on these agricultural entomologists with their murderous designs. To hear them talk, one would think that the insect has no right to live. Incapable of behaving like a ferocious land-owner who becomes filled with thoughts of massacre at the sight of a maggoty plum, I, more kindly, abandon my few rows of peas and beans to the Cicadella: she will leave me my share, I am convinced.

Besides, the insignificant ones of the earth are not the least rich in talent, in an originality of invention which will teach us much concerning the infinite variety of instinct. The Cicadella, in particular, possesses her recipes for aerated waters. Let us ask her by what process she succeeds in giving such a fine head of froth to her product, for the books that talk about boiling cauldrons and Cuckoo-spit are silent on this subject, the only one worthy of narration.

The foamy mass has no very definite shape and is hardly larger than a hazel-nut. It is remarkably persistent even when the insect is not working at it any longer. Deprived of its manufacturer, who would not fail to keep it going, and placed on a watch-glass, it lasts for more than twenty-four hours without evaporating or losing its bubbles. This persistency is striking, compared with the rapidity with which soapsuds, for instance, disappear.

Prolonged duration of the foam is necessary to the Cicadella, who would exhaust herself in the constant renewal of her products if her work were ordinary froth. Once the effervescent covering is obtained, it is essential that the insect should rest for a time, with no other task than to drink its fill and grow. And so the moisture converted into froth possesses a certain stickiness, conducive to longevity. It is slightly oily and trickles under one's finger like a weak solution of gum.

The bubbles are small and even, being all of the same dimensions. You can see that they have been scrupulously gauged, one by one; you suspect the presence of a graduated tube. Like our chemists and druggists, the insect must have its drop-measures.

A single Cicadella is usually crouching invisible in the depths of the foam; sometimes there are two or three or more. In such cases, it is a fortuitous association, the fabrics of the several workers being so close together that they merge into one common edifice.

Let us see the work begin and, with the aid of a magnifying-glass, follow the creature's proceedings. With her sucker inserted up to the hilt and her six short legs firmly fixed, the Cicadella remains motionless, flat on her stomach on the long-suffering leaf. You expect to see froth issuing from the edge of the well, effervescing under the action of the insect's implement, whose lancets, ascending and descending in turns and rubbing against each other like those of the Cicada, ought to make the sap foam as it is forced out. The froth, so it would seem, must come ready-made from the puncture. That is what the current descriptions of the Cicadella tell us; that was how I myself pictured it on the authority of the writers. All this is a huge mistake: the real thing is much more ingenious. It is a very clear liquid that comes up from the well, with no more trace of foam than in a dewdrop. Even so the Cicada, who possesses similar tools, makes the spot at which she slakes her thirst give forth a limpid fluid, with not a vestige of froth to it. Therefore, notwithstanding its dexterity in sucking up liquids, the Cicadella's mouth-apparatus has nothing to do with the manufacture of the foamy mattress. It supplies the raw material; another implement works it up. What implement? Have patience and we shall see.

The clear liquid rises imperceptibly and glides under the insect, which at last is half inundated. The work begins again without delay. To make white of egg into a froth we have two methods: we can whip it, thus dividing the sticky fluid into thin flakes and causing it to take in air in a network of cells; or we can blow into it and so inject air-bubbles right into the mass. Of these two methods, the Cicadella employs the second, which is less violent and more elegant. She blows her froth.

But how is the blowing done? The insect seems incapable of it, being devoid of any air-mechanism similar to that of the lungs. To breathe with tracheæ and to blow like a bellows are incompatible actions.

Agreed; but be sure that, if the insect needs a blast of air for its manufactures, the blowing-machine will be there, most ingeniously contrived. This machine the Cicadella possesses at the tip of her abdomen, at the end of the intestine. Here, split lengthwise in the shape of a Y, a little pocket opens and shuts in turns, a pocket whose two lips close hermetically when joined.

Having said this, let us watch the performance. The insect lifts the tip of its abdomen out of the bath in which it is swimming. The pocket opens, sucks in the air of the atmosphere till it is full, then closes and dives down, the richer by its prize. Inside the liquid, the apparatus contracts. The captive air escapes as from a nozzle and produces a first bubble of froth. Forthwith the air-pocket returns to the upper air, opens, takes in a fresh load and goes down again closed, to immerse itself once more and blow in its gas. A new bubble is produced.

And so it goes on with chronometrical regularity, from second to second, the blowing-machine swinging upwards to open its valve and fill itself with air, downwards to dive into the liquid and send out its gaseous contents. Such is the air-measurer, the drop-glass which accounts for the evenness of the frothy bubbles.

Ulysses, the favourite of the gods, received from the storm-dispenser, Æolus, bags in which the winds were confined. The carelessness of his crew, who untied the bags to find out what they contained, let loose a tempest which destroyed the fleet. I have seen those mythological wind-filled bags; I saw them years ago, when I was a child.

A peripatetic tinker, a son of Calabria, had set up between two stones the crucible in which a tin soup-tureen and plates were to be remelted. Æolus did the blowing, Æolus in the person of a little dark-skinned boy who, squatting on his heels, forced air towards the forge by alternately squeezing two goatskin bags, one on the right and one on the left. Thus must the prehistoric bronze-smelters have performed their task, they whose workshops and whose remains of copper-slag I find on the hills near my home: the blast of their furnaces was produced by these inflated skins.

The machine employed by my Æolus is pathetically simple. The hide of a goat, with the hair left on, is practically all that is necessary. It is a bag fastened at the bottom over a nozzle, open at the top and supplied, by way of lips, with two little boards which, when brought together, close up the whole apparatus. These two stiff lips are each furnished with a leather handle, one for the thumb, the other for the four remaining fingers. The hand opens; the lips of the bag part and it fills with air. The hand closes and brings the boards together; the air imprisoned in the compressed bag escapes by the nozzle. The alternate working of the two bags gives a continuous blast.

Apart from continuity, which is not a favourable condition when the gas has to be discharged in small bubbles, the Cicadella's bellows works like the Calabrian tinker's. It is a flexible pocket with stiff lips, which alternately part and unite, opening to let the air enter and closing to keep it imprisoned. The contraction of the sides takes the place of the shrinking of the bag and puffs out the gaseous contents when the pocket is immersed.

He certainly had a lucky inspiration who first thought of confining the wind in a bag, as mythology tells us that Æolus did. The goatskin turned into a bellows gave us our metals, the essential matter whereof our tools are made. Well, in this art of expelling air, an enormous source of progress, the Cicadella was the pioneer. She was blowing her froth before Tubalcain thought of urging the fire of his forge with a leather pouch. She was the first to invent bellows.

When, bubble by bubble, the foamy wrapper covers the insect to a height which the uplifted tip of her belly is unable to reach, it is no longer possible to take in air and the effervescence stops. Nevertheless, the gimlet that extracts the sap goes on working, for nourishment must be obtained. As a rule then, in the sloping part, the superfluous liquid, that which is not converted into foam, collects and forms a drop of perfectly clear liquid.

What does this limpid fluid lack in order to turn white and effervesce? Nothing but air blown into it, one would think. I am able to substitute my own devices for the Cicadella's syringe. I place between my lips a very slender glass tube and with delicate puffs send my breath into the drop of moisture. To my great surprise, it does not froth up. The result is just the same as that which I should have with plain water from the tap.

Instead of a plentiful, lasting, slow-subsiding foam, like that with which the insect covers itself, all that I obtain is a miserable ring of bubbles, which burst as soon as they appear. And I am equally unsuccessful with the liquid which the Cicadella collects under her abdomen at the start, before working her bellows. What is wrong in each case? The foamy product and its generating liquid shall tell us.

The first is oily to the touch, gummy and as fluid as, for instance, a weak solution of albumen would be; the second flows as readily as plain water. The Cicadella therefore does not draw from her well a liquid liable to effervesce merely by the action of the blow-pocket; she adds something to what oozes from the puncture, adds a viscous element which gives cohesion and makes frothing possible, even as a boy adds soap to the water which he blows into iridescent bubbles through a straw.

Where then does the insect keep its soap-works, its manufactory of the effervescent element? Evidently in the blow-pocket itself. It is here that the intestine ends and here that albuminous products, furnished either by the digestive canal or by special glands, can be expelled in infinitesimal doses. Each whiff sent out is thus accompanied by a trifle of adhesive matter, which dissolves in the water, making it sticky and enabling it to retain the captive air in permanent bubbles. The Cicadella covers herself with an icing of which her intestine is to some extent the manufacturer.

This method brings us back to the industry of the lily-dweller, the grub which makes itself a loathsome armour out of its excretions; but what a distance between the heap of ordure which it wears on its back and the Cicadella's aerated mattress!

Another fact, more difficult to explain, attracts our attention. A multitude of low-growing, herbaceous plants, whose sap starts flowing in April, suit the frothy insect, without distinction of species, genus or family. I could almost make a list of the non-ligneous vegetation of my neighbourhood by cataloguing the plants on which the little creature's foam is to be found in greater or lesser abundance. A few experiments will tell us how indifferent the Cicadella is to both the nature and the properties of the plant which she adopts as her home.

I pick the insect out of its froth with the tip of a hair-pencil and place it on some other plant, of an opposite flavour, letting the strong come after the mild, the spicy after the insipid, the bitter after the sweet. The new encampment is accepted without hesitation and soon covered with foam. For instance, a Cicadella taken from the bean, which has a neutral flavour, thrives excellently on the spurges, full of pungent milky sap, and particularly on *Euphorbia serrata,* the narrow notch-leaved spurge, which is one of her favourite dwelling-places. And she is equally satisfied when moved from the highly-spiced spurge to the comparatively flavourless bean.

This indifference is surprising when we reflect how scrupulously faithful other insects are to their plants. There are undoubtedly stomachs expressly made to drink corrosive and assimilate toxic matters. The caterpillar of *Acherontia atropos,* the Death's-head Hawk-moth, eats its fill of potato-leaves, which are seasoned with solanin; the caterpillar of the Spurge-moth browses in these parts on the upright red spurge (*Euphorbia characias*), whose milk produces much the same effect as red-hot iron on the tongue; but neither one nor the other would pass from these narcotics or these caustics to utterly insipid fare.

How does the Cicadella manage to feed on anything and everything, for she evidently obtains nourishment while putting a head on her liquid? I see her thrive, either of her own accord or by my devices, on the common buttercup (*Ranunculus acris*), which has a flavour un-equalled save by Cayenne pepper; on the Italian arum (*Arum italicum*), the veriest particle of whose leaves is enough to burn the lips; on the traveller's joy, or virgin's bower (*Clematis vitalba*), the famous beggars' herb, which reddens the skin and produces the sores in request among our sham cripples. After these highly-seasoned condiments, she will promptly accept the mild sainfoin, the scented savory, the bitter dande-lion, the sweet field eringo, in short, anything that I put before her, whether full-flavoured or tasteless.

As a matter of fact, this strange catholicity of diet might well be only apparent. When the Cicadella punctures this or that herb, of whatever species, all that she does is to extract an almost neutral liquid, just as the roots draw it from the soil; she does not admit to her fountain the fluids worked up into essential principles. The liquid that trickles forth under the insect's gimlet and forms a bead at the bottom of the foamy mass is perfectly clear.

I have gathered this drop on the spurge, the arum, the clematis and the buttercup. I expected to find a fire-water, pungent as the sap of those different plants. Well, it is nothing of the kind; it lacks all savour; it is water or little more. And this insipid stuff has issued from a reser-voir of vitriol.

If I prick the spurge with a fine needle, that which rises from the puncture is a white, milky drop, tasting horribly bitter. When the Cicadella pushes in her drill, a clear, flavourless fluid oozes out. The two operations seem to be directed towards different sources.

How does she manage to draw a liquid that is clear and harmless from the same barrel whence my needle brings up something milky and burning? Can the Cicadella, with her instrument, that incom-parable alembic, divide the fierce fluid into two, admitting the neutral and rejecting the peppery? Can she be drawing on certain vessels whose sap, not yet elaborated, has not acquired its final virulence? The delicate vegetable anatomy is helpless in the presence of the tiny creature's pump. I give up the problem.

When the Cicadella is exploring the spurge, as frequently happens, she has a serious reason for not admitting to her fountain all that would be yielded by simple bleeding, such as my needle would produce. The milky juice of the plant would be fatal to her.

I gather a drop or two of the liquid that trickles from a cut stalk and instal a Cicadella in it. The insect is not comfortable: I can see this by its efforts to escape. My hair-pencil pushes the fugitive back into the pool of milk, rich in dissolved rubber. Soon this rubber settles into clots similar to crumbs of cheese; the insect's legs become clad in gaiters that seem made of casein; a coating of gum obstructs the breathing-valves; possibly also the extremely delicate skin is hurt by the blistering qualities of the milky sap. If kept for some time in that environment, the Cicadella dies.

Even so would she die if her gimlet, working simply as a needle, brought the milk of the spurge to the surface. A sifting takes place then, which allows almost pure water to issue from the source that gives the wherewithal for making the froth. A subtle exhaustion-process, whose mechanism is hidden from our curiosity, a piston-play of unrivalled delicacy, effects this marvellous work of purification.

Water is always water, whether it come from the stagnant pool or the clear stream, from a poisonous liquid or a healing infusion; and it possesses the same properties, when it is rid of its impurities by distillation. In like manner, the sap, whether furnished by the spurge or the bean, the clematis or the sainfoin, the buttercup or the borage, is of the same watery nature when the Cicadella's syphon, by a reducing-process which would be the envy of our stills, has deprived it of its peculiar properties, which vary so greatly in different plants.

This would explain how the insect makes its froth rise on the first plant that it comes across. Everything suits it, because its apparatus reduces any sap to the condition of plain water. The inimitable well-sinker is able to produce the limpid from the cloudy and the harmless from the toxic.

It may possibly happen that the insect's well supplies water that is not quite pure. If left to evaporate in a watch-glass, the clear drop that trickles from the mass of foam yields a thin white residue, which dissolves by effervescence in nitric acid. This residue might well be carbonate of potash. I also suspect the presence of traces of albumen. Obviously, the Cicadella finds something to feed on at the bottom of the puncture. Now what does she consume? To all appearances, something with an albuminous basis, for the pigmy herself is, for the most part, but a grain of similar matter. This element is plentiful in all plants; and it is probable that the insect uses it lavishly to make up for the expenditure of gum needed for the formation of froth. Some albuminous product, perfected in the digestive canal and discharged by the intestine as and when the blow-pocket expels its bubble of air, might well give the liquid the power of swelling into a foam that lasts for a long time.

If we ask ourselves what advantage the Cicadella derives from her mass of froth, a very excellent answer is at once suggested: the insect keeps itself cool under that shelter, hides itself from the eyes of its persecutors and is protected against the rays of the sun and the attacks of parasites.

The Lily-beetle makes a similar use of the mantle of her own dirt; but she, most unhappily for herself, flings off her nasty cloak and descends naked from the plant to the ground, where she has to bury herself to slaver her cocoon. At this critical moment, the Flies lie in wait for her and entrust her with their eggs, the germs of parasites which will eat into her body.

The Cicadella is better-advised and altogether escapes the dangers attendant on a removal. Subject to certain summary changes which never interrupt her activity, she assumes the adult form in the very heart of her bastion, under the shelter of a viscous rampart capable of repelling any assailant. Here she enjoys perfect security when the difficult hour has come for tearing off her old skin and putting on another, brand-new and more decorative; here she finds profound peace for her excoriation and for the display of the attire of a riper age.

The insect does not leave its cool covering until it is grown up, when it appears in the form of a pretty little, brown-striped Cicadella. It is then able to take enormous and sudden leaps, which carry it far from the aggressor; and it leads an easy life, untroubled by the foe.

Looked upon as a system of defence, the frothy stronghold is indeed a magnificent invention, much superior to the squalid work of the invader of the lily. And, strange to say, the system has no imitators among the genera most nearly allied to the froth-blower.

In her larval form, the Asparagus-beetle is victimized by the Fly because she does not follow the example of her cousin, the Lily-beetle, and clothe herself in her own droppings. Even so, on the grass, on the trees displaying their tender leaves, other Cicadellæ abound, no less exposed to danger from the Warbler seeking a succulent morsel for his little ones; and, as they draw out the sap through the punctures made by their suckers, not one of them thinks of making it effervesce. Yet they too possess the elevator-pump, which they all work in the same manner; only they do not know how to turn the end of their intestine into a bellows. Why not? Because instincts are not to be acquired. They are primordial aptitudes, bestowed here and denied there; time cannot awaken them by a slow incubation, nor are they decreed by any similarity of organization.

THE POND

« 28 »

Until he found a haven at Sérignan, Fabre always thought of Saint-Léons, his birthplace and the scene of his happiest years, as the portion of the Earth he loved best. He said, for many years, that he wanted to be buried there when he died. His mind often reverted to the days of his earliest childhood and such recollections produced some of his most engaging pictures, including this story of the discoveries of a small boy taken from THE LIFE OF THE FLY. *The war that the tallow-factory owner had just returned from was the War of 1830 between France and Algiers.*

The pond, the delight of my early childhood, is still a sight whereof my old eyes never tire. What animation in that verdant world! On the warm mud of the edges, the Frog's little Tadpole basks and frisks in its black legions; down in the water, the orange-bellied Newt steers his way slowly with the broad rudder of his flat tail; among the reeds are stationed the flotillas of the Caddis-worms, half-protruding from their tubes, which are now a tiny bit of stick and again a turret of little shells.

In the deep places, the Water-beetle dives, carrying with him his reserves of breath: an air-bubble at the tip of the wing-cases and, under the chest, a film of gas that gleams like a silver breastplate; on the surface, the ballet of those shimmering pearls, the Whirligigs, turns and twists about; hard by there skims the insubmersible troop of the Pond-skaters, who glide along with side-strokes similar to those which the cobbler makes when sewing.

Here are the Water-boatmen, who swim on their backs with two oars spread cross-wise, and the flat Water-scorpions; here, squalidly clad in mud, is the grub of the largest of our Dragon-flies, so curious because of its manner of progression: it fills its hinder-parts, a yawning

218

funnel, with water, spirts it out again and advances just so far as the recoil of its hydraulic cannon.

The Molluscs abound, a peaceful tribe. At the bottom, the plump River-snails discreetly raise their lid, opening ever so little the shutters of their dwelling; on the level of the water, in the glades of the aquatic garden, the Pond-snails—Physa, Limnæa and Planorbis—take the air. Dark Leeches writhe upon their prey, a chunk of Earth-worm; thousands of tiny, reddish grubs, future Mosquitoes, go spinning around and twist and curve like so many graceful Dolphins.

Yes, a stagnant pool, though but a few feet wide, hatched by the sun, is an immense world, an inexhaustible mine of observation to the studious man and a marvel to the child who, tired of his paper boat, diverts his eyes and thoughts a little with what is happening in the water. Let me tell what I remember of my first pond, at a time when ideas began to dawn in my seven-year-old brain.

How shall a man earn his living in my poor native village, with its inclement weather and its niggardly soil? The owner of a few acres of grazing-land rears sheep. In the best parts, he scrapes the soil with the swing-plough; he flattens it into terraces banked by walls of broken stones. Pannierfuls of dung are carried up on donkey-back from the cowshed. Then, in due season, comes the excellent potato, which, boiled and served hot in a basket of plaited straw, is the chief stand-by in winter.

Should the crop exceed the needs of the household, the surplus goes to feed a pig, that precious beast, a treasure of bacon and ham. The ewes supply butter and curds; the garden boasts cabbages, turnips and even a few hives in a sheltered corner. With wealth like that one can look fate in the face.

But we, we have nothing, nothing but the little house inherited by my mother and its adjoining patch of garden. The meagre resources of the family are coming to an end. It is time to see to it and that quickly. What is to be done? That is the stern question which father and mother sat debating one evening.

Hop-o'-my-Thumb, hiding under the woodcutter's stool, listened to his parents overcome by want. I also, pretending to sleep, with my elbows on the table, listen not to blood-curdling designs, but to grand plans that set my heart rejoicing. This is how the matter stands: at the bottom of the village, near the church, at the spot where the water of the large roofed spring escapes from its underground weir and joins the brook in the valley, an enterprising man, back from the war, has set up a small tallow-factory. He sells the scrapings of his pans, the burnt fat, reeking of candle-grease, at a low price. He proclaims these wares to be excellent for fattening ducks.

"Suppose we bred some ducks," says mother. "They sell very well in town. Henri would mind them and take them down to the brook."

"Very well," says father, "let's breed some ducks. There may be difficulties in the way; but we'll have a try."

That night, I had dreams of paradise: I was with my ducklings, clad in their yellow suits; I took them to the pond, I watched them have their bath, I brought them back again, carrying the more tired ones in a basket.

A month or two after, the little birds of my dreams were a reality. There were twenty-four of them. They had been hatched by two hens, of whom one, the big, black one, was an inmate of the house, while the other was borrowed from a neighbour.

To bring them up, the former is sufficient, so careful is she of her adopted family. At first, everything goes perfectly: a tub with two fingers' depth of water serves as a pond. On sunny days, the ducklings bathe in it under the anxious eye of the hen.

A fortnight later, the tub is no longer enough. It contains neither cresses crammed with tiny Shellfish nor Worms and Tadpoles, dainty morsels both. The time has come for dives and hunts amid the tangle of the water-weeds; and for us the day of trouble has also come. True, the miller, down by the brook, has fine ducks, easy and cheap to bring up; the tallow-smelter, who has extolled his burnt fat so loudly, has some as well, for he has the advantage of the waste water from the spring at the bottom of the village; but how are we, right up there, at the top, to procure aquatic sports for our broods? In summer, we have hardly water to drink!

Near the house, in a freestone recess, a scanty source trickles into a basin made in the rock. Four or five families have, like ourselves, to draw their water there with copper pails. By the time that the schoolmaster's donkey has slaked her thirst and the neighbours have taken their provision for the day, the basin is dry. We have to wait for four-and-twenty hours for it to fill. No, this is not the hole in which the ducks would delight nor indeed in which they would be tolerated.

There remains the brook. To go down to it with the troop of ducklings is fraught with danger. On the way through the village, we might meet cats, bold ravishers of small poultry; some surly mongrel might frighten and scatter the little band; and it would be a hard puzzle to collect it in its entirety. We must avoid the traffic and take refuge in peaceful and sequestered spots.

On the hills, the path that climbs behind the château soon takes a sudden turn and widens into a small plain beside the meadows. It skirts a rocky slope whence trickles, level with the ground, a streamlet, forming a pond of some size. Here profound solitude reigns all day long. The ducklings will be well off; and the journey can be made in peace by a deserted foot-path.

You, little man, shall take them to that delectable spot. What a day it was that marked my first appearance as a herdsman of ducks! Why must there be a jar to the even tenor of such joys? The too-frequent encounter of my tender skin with the hard ground had given me a large and painful blister on the heel. Had I wanted to put on the shoes stowed away in the cupboard for Sundays and holidays, I could not. There was nothing for it but to go barefoot over the broken stones, dragging my leg and carrying high the injured heel.

Let us make a start, hobbling along, switch in hand, behind the ducks. They too, poor little things, have sensitive soles to their feet; they limp, they quack with fatigue. They would refuse to go any farther if I did not, from time to time, call a halt under the shelter of an ash.

We are there at last. The place could not be better for my birdlets; shallow, tepid water, interspersed with muddy knolls and green eyots. The diversions of the bath begin forthwith. The ducklings clap their beaks and rummage here, there and everywhere; they sift each mouthful, rejecting the clear water and retaining the good bits. In the deeper parts, they point their sterns into the air and stick their heads under water. They are happy; and it is a blessed thing to see them at work. We will let them be. It is my turn to enjoy the pond.

What is this? On the mud lie some loose, knotted, soot-coloured cords. One could take them for threads of wool like those which you pull out of an old ravelly stocking. Can some shepherdess, knitting a black sock and finding her work turn out badly, have begun all over again and, in her impatience, have thrown down the wool with all the dropped stitches? It really looks like it.

I take up one of those cords in my hand. It is sticky and extremely slack; the thing slips through the fingers before they can catch hold of it. A few of the knots burst and shed their contents. What comes out is a black globule, the size of a pin's head, followed by a flat tail. I recognize, on a very small scale, a familiar object: the Tadpole, the Frog's baby. I have seen enough. Let us leave the knotted cords alone.

The next creatures please me better. They spin round on the surface of the water and their black backs gleam in the sun. If I lift a hand to seize them, that moment they disappear, I know not where. It's a pity: I should have much liked to see them closer and to make them wriggle in a little bowl which I should have put ready for them.

Let us look at the bottom of the water, pulling aside those bunches of green string whence beads of air are rising and gathering into foam. There is something of everything underneath. I see pretty shells with compact whorls, flat as beans; I notice little worms carrying tufts and feathers; I make out some with flabby fins constantly flapping on their backs. What are they all doing there? What are their names? I do not know. And I stare at them for ever so long, held by the incomprehensible mystery of the waters.

At the place where the pond dribbles into the adjoining field are some alder-trees; and here I make a glorious find. It is a Scarab—not a very large one, oh no! He is smaller than a cherry-stone, but of an unutterable blue. The angels in paradise must wear dresses of that colour. I put the glorious one inside an empty snail-shell, which I plug up with a leaf. I shall admire that living jewel at my leisure, when I get back. Other distractions summon me away.

The spring that feeds the pond trickles from the rock, cold and clear. The water first collects into a cup, the size of the hollow of one's two hands, and then runs over in a stream. These falls call for a mill: that goes without saying. Two bits of straw, artistically crossed upon an axis, provide the machinery; some flat stones set on edge afford supports. It is a great success: the mill turns admirably. My triumph would be complete, could I but share it. For want of other playmates, I invite the ducks.

Everything palls in this poor world of ours, even a mill made of two straws. Let us think of something else: let us contrive a dam to hold back the waters and form a pool. There is no lack of stones for the brick-work. I pick the most suitable; I break the larger ones. And, while collecting these blocks, suddenly I forget all about the dam which I meant to build.

On one of the broken stones, in a cavity large enough for me to put my fist in, something gleams like glass. The hollow is lined with facets gathered in sixes which flash and glitter in the sun. I have seen something like this in church, on the great saints'-days, when the light of the candles in the big chandelier kindles the stars in its hanging crystal.

We children, lying, in summer, on the straw of the threshing-floor, have told one another stories of the treasures which a dragon guards underground. Those treasures now return to my mind: the names of precious stones ring out uncertainly but gloriously in my memory. I think of the king's crown, of the princesses' necklaces. In breaking stones, can I have found, but on a much richer scale, the thing that shines quite small in my mother's ring? I want more such.

The dragon of the subterranean treasures treats me generously. He gives me his diamonds in such quantities that soon I possess a heap of broken stones sparkling with magnificent clusters. He does more: he gives me his gold. The trickle of water from the rock falls on a bed of fine sand which it swirls into bubbles. If I bent over towards the light, I see something like gold-filings whirling where the fall touches the bottom. Is it really the famous metal of which twenty-franc pieces, so rare with us at home, are made? One would think so, from the glitter.

I take a pinch of sand and place it in my palm. The brilliant particles are numerous, but so small that I have to pick them up with a straw moistened in my mouth. Let us drop this: they are too tiny and too bothersome to collect. The big, valuable lumps must be farther on, in the thickness of the rock. We'll come back later; we'll blast the mountain.

I break more stones. Oh, what a queer thing has just come loose, all in one piece! It is turned spiral-wise, like certain flat Snails that come out of the cracks of old walls in rainy weather. With its gnarled sides, it looks like a little ram's-horn. Shell or horn, it is very curious. How do things like that find their way into the stone?

Treasures and curiosities make my pockets bulge with pebbles. It is late and the little ducklings have had all they want to eat. Come along, youngsters, let's go home. My blistered heel is forgotten in my excitement.

The walk back is a delight. A voice sings in my ear, an untranslatable voice, softer than any language and bewildering as a dream. It speaks to me for the first time of the mysteries of the pond; it glorifies the heavenly insect which I hear moving in the empty snail-shell, its temporary cage; it whispers the secrets of the rock, the gold-filings, the faceted jewels, the ram's-horn turned to stone.

Poor simpleton, smother your joy! I arrive. My parents catch sight of my bulging pockets, with their disgraceful load of stones. The cloth has given way under the rough and heavy burden.

"You rascal!" says father, at sight of the damage. "I send you to mind the ducks and you amuse yourself picking up stones, as though there weren't enough of them all round the house! Make haste and throw them away!"

Broken-hearted, I obey. Diamonds, gold-dust, petrified ram's-horn, heavenly Beetle are all flung on a rubbish-heap outside the door.

Mother bewails her lot:

"A nice thing, bringing up children to see them turn out so badly! You'll bring me to my grave. Green stuff I don't mind: it does for the rabbits. But stones, which ruin your pockets; poisonous animals, which'll sting your hand: what good are they to you, silly? There's no doubt about it: some one has thrown a spell over you!"

Yes, my poor mother, you were right, in your simplicity: a spell had been cast upon me . . .

COURTSHIP OF THE SCORPION

« 29 »

Like the Praying Mantis and numerous species of spiders, the female scorpion devours her husband. Fabre's investigations of this phase of the life-story of the creature extended, as his investigations often did, over a period of many years. His step-by-step advance in understanding, as recorded in his notes set down at the time, is given here. The source of this selection is THE LIFE OF THE SCORPION.

In April, when the Swallow returns to us and the Cuckoo sounds his first note, a revolution takes place among my hitherto peaceable Scorpions. Several whom I have established in the colony in the enclosure, leave their shelter at nightfall, go wandering about and do not return to their homes. A more serious business: often, under the same stone, are two Scorpions of whom one is in the act of devouring the other. Is this a case of brigandage among creatures of the same order, who, falling into vagabond ways when the fine weather sets in thoughtlessly enter their neighbours' houses and there meet with their undoing unless they be the stronger? One would almost think it, so quickly is the intruder eaten up, for days at a time and in small mouthfuls, even as the usual game would be.

Now here is something to give us a hint. The Scorpions devoured are invariably of middling size. Their lighter colouring, their less protuberant bellies, mark them as males, always males. The others, larger, more paunchy and a little darker in shade, do not end in this unhappy fashion. So these are probably not brawls between neighbours who, jealous of their solitude, would soon settle the hash of any visitor and eat him afterwards, a drastic method of putting a stop to further indiscretions; they are rather nuptial rites, tragically performed by the matron after pairing. To determine how much ground there is for this suspicion is beyond my powers until next year: I am still too badly equipped.

Spring returns once more. I have prepared the large glass cage in advance and stocked it with twenty-five inhabitants, each with his bit of crockery. From mid-April onwards, every evening, when it grows dark, between seven and nine o'clock, great animation reigns in the crystal palace. That which seemed deserted by day now becomes a scene of festivity. As soon as supper is finished, the whole household runs out to look on. A lantern hung outside the panes allows us to follow events.

It is our distraction after the worries of the day; it is our play-house. In this theatre for simple folk, the performances are so highly interesting that, the moment the lantern is lighted, all of us, great and small alike, come and take our places in the stalls; all, down to Tom, the House-dog. Tom, it is true, indifferent to Scorpion affairs, like the true philosopher that he is, lies at our feet and dozes, but only with one eye, keeping the other always open on his friends the children.

Let me try to give the reader an idea of what happens. A numerous assembly soon gathers near the glass panes in the region discreetly lit by the lanterns. Every elsewhere, here, there, single Scorpions walk about and, attracted by the light, leave the shade and hasten to the illuminated festival. The very Moths betray no greater eagerness to flutter to the rays of our lamps. The newcomers mingle with the crowd, while others, tired of their pastimes, withdraw into the shade, snatch a few moments' rest and then impetuously return upon the scene.

These hideous devotees of gaiety provide a dance that is not wholly devoid of charm. Some come from afar: solemnly they emerge from the shadow; then, suddenly, with a rush as swift and easy as a slide, they join the crowd, in the light. Their agility reminds one of Mice scurrying along with their tiny steps. They seek one another and fly precipitately the moment they touch, as though they had mutually burnt their fingers. Others, after tumbling about a little with their play-fellows, make off hurriedly wildly. They take fresh courage in the dark and return.

At times, there is a violent tumult: a confused mass of swarming legs, snapping claws, tails curving and clashing, threatening or fondling, it is hard to say which. In this affray, under favourable conditions, twin specks of light flare and shine like carbuncles. One would take them for eyes that emit flashing glances; in reality they are two polished, reflecting facets, which occupy the front of the head. All, large and small alike, take part in the brawl; it might be a battle to the death, a general massacre; and it is just a wanton frolic. Even so do kittens bemaul each other. Soon, the group disperses; all make off in all sorts of directions, without a scratch, without a sprain.

Behold the fugitives collecting once more beneath the lantern. They pass and pass again; they come and go, often meeting front to front. He who is in the greatest hurry walks over the back of the other, who lets him have his way without any protest but a movement of the body. It is no time for blows: at most, two Scorpions meeting will exchange a cuff, that is to say, a rap of the caudal staff. In their community, this friendly thump, in which the point of the sting plays no part, is a sort of a fisticuff in frequent use. There are better things than entangled legs and brandished tails; there are sometimes poses of the highest originality. Face to face, with claws drawn back, two wrestlers proceed to stand on their heads like acrobats, that is to say, resting only on the forequarters, they raise the whole hinder portion of the body, so much so that the chest displays the four little lung pockets uncovered. Then the tails, held vertically erect in a straight line, exchange mutual rubs, gliding one over the other, while their extremities are hooked together and repeatedly fastened and unfastened. Suddenly, the friendly pyramid falls to pieces and each runs off hurriedly, without ceremony.

What were these two wrestlers trying to do, in their eccentric posture? Was it a set-to between two rivals? It would seem not, so peaceful is the encounter. My subsequent observations were to tell me that this was the mutual teasing of a betrothed couple. To declare his flame, the Scorpion stands on his head.

To continue as I have begun and give a homogeneous picture of the thousand tiny particulars gathered day by day would have its advantages: the story would sooner be told; but, at the same time deprived of its details, which vary greatly between one observation and the next and are difficult to piece together, it would be less interesting. Nothing must be neglected in the relation of manners so strange and as yet so little known. At the risk of repeating one's self here and there, it is preferable to adhere to chronological order and to tell the story by fragments, as one's observations reveal fresh facts. Order will emerge from this disorder; for each of the more remarkable evenings supplies some feature that corroborates and completes those which go before. I will therefore continue my narration in the form of a diary.

25th April, 1904.—Hullo! What is this, something I have not yet seen? My eyes, ever on the watch, look upon the affair for the first time. Two

Scorpions face each other, with claws outstretched and fingers clasped. It is a question of a friendly grasp of the hand and not the prelude to a battle, for the two partners are behaving to each other in the most peaceful way. There is one of either sex. One is paunchy and browner than the other: this is the female; the other is comparatively slim and pale: this is the male. With their tails prettily curled, the couple stroll with measured steps along the pane. The male is ahead and walks backwards, without jolt or jerk, without any resistance to overcome. The female follows obediently, clasped by her finger-tips and face to face with her leader.

The stroll is interrupted by halts that do not affect the method of conjunction; it is resumed, now here, now there, from end to end of the enclosure. Nothing shows the object which the strollers have in view. They loiter, they dawdle, they most certainly exchange ogling glances. Even so in my village, on Sundays, after vespers, do the youth of both sexes saunter along the hedges, every Jack with his Jill.

Often they tack about. It is always the male who decides which fresh direction the pair shall take. Without releasing her hands, he turns gracefully to the left or right about and places himself side by side with his companion. Then, for a moment, with tail laid flat, he strokes her spine. The other stands motionless, impassive.

For over an hour, without tiring, I watch these interminable comings and goings. A part of the household lends me its eyes in the presence of the strange sight which no one in the world has yet seen, at least with a vision capable of observing. In spite of the lateness of the hour, which upsets all our habits, our attention is concentrated and no essential thing escapes us.

At last, about ten o'clock, something happens. The male has hit upon a potsherd whose shelter seems to suit him. He releases his companion with one hand, with one alone, and continuing to hold her with the other, he scratches with his legs and sweeps with his tail. A grotto opens. He enters and, slowly, without violence, drags the patient Scorpioness after him. Soon both have disappeared. A plug of sand closes the dwelling. The couple are at home.

To disturb them would be a blunder: I should be interfering too soon, at an inopportune moment, if I tried at once to see what was happening below. The preliminary stages may last for the best part of the night; and it does not do for me, who have turned eighty, to sit up so late. I feel my legs giving way; and my eyes seem full of sand.

All night long I dream of Scorpions. They crawl under my bedclothes, they pass over my face; and I am not particularly excited, so many curious things do I see in my imagination. The next morning, at daybreak, I lift the stoneware. The female is alone. Of the male there is no trace, either in the home or in the neighbourhood. First disappointment, to be followed by many others.

10th May.—It is nearly seven o'clock in the evening; the sky is overcast with signs of an approaching shower. Under one of the potsherds is a motionless couple, face to face, with linked fingers. Cautiously I raise the potsherd and leave the occupants uncovered, so as to study the consequences of the interview at my ease. The darkness of the night falls and nothing, it seems to me, will disturb the calm of the home deprived of its roof. A sharp shower compels me to retire. They, under the lid of the cage, have no need to take shelter against the rain. What will they do, left to their business as they are but deprived of a canopy to their alcove?

An hour later, the rain ceases and I return to my Scorpions. They are gone. They have taken up their abode under a neighbouring tile. Still with their fingers linked, the female is outside and the male indoors, preparing the home. At intervals of ten minutes, the members of my family relieve one another, so as not to lose the exact moment of the pairing, which appears to be imminent. Wasted pains: at eight o'clock, it being now quite dark, the couple, dissatisfied with the spot, set out on a fresh ramble, hand in hand, and go prospecting elsewhere. The male, walking backwards, leads the way, chooses the dwelling as he pleases; the female follows with docility. It is an exact repetition of what I saw on the 25th of April.

At last a tile is found to suit them. The male goes in first but this time neither hand releases his companion for a moment. The nuptial chamber is prepared with a few sweeps of the tail. Gently drawn towards him, the Scorpioness enters in the wake of her guide.

I visit them a couple of hours later, thinking that I've given them time enough to finish their preparations. I lift the potsherd. They are there in the same posture, face to face and hand in hand. I shall see no more to-day.

The next day, nothing new either. Each sits confronting the other, meditatively. Without stirring a limb, the gossips, holding each other by the finger-tips, continue their endless interview under the tile. In the evening, at sunset, after sitting linked together for four-and-twenty hours, the couple separate. He goes away from the tile, she remains; and matters have not advanced by an inch.

This observation gives us two facts to remember. After the stroll to celebrate the betrothal, the couple need the mystery and quiet of a shelter. Never would the nuptials be consummated in the open air, amid the bustling crowd, in sight of all. Remove the roof of the house, by night or day, with all possible discretion; and the husband and wife, who seem absorbed in meditation, march off in search of another spot. Also, the sojourn under the cover of a stone is a long one: we have just seen it spun out to twenty-four hours and even then without a decisive result.

12th May.—What will this evening's sitting teach us? The weather is calm and hot, favourable to nocturnal pastimes. A couple has been formed: how things began I do not know. This time the male is greatly inferior to his corpulent mate. Nevertheless, the skinny wight performs his duty gallantly. Walking backwards, according to rule, with his tail rolled trumpetwise, he marches the fat Scorpioness around the glass ramparts. After one circuit follows another, sometimes in the same, sometimes in the opposite direction.

Pauses are frequent. Then the foreheads touch, bend a little to left and right, as if the two were whispering in each other's ears. The little fore-legs flutter in feverish caresses. What are they saying to each other? How shall we translate their silent epithalamium into words?

The whole household turns out to see this curious team, which our presence in no way disturbs. The pair are pronounced to be "pretty"; and the expression is not exaggerated. Semitranslucent and shining in the light of the lantern, they seem carved out of a block of amber. Their arms outstretched, their tails rolled into graceful spirals, they wander on with a slow movement and with measured tread.

Nothing puts them out. Should some vagabond, taking the evening air and keeping to the wall like themselves, meet them on their way, he stands aside—for he understands these delicate matters—and leaves them a free passage. Lastly, the shelter of a tile receives the strolling pair, the male entering first and backwards: that goes without saying. It is nine o'clock.

The idyll of the evening is followed, during the night, by a hideous tragedy. Next morning, we find the Scorpioness under the potsherd of the previous day. The little male is by her side, but slain, and more or less devoured. He lacks the head, a claw, a pair of legs. I place the corpse in the open, on the threshold of the home. All day long, the recluse does not touch it. When night returns, she goes out and, meeting the deceased on her passage, carries him off to a distance to give him a decent funeral, that is to finish eating him.

This act of cannibalism agrees with what the open-air colony showed me last year. From time to time, I would find, under the stones, a pot-bellied female making a comfortable ritual meal off her companion of the night. I suspected that the male, if he did not break loose in time, once his functions were fulfilled, was devoured, wholly or partly, according to the matron's appetite. I now have the certain proof before my eyes. Yesterday, I saw the couple enter their home after their usual preliminary, the stroll; and, this morning, under the same tile, at the moment of my visit, the bride is consuming her mate.

THE BURYING-BEETLE

« 30 »

One of the remarkable features of Fabre's writing is his ability to make the repulsive interesting. This study of the activity of those sanitary workers of the fields, the Burying-Beetles—selected from THE GLOW-WORM AND OTHER BEETLES—*is a case in point. Various species of Necrophorus burying beetles are found in the United States; also those of Silpha, the carrion beetle. The Dermestes beetles are minute insects that feed upon dried animal matter, often riddling the pinned specimens of neglected insect-collections. The Saprini are carnivorous beetles of small size. More than 3,000 kinds of Staphylinus, or Rove Beetles, have been described from the United States.*

Beside the footpath in April lies the Mole, disembowelled by the peasant's spade; at the foot of the hedge the pitiless urchin has stoned to death the Lizard, who was about to don his green, pearl-embellished costume. The passer-by has thought it a meritorious deed to crush beneath his heel the chance-met Adder; and a gust of wind has thrown a tiny unfledged bird from its nest. What will become of these little bodies and so many other pitiful remnants of life? They will not long offend our sense of sight and smell. The sanitary officers of the fields are legion.

An eager freebooter, ready for any task, the Ant is the first to come hastening and begin, particle by particle, to dissect the corpse. Soon the odour attracts the Fly, the genitrix of the odious maggot. At the same time, the flattened Silpha, the glistening, slow-trotting Cellar-beetle, the Dermestes, powdered with snow upon the abdomen, and the slender Staphylinus, all, whence coming no one knows, hurry hither in squads, with never-wearied zeal, investigating, probing and draining the infection.

What a spectacle, in the spring, beneath a dead Mole! The horror of this laboratory is a beautiful sight for one who is able to observe and to meditate. Let us overcome our disgust; let us turn over the unclean refuse with our foot. What a swarming there is beneath it, what a tumult of busy workers! The Silphæ, with wing-cases wide and dark, as though in mourning, flee distraught, hiding in the cracks in the soil; the Saprini, of polished ebony which mirrors the sunlight, jog hastily off, deserting their workshop; the Dermestes, of whom one wears a fawn-coloured tippet flecked with white, seek to fly away, but, tipsy with the putrid nectar, tumble over and reveal the immaculate whiteness of their bellies, which forms a violent contrast with the gloom of the rest of their attire.

What were they doing there, all these feverish workers? They were making a clearance of death on behalf of life. Transcendent alchemists, they were transforming that horrible putrescence into a living and in-offensive product. They were draining the dangerous corpse to the point of rendering it as dry and sonorous as the remains of an old slipper hardened on the refuse-heap by the frosts of winter and the heats of summer. They were working their hardest to render the carrion innocuous.

Others will soon put in their appearance, smaller creatures and more patient, who will take over the relic and exploit it ligament by ligament, bone by bone, hair by hair, until the whole has been restored to the treasury of life. All honour to these purifiers! Let us put back the Mole and go our way.

Some other victim of the agricultural labours of spring, a Shrew-mouse, Field-mouse, Mole, Frog, Adder, or Lizard, will provide us with the most vigorous and famous of these expurgators of the soil. This is the Burying-beetle, the Necrophorus, so different from the cadaveric mob in dress and habits. In honour of his exalted functions he exhales an odour of musk; he bears a red tuft at the tip of his antennæ; his breast is covered with nankeen; and across his wing-cases he wears a double, scalloped scarf of vermillion. An elegant, almost sumptuous costume, very superior to that of the others, but yet lugubrious, as be-fits your undertaker's man.

He is no anatomical dissector, cutting his subject open, carving its flesh with the scalpel of his mandibles; he is literally a grave-digger, a sexton. While the others—Silphæ, Dermestes, Cellar-beetles—gorge themselves with the exploited flesh, without, of course, forgetting the interests of the family, he, a frugal eater, hardly touches his find on his own account. He buries it entire, on the spot, in a cellar where the thing, duly ripened, will form the diet of his larvæ. He buries it in order to establish his progeny.

This hoarder of dead bodies, with his stiff and almost heavy movements, is astonishingly quick at storing away wreckage. In a shift of a few hours, a comparatively enormous animal, a Mole, for instance, disappears, engulfed by the earth. The others leave the dried, emptied carcass to the air, the sport of the winds for months on end; he, treating it as a whole, makes a clean job of things at once. No visible trace of his work remains but a tiny hillock, a burial-mound, a tumulus.

With his expeditious method, the Necrophorus is the first of the little purifiers of the fields. He is also one of the most celebrated of insects in respect of his psychical capacities. This undertaker is endowed, they say, with intellectual faculties approaching to reason, such as are not possessed by the most gifted of the Bees and Wasps, the collectors of honey or game. He is honoured by the two following anecdotes, which I quote from Lacordaire's *Introduction a l'entomologie,* the only general treatise at my disposal:

"Clairville," says the author, "reports that he saw a *Necrophorus vespillo,* who, wishing to bury a dead Mouse and finding the soil on which the body lay too hard, went to dig a hole at some distance, in soil more easily displaced. This operation completed, he attempted to bury the Mouse in the cavity, but, not succeeding, he flew away and returned a few moments later, accompanied by four of his fellows, who assisted him to move the Mouse and bury it."

In such actions, Lacordaire adds, we cannot refuse to admit the intervention of reason.

"The following case," he continues, "recorded by Gleditsch, has also every indication of the intervention of reason. One of his friends, wishing to desiccate a Frog, placed it on the top of a stick thrust into the ground, in order to make sure that the Necrophori should not come and carry it off. But this precaution was of no effect; the insects, being unable to reach the Frog, dug under the stick and, having caused it to fall, buried it as well as the body."

To grant, in the intellect of the insect, a lucid understanding of the relations between cause and effect, between the end and the means, is to make a statement of serious import. I know of scarcely any more suited to the philosophical brutalities of my time. But are these two anecdotes really true? Do they involve the consequences deduced from them? Are not those who accept them as sound evidence just a little too simple?

To be sure, simplicity is needed in entomology. Without a good dose of this quality, a mental defect in the eyes of practical folk, who would busy himself with the lesser creatures? Yes, let us be simple, without being childishly credulous. Before making insects reason, let us reason a little ourselves; let us, above all, consult the experimental test. A fact gathered at random, without criticism, cannot establish a law.

I do not propose, O valiant grave-diggers, to depreciate your merits; such is far from being my intention. I have that in my notes, on the other hand, which will do you more honour than the story of the gibbet and the Frog; I have gleaned, for your benefit, examples of prowess which will shed a new lustre upon your reputation.

No, my intention is not to belittle your renown. Besides, it is not the business of impartial history to maintain a given thesis; it follows facts. I wish simply to question you upon the power of logic attributed to you. Do you or do you not enjoy gleams of reason? Have you within you the humble germ of human thought? That is the problem before us.

To solve it we will not rely upon the accidents which good fortune may now and again procure for us. We must employ the breeding-cage, which will permit of assiduous visits, continuous enquiry and a variety of artifices. But how to stock the cage? The land of the olive-tree is not rich in Necrophori. To my knowledge it possesses only a single species, *N. vestigator,* HERSCH.; and even this rival of the grave-diggers of the north is pretty scarce. The discovery of three or four in the spring was as much as my hunting-expeditions yielded in the old days. This time, if I do not resort to the ruses of the trapper, I shall obtain no more than that, whereas I stand in need of at least a dozen.

These ruses are very simple. To go in search of the sexton, who exists only here and there in the country-side, would be nearly always a waste of time; the favourable month, April, would be past before my cage was suitably stocked. To run after him is to trust too much to accident; so we will make him come to us by scattering in the orchard an abundant collection of dead Moles. To this carrion, ripened by the sun, the insect will not fail to hasten from the various points of the horizon, so accomplished is he in detecting such a delicacy.

I make an arrangement with a gardener in the neighbourhood, who, two or three times a week, makes up for the penury of my two acres of stony ground by providing me with vegetables raised in a better soil. I explain to him my urgent need of Moles in unlimited numbers. Battling daily with trap and spade against the importunate excavator who uproots his crops, he is in a better position than any one to procure for me what I regard for the moment as more precious than his bunches of asparagus or his white-heart cabbages.

The worthy man at first laughs at my request, being greatly surprised by the importance which I attribute to the abhorrent animal, the *Darboun;* but at last he consents, not without a suspicion at the back of his mind that I am going to make myself a gorgeous winter waist-coat with the soft, velvety skins of the Moles. A thing like that must be good for pains in the back. Very well. We settle the matter. The essential thing is that the *Darbouns* reach me.

They reach me punctually, by twos, by threes, by fours, packed in a few cabbage-leaves, at the bottom of the gardener's basket. The excellent fellow who lent himself with such good grace to my strange wishes will never guess how much comparative psychology will owe him! In a few days I was the possessor of thirty Moles, which were scattered here and there, as they reached me, in bare spots of the orchard, among the rosemary-bushes, the strawberry-trees and the lavender-beds.

Now it only remained to wait and to examine, several times a day, the under-side of my little corpses, a disgusting task which any one would avoid whose veins were not filled with the sacred fire of enthusiasm. Only little Paul, of all the household, lent me the aid of his nimble hand to seize the fugitives. I have already said that the entomologist needs simplicity of mind. In this important business of the Necrophori, my assistants were a small boy and an illiterate.

Little Paul's visits alternating with mine, we had not long to wait. The four winds of heaven bore forth in all directions the odour of the carrion; and the undertakers hurried up, so that the experiments, begun with four subjects, were continued with fourteen, a number not attained during the whole of my previous searches, which were unpremeditated and in which no bait was used as decoy. My trapper's ruse was completely successful.

Before I report the results obtained in the cage, let us stop for a moment to consider the normal conditions of the labours that fall to the lot of the Necrophori. The Beetle does not select his head of game, choosing one in proportion to his strength, as do the Hunting Wasps; he accepts what chance offers. Among his finds some are small, such as

the Shrew-mouse; some medium-sized, such as the Field-mouse; some
enormous, such as the Mole, the Sewer-rat and the Snake, any of which
exceeds the digging-powers of a single sexton. In the majority of cases,
transportation is impossible, so greatly disproportioned is the burden to
the motive-power. A slight displacement, caused by the effort of the insects' backs, is all that can possibly be effected.

Ammophila and Cerceris, Sphex and Pompilus excavate their burrows
wherever they please; they carry their prey on the wing, or, if too heavy,
drag it afoot. The Necrophorus knows no such facilities in his task. Incapable of carting the monstrous corpse, no matter where encountered,
he is forced to dig the grave where the body lies.

This obligatory place of sepulture may be in stony soil or in shifting
sand; it may occupy this or that bare spot, or some other where the grass,
especially the couch-grass, plunges into the ground its inextricable network of little cords. There is a great probability, too, that a bristle of
stunted brambles may be supporting the body at some inches above the
soil. Slung by the labourer's spade, which has just broken his back,
the Mole falls here, there, anywhere, at random; and where the body
falls, no matter what the obstacles, provided that they be not insurmountable, there the undertaker must utilize it.

The difficulties of inhumation are capable of such variety as causes
us already to foresee that the Necrophorus cannot employ fixed methods
in performing his task. Exposed to fortuitous hazards, he must be able
to modify his tactics within the limits of his modest discernment. To
saw, to break, to disentangle, to lift, to shake, to displace: these are so
many means which are indispensable to the grave-digger in a predicament. Deprived of these resources, reduced to uniformity of procedure,
the insect would be incapable of pursuing its calling.

We see at once how imprudent it would be to draw conclusions from
an isolated case in which rational co-ordination or premeditated intention might appear to play its part. Every instinctive action no doubt has
its motive; but does the animal in the first place judge whether the action
is opportune? Let us begin by a careful consideration of the creature's
labours; let us support each piece of evidence by others; and then we
shall perhaps be able to answer the question.

First of all, a word as to diet. A general scavenger, the Burying-beetle
refuses no sort of cadaveric putrescence. All is good to his senses,
feathered game or furry, provided that the burden do not exceed his
strength. He exploits the batrachian or the reptile with no less animation. He accepts without hesitation extraordinary finds, probably un-

known to his race, as witness a certain Gold-fish, a red Chinese Carp, whose body, placed in one of my cages, was forthwith considered an excellent tit-bit and buried according to the rules. Nor is butcher's meat despised. A mutton-cutlet, a strip of beef-steak, in the right stage of maturity, disappeared beneath the soil, receiving the same attentions as those lavished on the Mole or the Mouse. In short, the Necrophorus has no exclusive preferences; anything putrid he conveys underground.

The maintenance of his industry, therefore, presents no sort of difficulty. If one kind of game be lacking, some other, the first to hand, will very well replace it. Nor is there much trouble in fixing the site of his industry. A capacious wire-gauze cover, resting on an earthen pan filled to the brim with fresh, heaped sand, is sufficient. To obviate criminal attempts on the part of the Cats, whom the game would not fail to tempt, the cage is installed in a closed glass-house, which in winter shelters the plants and in summer serves as an entomological laboratory.

Now to work. The Mole lies in the centre of the enclosure. The soil, easily shifted and homogeneous, realizes the best conditions for comfortable work. Four Necrophori, three males and a female, are there with the body. They remain invisible, hidden beneath the carcase, which from time to time seems to return to life, shaken from end to end by the backs of the workers. An observer not in the secret would be somewhat astonished to see the dead creature move. From time to time, one of the sextons, almost always a male, comes out and walks round the animal, which he explores, probing its velvet coat. He hurriedly returns, appears again, once more investigates and creeps back under the corpse.

The tremors become more pronounced; the carcase oscillates, while a cushion of sand, pushed out from below, grows up all around it. The Mole, by reason of his own weight and the efforts of the grave-diggers, who are labouring at their task underneath, gradually sinks, for lack of support, into the undermined soil.

Presently the sand which has been pushed out quivers under the thrust of the invisible miners, slips into the pit and covers the interred Mole. It is a clandestine burial. The body seems to disappear of itself, as though engulfed by a fluid medium. For a long time yet, until the depth is regarded as sufficient, the body will continue to descend.

It is, on the whole, a very simple operation. As the diggers below deepen the cavity into which the corpse, shaken and tugged above, sinks without the direct intervention of the sextons, the grave fills of itself by the mere slipping of the soil. Stout shovels at the tips of their claws, powerful backs, capable of creating a little earthquake: the diggers need nothing more for the practice of their profession. Let us add—for this is an essential point—the art of continually jerking the body, so as to pack it into a lesser volume and make it glide through difficult passages. We shall soon see that this art plays a leading part in the industry of the Necrophori.

Although he has disappeared, the Mole is still far from having reached his destination. Let us leave the undertakers to finish their job. What they are now doing below ground is a continuation of what they did on the surface and would teach us nothing new. We will wait for two or three days.

The moment has come. Let us inform ourselves of what is happening down there. Let us visit the place of corruption. I shall never invite anybody to the exhumation. Of those about me, only little Paul has the courage to assist me.

The Mole is a Mole no longer, but a greenish horror, putrid, hairless, shrunk into a sort of fat, greasy rasher. The thing must have undergone careful manipulation to be thus condensed into a small volume, like a fowl in the hands of the cook, and, above all, to be so completely deprived of its furry coat. Is this culinary procedure undertaken in respect of the larvæ, which might be incommoded by the fur? Or is it just a casual result, a mere loss of hair due to putridity? I am not certain. But it is always the case that these exhumations, from first to last, have revealed the furry game furless and the feathered game featherless, except for the pinion- and tail-feathers. Reptiles and fish, on the other hand, retain their scales.

Let us return to the unrecognizable thing that was once a Mole. The tit-bit lies in a spacious crypt, with firm walls, a regular workshop, worthy of being the bake-house of a Copris. Except for the fur, which lies scattered about in flocks, it is intact. The grave-diggers have not eaten into it: it is the patrimony of the sons, not the provision of the parents, who, to sustain themselves, levy at most a few mouthfuls of the ooze of putrid humours.

Beside the dish which they are kneading and protecting are two Necrophori; a couple, no more. Four collaborated in the burial. What has become of the other two, both males? I find them hidden in the soil, at a distance, almost on the surface.

This observation is not an isolated one. Whenever I am present at a funeral undertaken by a squad in which the males, zealous one and all, predominate, I find presently, when the burial is completed, only one couple in the mortuary cellar. After lending their assistance, the rest have discreetly retired.

These grave-diggers, in truth, are remarkable fathers. They have nothing of the happy-go-lucky paternal carelessness that is the general rule among insects, which pester the mother for a moment with their attentions and then leave her to care for the offspring! But those who would be idlers in the other castes here labour valiantly, now in the interest of their own family, now in that of another's, without distinction. If a couple is in difficulties, helpers arrive, attracted by the odour of carrion; anxious to serve a lady, they creep under the body, work at it with back and claw, bury it and then go their ways, leaving the master and mistress of the house to their happiness.

For some time longer these two manipulate the morsel in concert, stripping it of fur or feather, trussing it and allowing it to simmer to the grub's taste. When everything is in order, the couple go forth, dissolving their partnership; and each, following his fancy, begins again elsewhere, even if only as a mere auxiliary.

Twice and no oftener hitherto have I found the father preoccupied by the future of his sons and labouring in order to leave them rich: it happens with certain dung-workers and with the Necrophori, who bury dead bodies. Scavengers and undertakers both have exemplary morals. Who would look for virtue in such a quarter?

What follows—the larval existence and the metamorphosis—is a secondary and, for that matter, a familiar detail. It is a dry subject and I will deal with it briefly. At the end of May, I exhume a Brown Rat, buried by the grave-diggers a fortnight earlier. Transformed into a black, sticky mass, the horrible dish provides me with fifteen larvæ already, for the most part, of the normal size. A few adults, unquestionably connections of the brood, are also swarming amid the putrescence. The laying-time is over now and victuals are plentiful. Having nothing else to do, the foster-parents have sat down to the feast with the nurslings.

The undertakers are quick at rearing a family. It is at most a fortnight since the Rat was laid in the earth; and here already is a vigorous population on the verge of the metamorphosis. This precocity amazes me. It would seem as though carrion liquefaction, deadly to any other stomach, were in this case a food productive of special energy, which stimulates the organism and accelerates its growth, so that the fare may be consumed before its approaching conversion into mould. Living chemistry makes haste to outstrip the ultimate reactions of mineral chemistry.

White, naked, blind, possessing the customary attributes of life spent in the dark, the larva, with its tapering outline, is slightly reminiscent of the Ground-beetles'. The mandibles are black and powerful and make excellent dissecting-scissors. The limbs are short, but capable of a quick, toddling gait. The segments of the abdomen are clad on the upper surface in a narrow red plate, armed with four little spikes, whose office apparently is to furnish points of support when the larva quits the natal dwelling and dives into the soil, there to undergo the transformation. The thoracic segments are provided with wider plates, but unarmed.

The adults discovered in the company of their larval family, in this putrescence which was a Rat, are all abominably verminous. So shiny and neat in their attire, when at work under the first Moles of April, the Necrophori, when June approaches, become odious to look upon. A layer of parasites envelops them; insinuating itself into the joints, it forms an almost continuous crust. The insect presents a misshapen appearance under this overcoat of vermin, which my hair-pencil can hardly brush aside. Driven off the belly, the horde runs round the sufferer, perches on his back and refuses to let go.

I recognize the Beetle's Gamasus, the Tick who so often soils the ventral amethyst of our Geotrupes. No, life's prizes do not go to the useful. Necrophori and Geotrupes devote themselves to the general health; and these two corporations, so interesting in their hygienic functions, so remarkable for their domestic morals, fall victims to the vermin of poverty. Alas, of this discrepancy between the services rendered and the harshness of life there are many other examples outside the world of scavengers and undertakers!

The Burying-beetles display an exemplary domestic morality, but it does not continue till the end. In the first fortnight of June, the family being sufficiently provided, the sextons strike work and my cages are deserted on the surface, in spite of new arrivals of Mice and Sparrows. From time to time, some grave-digger leaves the subsoil and comes crawling languidly into the fresh air.

Another rather curious fact now attracts my attention. All those who climb up from underground are maimed, with limbs amputated at the joints, some higher up, some lower down. I see one cripple who has only one leg left entire. With this odd limb and the stumps of the others, lamentably tattered, scaly with vermin, he rows, as it were, over the sheet of dust. A comrade emerges, better off for legs, who finishes the invalid and cleans out his abdomen. Thus do my thirteen remaining Necrophori end their days, half-devoured by their companions, or at least shorn of several limbs. The pacific relations of the outset are succeeded by cannibalism.

History tells us that certain peoples, the Massagetæ and others, used to kill off their old men to save them from senile misery. The fatal blow on the hoary skull was in their eyes an act of filial piety. The Necrophori have their share of these ancient barbarities. Full of days and henceforth useless, dragging out a weary existence, they mutually exterminate one another. Why prolong the agony of the impotent and the imbecile?

The Massagetæ might plead, as an excuse for their atrocious custom, a dearth of provisions, which is an evil counsellor; not so the Necrophori, for, thanks to my generosity, victuals are more than plentiful, both beneath the soil and on the surface. Famine plays no part in this slaughter. What we see is an aberration due to exhaustion, the morbid fury of a life on the point of extinction. As is generally the case, work bestows a peaceable disposition on the grave-digger, while inaction inspires him with perverted tastes. Having nothing left to do, he breaks his kinsman's limbs and eats him up, heedless of being maimed or eaten himself. It is the final deliverance of verminous old age.

This murderous frenzy, breaking out late in life, is not peculiar to the Necrophorus. I have described elsewhere the perversity of the Osmia, so placid in the beginning. Feeling her ovaries exhausted, she smashes her neighbours' cells and even her own; she scatters the dusty honey, rips open the egg, eats it. The Mantis devours the lovers who have played their parts; the mother Decticus willingly nibbles a thigh of her decrepit husband; the merry Crickets, once the eggs are laid in the ground, indulge in tragic domestic quarrels and with not the least compunction slash open one another's bellies. When the cares of the family are finished, the joys of life are finished likewise. The insect then sometimes becomes depraved; and its disordered mechanism ends in aberrations.

The larva has nothing striking to show in the way of industry. When it has fattened to the desired extent, it leaves the charnel-house of the natal crypt and descends into the earth, far from the putrefaction. Here, working with its legs and its dorsal armour, it presses back the sand around it and makes itself a close cabin wherein to rest for the metamorphosis. When the lodge is ready and the torpor of the approaching moult arrives, it lies inert; but, at the least alarm, it comes to life and turns round on its axis.

Even so do several nymphs spin round and round when disturbed, notably that of *Ægasomus scabricornis* which I have now before my eyes in July. It is always a fresh surprise to see these mummies suddenly throw off their immobility and gyrate on their own axis with a mechanism whose secret deserves to be fathomed. The science of rational mechanics might find something here to whet its finest theories upon. The strength and litheness of a clown cannot compare with those of this budding flesh, this hardly coagulated glair.

Once isolated in its cell, the larva of the Necrophorus becomes a nymph in ten days or so. I lack the evidence furnished by direct observation, but the story is completed of itself. The Necrophorus must assume the adult form in the course of the summer; like the Dung-beetle, he must enjoy in the autumn a few days of revelry free from family cares. Then, when the cold weather draws near, he goes to earth in his winter quarters, whence he emerges as soon as spring arrives.

EXPERIMENTS WITH BURYING-BEETLES

« 31 »

Also from THE GLOW-WORM AND OTHER BEETLES, *this account of Fabre's researches with the long-suffering beetles of his harmas is one of his most celebrated chapters. Like a scientific detective story, it unfolds engrossingly and logically, step by step, link by link. The Copres, Sisyphi and Gymnopleuri, mentioned by Fabre, are all dung beetles. Audubon is, of course, John James Audubon (1780–1851), the great American bird-painter and ornithologist. Fabre admired Audubon greatly and once said he found in him the man most nearly akin to himself in mind and temperament.*

Let us come to the feats of reason which have earned for the Necrophorus the best part of his fame and, to begin with, submit the case related by Clairville, that of the too hard soil and the call for assistance, to the test of experiment.

With this object I pave the centre of the space beneath the cover, flush with the soil, with a brick, which I sprinkle with a thin layer of sand. This will be the soil that cannot be dug. All around it, for some distance and on the same level, lies the loose soil, which is easy to delve.

In order to approach the conditions of the anecdote, I must have a Mouse; with a Mole, a heavy mass, the removal would perhaps present too much difficulty. To obtain one, I place my friends and neighbours under requisition; they laugh at my whim but none the less proffer their traps. Yet, the moment a very common thing is needed, it becomes rare. Defying decency in his speech, after the manner of his ancestors' Latin, the Provençal says, but even more crudely than in my translation:

"If you look for dung, the Donkeys become constipated!"

At last I possess the Mouse of my dreams! She comes to me from that refuge, furnished with a truss of straw, in which official charity grants a day's hospitality to the pauper wandering over the face of the fertile earth, from that municipal hostel whence one inevitably issues covered with Lice. O Réaumur, who used to invite marchionesses to see your caterpillars change their skins, what would you have said of a future disciple conversant with such squalor as this? Perhaps it is well that we should not be ignorant of it, so that we may have compassion with that of the beast.

The Mouse so greatly desired is mine. I place her upon the centre of the brick. The grave-diggers under the wire cover are now seven in number, including three females. All have gone to earth; some are inactive, close to the surface; the rest are busy in their crypts. The presence of the fresh corpse is soon perceived. About seven o'clock in the morning, three Necrophori come hurrying up, two males and a female. They slip under the Mouse, who moves in jerks, a sign of the efforts of the burying-party. An attempt is made to dig into the layer of sand which hides the brick, so that a bank of rubbish accumulates round the body.

For a couple of hours the jerks continue without results. I profit by the circumstance to learn the manner in which the work is performed. The bare brick allows me to see what the excavated soil would conceal from me. When it is necessary to move the body, the Beetle turns over; with his six claws he grips the hair of the dead animal, props himself upon his back and pushes, using his forehead and the tip of his abdomen as a lever. When he wants to dig, he resumes the normal position. So, turn and turn about, the sexton strives, now with his legs in the air, when it is a question of shifting the body or dragging it lower down; now with his feet on the ground, when it is necessary to enlarge the grave.

The point at which the Mouse lies is finally recognized as unassailable. A male appears in the open. He explores the corpse, goes round it, scratches a little at random. He goes back; and immediately the dead body rocks. Is he advising his collaborators of what he has discovered? Is he arranging the work with a view to their establishing themselves elsewhere, on propitious soil?

The facts are far from confirming this idea. When he shakes the body, the others imitate him and push, but without combining their efforts in a given direction, for, after advancing a little towards the edge of the brick, the burden goes back again, returning to the point of departure. In the absence of a concerted understanding, their efforts of leverage are wasted. Nearly three hours are occupied by oscillations which mutually annul one another. The Mouse does not cross the little sand-hill heaped about her by the rakes of the workers.

For the second time, a male appears and makes a round of exploration. A boring is effected in loose earth, close beside the brick. This is a trial excavation, to learn the nature of the soil, a narrow well, of no great depth, into which the insect plunges to half its length. The well-sinker returns to the other workers, who arch their backs, and the load progresses a finger's-breadth towards the point recognized as favourable. Have we done the trick this time? No, for after a while the Mouse recoils. There is no progress towards a solution of the difficulty.

Now two males come out in search of information, each of his own accord. Instead of stopping at the point already sounded, a point most judiciously chosen, it seemed, on account of its proximity, which would save laborious carting, they precipitately scour the whole area of the cage, trying the soil on this side and on that and ploughing superficial furrows in it. They get as far from the brick as the limits of the enclosure permit.

They dig, by preference, against the base of the cover; here they make several borings, without any reason, so far as I can see, the bed of soil being everywhere equally assailable away from the brick; the first point sounded is abandoned for a second, which is rejected in its turn. A third and fourth are tried; then another. At the sixth point the choice is made. In all these cases the excavation is by no means a grave destined to receive the Mouse, but a mere trial boring, of inconsiderable depth and of the diameter of the digger's body.

Back again to the Mouse, who suddenly shakes, swings, advances, recoils, first in one direction, then in another, until in the end the hillock of sand is crossed. Now we are free of the brick and on excellent soil. Little by little the load advances. This is no cartage by a team hauling in the open, but a jerky removal, the work of invisible levers. The body seems to shift of its own accord.

This time, after all those hesitations, the efforts are concerted; at least, the load reaches the region sounded far more rapidly than I expected. Then begins the burial, according to the usual method. It is one o'clock. It has taken the Necrophori half-way round the clock to ascertain the condition of the locality and to displace the Mouse.

In this experiment it appears, in the first place, that the males play a major part in the affairs of the household. Better-equipped, perhaps, than their mates, they make investigations when a difficulty occurs;

they inspect the soil, recognize whence the check arises and choose the spot at which the grave shall be dug. In the lengthy experiment of the brick, the two males alone explored the surroundings and set to work to solve the difficulty. Trusting her assistants, the female, motionless beneath the Mouse, awaited the result of their enquiries. The tests which are to follow will confirm the merits of these valiant auxiliaries.

In the second place, the points where the Mouse lies being recognized as presenting an insurmountable resistance, there is no grave dug in advance, a little farther off, in the loose soil. All the attempts are limited, I repeat, to shallow soundings, which inform the insect of the possibility of inhumation.

It is absolute nonsense to speak of their first preparing the grave to which the body will afterwards be carted. In order to excavate the soil, our sextons have to feel the weight of their dead upon their backs. They work only when stimulated by the contact of its fur. Never, never in this world, do they venture to dig a grave unless the body to be buried already occupies the site of the cavity. This is absolutely confirmed by my two months and more of daily observations.

The rest of Clairville's anecdote bears examination no better. We are told that the Necrophorus in difficulties goes in search of assistance and returns with companions who assist him to bury the Mouse. This, in another form, is the edifying story of the Sacred Beetle whose pellet has rolled into a rut. Powerless to withdraw his booty from the abyss, the wily Dung-beetle summons three or four of his neighbours, who kindly pull out the pellet and return to their labours when the work of salvage is done.

The ill-interpreted exploit of the thieving pill-roller sets me on my guard against that of the undertaker. Shall I be too particular if I ask what precautions the observer took to recognize the owner of the Mouse on his return, when he reappears, as we are told, with four assistants? What sign denotes that one of the five who was able, in so rational a manner, to call for help? Can we even be sure that the one to disappear returns and forms one of the band? There is nothing to tell us so; and this was the essential point which a sterling observer was bound not to neglect. Were they not rather five chance Necrophori who, guided by the smell, without any previous understanding, hastened to the abandoned Mouse to exploit her on their own account? I incline to this opinion, the likeliest of all in the absence of exact information.

Probability becomes certainty if we check the fact by experiment. The test with the brick already tells us something. For six hours my three specimens exhausted themselves in efforts before they succeed in removing their booty and placing it on practicable soil. In this long and heavy job, helpful neighbours would have been most welcome. Four other Necrophori, buried here and there under a little sand, comrades and acquaintances, fellow-workers of the day before, were occupying the same cage; and not one of the busy ones thought of calling on them to assist. Despite their extreme embarrassment, the owners of the Mouse accomplished their task to the end, without the least help, though this could have been so easily requisitioned.

Being three, one might say, they deemed themselves strong enough; they needed no one else to lend them a hand. The objection does not hold good. On many occasions and under conditions even more difficult than those presented by a hard soil, I have again and again seen isolated Necrophori wearing themselves out against my artifices; yet not once did they leave their workshop to recruit helpers. Collaborators, it is true, often arrive, but they are summoned by their sense of smell, not by the first occupant. They are fortuitous helpers; they are never called in. They are received without strife but also without gratitude. They are not summoned; they are tolerated.

In the glazed shelter where I keep the cage I happened to catch one of these chance assistants in the act. Passing that way in the night and scenting dead flesh, he had entered where none of his kind had yet penetrated of his own accord. I surprised him on the dome of the cover. If the wire had not prevented him, he would have set to work incontinently, in company with the rest. Had my captives invited this one? Assuredly not. Heedless of others' efforts, he hastened up, attracted by the odour of the Mole. So it was with those whose obliging assistance is extolled. I repeat, in respect of their imaginary prowess, what I have said elsewhere of the Sacred Beetle's: it is a child's story, worthy to rank with any fairy-tale for the amusement of the simple.

A hard soil, necessitating the removal of the body, is not the only difficulty with which the Necrophori are acquainted. Frequently, perhaps more often than not, the ground is covered with grass, above all with couch-grass, whose tenacious rootlets form an inextricable network below the surface. To dig in the interstices is possible, but to drag the dead animal through them is another matter: the meshes of the net are too close to give it passage. Will the grave-digger find himself helpless against such an obstacle, which must be an extremely common one? That could not be.

Exposed to this or that habitual impediment in the exercise of its calling, the animal is always equipped accordingly; otherwise its profession would be impracticable. No end is attained without the necessary means and aptitudes. Besides that of the excavator, the Necrophorus certainly possesses another art: the art of breaking the cables, the roots, the stolons, the slender rhizomes which check the body's descent into the grave. To the work of the shovel and the pick must be added that of the shears. All this is perfectly logical and may be clearly foreseen. Nevertheless, let us call in experiment, the best of witnesses.

I borrow from the kitchen-range an iron trivet whose legs will supply a solid foundation for the engine which I am devising. This is a coarse network made of strips of raffia, a fairly accurate imitation of that of the couch-grass. The very irregular meshes are nowhere wide enough to admit of the passage of the creature to be buried, which this time is a Mole. The machine is planted by its three feet in the soil of the cage, level with the surface. A little sand conceals the ropes. The Mole is placed in the centre; and my bands of sextons are let loose upon the body.

The burial is performed without a hitch in the course of an afternoon. The raffia hammock, almost the equivalent of the natural network of the couch-grass, scarcely disturbs the burying-process. Matters do not proceed quite so quickly; and that is all. No attempt is made to shift the Mole, who sinks into the ground where he lies. When the operation is finished, I remove the trivet. The network is broken at the spot where the corpse was lying. A few strips have been gnawed through; a small number, only as many as were strictly necessary to permit the passage of the body.

Well done, my undertakers! I expected no less of your skill and tact. You foiled the experimenter's wiles by employing the resources which you use against natural obstacles. With mandibles for shears, you patiently cut my strings as you would have gnawed the threads of the grass-roots. This is meritorious, if not deserving of exceptional glorification. The shallowest of the insects that work in earth would have done as much if subjected to similar conditions.

Let us ascend a stage in the series of difficulties. The Mole is now fixed by a strap of raffia fore and aft to a light horizontal cross-bar resting on two firmly-planted forks. It is like a joint of venison on the spit, eccentrically fastened. The dead animal touches the ground throughout the length of its body.

The Necrophori disappear under the corpse and, feeling the contact of its fur, begin to dig. The grave grows deeper and an empty space appears; but the coveted object does not descend, retained as it is by the cross-bar which the two forks keep in place. The digging slackens, the hesitations become prolonged.

However, one of the grave-diggers climbs to the surface, wanders over the Mole, inspects him and ends by perceiving the strap at the back. He gnaws and ravels it tenaciously. I hear the click of the shears that completes the rupture. Crack! The thing is done. Dragged down by his own weight, the Mole sinks into the grave, but slantwise, with his head still outside, kept in place by the second strap.

The Beetles proceed with the burial of the hinder part of the Mole; they twitch and jerk it now in this direction, now in that. Nothing comes of it; the thing refuses to give. A fresh sortie is made by one of them, to find out what is happening overhead. The second strap is perceived, is severed in turn; and henceforth the work goes on as well as could be wished.

My compliments, perspicacious cable-cutters! But I must not exaggerate. The Mole's straps were for you the little cords with which you are so familiar in turfy soil. You broke them, as well as the hammock of the previous experiment, just as you sever with the blades of your shears any natural thread stretching across your catacombs. It is an indispensable trick of your trade. If you had had to learn it by experience, to think it out before practising it, your race would have disappeared, killed by the hesitations of its apprenticeship, for the spots prolific of Moles, Frogs, Lizards and other viands to your taste are usually covered with grass.

You are capable of much better things still; but, before setting forth these, let us examine the case when the ground bristles with slender brushwood, which holds the corpse at a short distance from the ground. Will the find thus hanging where it chances to fall remain unemployed? Will the Necrophori pass on, indifferent to the superb morsel which they see and smell a few inches above their heads, or will they make it drop from its gibbet?

Game does not abound to such a point that it can be despised if a few efforts will obtain it. Before I see the thing happen, I am persuaded that it will fall, that the Necrophori, often confronted with the difficulties of a body not lying on the soil, must possess the instinct to shake it to the ground. The fortuitous support of a few bits of stubble, of a few interlaced twigs, so common in the fields, cannot put them off. The drop of the suspended body, if placed too high, must certainly form part of their instinctive methods. For the rest, let us watch them at work.

I plant in the sand of the cage a meagre tuft of thyme. The shrub is at most some four inches in height. In the branches I place a Mouse, entangling the tail, the paws and the neck among the twigs to increase the difficulty. The population of the cage now consists of fourteen Necrophori and will remain the same until the close of my investigations. Of course they do not all take part simultaneously in the day's work: the majority remain underground, dozing or occupied in setting their cellars in order. Sometimes only one, often two, three or four, rarely more, busy themselves with the corpse which I offer them. To-day, two hasten to the Mouse, who is soon perceived overhead on the tuft of thyme.

They gain the top of the plant by way of the trelliswork of the cage. Here are repeated, with increased hesitation, due to the inconvenient nature of the support, the tactics employed to remove the body when the soil is unfavourable. The insect props itself against a branch, thrusting alternately with back and claws, jerking and shaking vigorously until the point whereat it is working is freed from its fetters. In one brief shift, by dint of heaving their backs, the two collaborators extricate the body from the tangle. Yet another shake; and the Mouse is down. The burial follows.

There is nothing new in this experiment: the find has been treated just as though it lay on soil unsuitable for burial. The fall is the result of an attempt to transport the load.

The time has come to set up the Frog's gibbet made famous by Gleditsch. The batrachian is not indispensable; a Mole will serve as well or even better. With a ligament of raffia I fix him, by his hind-legs, to a twig which I plant vertically in the ground, inserting it to no great depth. The creature hangs plumb against the gibbet, its head and shoulders making ample contact with the soil.

The grave-diggers set to work beneath the part which lies along the ground, at the very foot of the stake; they dig a funnel into which the Mole's muzzle, head and neck sink little by little. The gibbet becomes uprooted as they descend and ends by falling, dragged over by the weight of its heavy burden. I am assisting at the spectacle of the overturned stake, one of the most astonishing feats of reason with which the insect has ever been credited.

This, for one who is considering the problem of instinct, is an exciting moment. But let us beware of forming conclusions just yet; we might be in too great a hurry. Let us first ask ourselves whether the fall of the stake was intentional or accidental. Did the Necrophori lay it bare with the express purpose of making it fall? Or did they, on the contrary, dig at its base solely in order to bury that part of the Mole which lay on the ground? That is the question, which, for the rest, is very easy to answer.

The experiment is repeated; but this time the gibbet is slanting and the Mole, hanging in a vertical position, touches the ground at a couple of inches from the base of the apparatus. Under these conditions, absolutely no attempt is made to overthrow it. Not the least scrape of a claw is delivered at the foot of the gibbet. The entire work of excavation is performed at a distance, under the body, whose shoulders are lying on the ground. Here and here only a hole is dug to receive the front of the body, the part accessible to the sextons.

A difference of an inch in the position of the suspended animal destroys the famous legend. Even so, many a time, the most elementary sieve, handled with a little logic, is enough to winnow a confused mass of statements and to release the good grain of truth.

Yet another shake of this sieve. The gibbet is slanting or perpendicular, no matter which; but the Mole, fixed by his hind-legs to the top of the twig, does not touch the soil; he hangs a few fingers'-breadths from the ground, out of the sextons' reach.

What will they do now? Will they scrape at the foot of the gibbet in order to overturn it? By no means; and the ingenuous observer who looked for such tactics would be greatly disappointed. No attention is paid to the base of the support. It is not vouchsafed even a stroke of the rake. Nothing is done to overturn it, nothing, absolutely nothing! It is by other methods that the Burying-beetles obtain the Mole.

These decisive experiments, repeated under many different forms, prove that never, never in this world, do the Necrophori dig, or even give a superficial scrape, at the foot of the gallows, unless the hanging body touch the ground at that point. And, in the latter case, if the twig should happen to fall, this is in no way an intentional result, but a mere fortuitous effect of the burial already commenced.

What, then, did the man with the Frog, of whom Gleditsch tells us, really see? If his stick was overturned, the body placed to dry beyond the assaults of the Necrophori must certainly have touched the soil: a strange precaution against robbers and damp! We may well attribute more foresight to the preparer of dried Frogs and allow him to hang his animal a few inches off the ground. In that case, as all my experiments emphatically declare, the fall of the stake undermined by the sextons is a pure matter of imagination.

Yet another of the fine arguments in favour of the reasoning-power of insects flies from the light of investigation and founders in the slough of error! I wonder at your simple faith, O masters who take seriously the statements of chance-met observers, richer in imagination than in veracity; I wonder at your credulous zeal, when, without criticism, you build up your theories on such absurdities!

Let us continue. The stake is henceforth planted perpendicularly, but the body hanging on it does not reach the base: a condition enough to ensure that there will never be any digging at this point. I make use of a Mouse, who, by reason of her light weight, will lend herself better to the insect's manœuvres. The dead animal is fixed by the hind-legs to the top of the apparatus with a raffia strap. It hangs plumb, touching the stick.

Soon two Necrophori have discovered the morsel. They climb the greased pole; they explore the prize, poking their foreheads into its fur. It is recognized as an excellent find. To work, therefore. Here we have again, but under more difficult conditions, the tactics employed when it was necessary to displace the unfavourably situated body: the two collaborators slip between the Mouse and the stake and, taking a grip of the twig and exerting a leverage with their backs, they jerk and shake the corpse, which sways, twirls about, swings away from the stake and swings back again. All the morning is passed in vain attempts, interrupted by explorations on the animal's body.

In the afternoon, the cause of the check is at last recognized; not very clearly, for the two obstinate gallow-robbers first attack the Mouse's hind-legs, a little way below the strap. They strip them bare, flay them and cut away the flesh about the foot. They have reached the bone, when one of them finds the string of raffia beneath his mandibles. This, to him, is a familiar thing, representing the grass-thread so frequent in burials in turfy soil. Tenaciously the shears gnaw at the bond; the fibrous fetter is broken; and the Mouse falls, to be buried soon after.

If it stood alone, this breaking of the suspending tie would be a magnificent performance; but considered in connection with the sum of the Beetle's customary labours it loses any far-reaching significance. Before attacking the strap, which was not concealed in any way, the insect exerted itself for a whole morning in shaking the body, its usual method. In the end, finding the cord, it broke it, as it would have broken a thread of couch-grass encountered underground.

Under the conditions devised for the Beetle, the use of the shears is the indispensable complement of the use of the shovel; and the modicum of discernment at his disposal is enough to inform him when it will be well to employ the clippers. He cuts what embarrasses him, with no more exercise of reason than he displays when lowering his dead Mouse underground. So little does he grasp the relation of cause and effect that he tries to break the bone of the leg before biting the raffia which is knotted close beside him. The difficult task is attempted before the extremely easy one.

Difficult, yes, but not impossible, provided that the Mouse be young. I begin over again with a strip of iron wire, on which the insect's shears cannot get a grip, and a tender Mousekin, half the size of an adult. This time a tibia is gnawed through, sawed in two by the Beetle's mandibles, near the spring of the heel. The detached leg leaves plenty of space for the other, which readily slips from the metal band; and the little corpse falls to the ground.

But, if the bone be too hard, if the prize suspended be a Mole, an adult Mouse or a Sparrow, the wire ligament opposes an insurmountable obstacle to the attempts of the Necrophori, who, for nearly a week, work at the hanging body, partly stripping it of fur or feather and dishevelling it until it forms a lamentable object, and at last abandon it when desiccation sets in. And yet a last resource remained, one as rational as infallible: to overthrow the stake. Of course, not one dreams of doing so.

For the last time let us change our artifices. The top of the gibbet consists of a little fork, with the prongs widely opened and measuring barely two-fifths of an inch in length. With a thread of hemp, less easily attacked than a strip of raffia, I bind the hind-legs of an adult Mouse together, a little above the heels; and I slip one of the prongs in between. To bring the thing down one has only to slide it a little way upwards; it is like a young Rabbit hanging in the window of a poulterer's shop.

Five Necrophori come to inspect what I have prepared. After much futile shaking, the tibiæ are attacked. This, it seems, is the method usually employed when the corpse is caught by one of its limbs in some narrow fork of a low-growing plant. While trying to saw through the bone—a heavy job this time—one of the workers slips between the shackled legs; in this position, he feels the furry touch of the Mouse against his chine. No more is needed to arouse his propensity to thrust with his back. With a few heaves of the lever the thing is done: the Mouse rises a little, slides over the supporting peg and falls to the ground.

Is this manœuvre really thought out? Has the insect indeed perceived, by the light of a flash of reason, that to make the morsel fall it was necessary to unhook it by sliding it along the peg? Has it actually perceived the mechanism of the hanging? I know some persons—indeed, I know many—who, in the presence of this magnificent result, would be satisfied without further investigation.

More difficult to convince, I modify the experiment before drawing a conclusion. I suspect that the Necrophorus, without in any way foreseeing the consequences of his action, heaved his back merely because he felt the animal's legs above him. With the system of suspension adopted, the push of the back, employed in all cases of difficulty, was brought to bear first upon the point of support; and the fall resulted from this happy coincidence. That point, which has to be slipped along the peg in order to unhook the object, ought really to be placed at a short distance from the Mouse, so that the Necrophori may no longer feel her directly on their backs when they push.

A wire binds together now the claws of a Sparrow, now the heels of a Mouse and is bent, three-quarters of an inch farther away, into a little ring, which slips very loosely over one of the prongs of the fork, a short, almost horizontal prong. The least push of this ring is enough to bring the hanging body to the ground; and because it stands out it lends itself excellently to the insect's methods. In short, the arrangement is the same as just now, with this difference, that the point of support is at a short distance from the animal hung up.

My trick, simple though it be, is quite successful. For a long time the body is repeatedly shaken, but in vain; the tibiæ, the hard claws refuse to yield to the patient saw. Sparrows and Mice grow dry and shrivel, unused, upon the gallows. My Necrophori, some sooner, some later, abandon the insoluble mechanical problem: to push, ever so little, the movable support and so to unhook the coveted carcase.

Curious reasoners, in faith! If, just now, they had a lucid idea of the mutual relations between the tied legs and the suspending peg; if they made the Mouse fall by a reasoned manœuvre, whence comes it that the present artifice, no less simple than the first, is to them an insurmountable obstacle? For days and days they work on the body, examining it from head to foot, without noticing the movable support, the cause of their mishap. In vain I prolong my watch; I never see a single one of them push the support with his foot or butt it with his head.

Their defeat is not due to lack of strength. Like the Geotrupes, they are vigorous excavators. When you grasp them firmly in your hand, they slip into the interstices of the fingers and plough up your skin so as to make you quickly loose your hold. With his head, a powerful ploughshare, the Beetle might very easily push the ring off its short support. He is not able to do so, because he does not think of it; he does not think of it, because he is devoid of the faculty attributed to him, in order to support their theories, by the dangerous generosity of the evolutionists.

Divine reason, sun of the intellect, what a clumsy slap in thy august countenance, when the glorifiers of the animal degrade thee with such denseness!

Let us now examine the mental obscurity of the Necrophori under another aspect. My captives are not so satisfied with their sumptuous lodging that they do not seek to escape, especially when there is a dearth of labour, that sovran consoler of the afflicted, man or beast. Interment within the wire cover palls upon them. So, when the Mole is buried and everything in order in the cellar, they stray uneasily over the trellised dome; they clamber up, come down, go up again and take to flight, a flight which instantly becomes a fall, owing to collision with the wire grating. They pick themselves up and begin all over again. The sky is splendid; the weather is hot, calm and propitious for those in search of the Lizard crushed beside the footpath. Perhaps the effluvia of the gamy tit-bit have reached them from afar, imperceptible to any other sense than that of the grave-diggers. My Necrophori therefore would be glad to get away.

Can they? Nothing would be easier, if a glimmer of reason were to aid them. Through the trelliswork, over which they have so often strayed, they have seen, outside, the free soil, the promised land which they want to reach. A hundred times if once have they dug at the foot of the rampart. There, in vertical wells, they take up their station, drowsing whole days on end while unemployed. If I give them a fresh Mole, they emerge from their retreat by the entrance-corridor and come to hide themselves beneath the belly of the beast. The burial over, they return, one here, one there, to the confines of the enclosure and disappear underground.

Well, in two and a half months of captivity, despite long stays at the base of the trellis, at a depth of three-quarters of an inch beneath the surface, it is rare indeed for a Necrophorus to succeed in circumventing the obstacle, in prolonging his excavation beneath the barrier, in digging an elbow and bringing it out on the other side, a trifling task for these vigorous creatures. Of fourteen only one succeeds in escaping.

A chance deliverance and not premeditated; for, if the happy event had been the result of a mental combination, the other prisoners, practically his equals in powers of perception, would all, from first to last, have discovered by rational means the elbowed path leading to the outer world; and the cage would promptly be deserted. The failure of the great majority proves that the single fugitive was simply digging at random. Circumstances favoured him; and that is all. We must not put it to his credit that he succeeded where all the others failed.

We must also beware of attributing to the Necrophori a duller understanding than is usual in insect psychology. I find the ineptness of the undertaker in all the Beetles reared under the wire cover, on the bed of sand into which the rim of the dome sinks a little way. With very rare exceptions, fortuitous accidents, not one thinks of circumventing the barrier by way of the base; not one manages to get outside by means of a slanting tunnel, not even though he be a miner by profession, as are the Dung-beetles *par excellence.* Captives under the wire dome and anxious to escape, Sacred Beetles, Geotrupes, Copres, Gymnopleuri, Sisyphi, all see about them the free space, the joys of the open sunlight; and not one thinks of going round under the rampart, which would present no difficulty to their pickaxes.

Even in the higher ranks of animality, examples of similar mental obfuscation are not lacking. Audubon tells us how, in his days, wild Turkeys were caught in North America. In a clearing known to be frequented by these birds, a great cage was constructed with stakes driven into the ground. In the centre of the enclosure opened a short tunnel, which dipped under the palisade and returned to the surface outside the cage by a gentle slope, which was open to the sky. The central opening, wide enough to give a bird free passage, occupied only a portion of the enclosure, leaving around it, against the circle of stakes, a wide unbroken zone. A few handfuls of maize were scattered in the interior of the trap, as well as round about it, and in particular along the sloping path, which passed under a sort of bridge and led to the centre of the contrivance. In short, the Turkey-trap presented an ever-open door. The bird found it in order to enter, but did not think of looking for it in order to go out.

According to the famous American ornithologist, the Turkeys, lured by the grains of maize, descended the insidious slope, entered the short underground passage and beheld, at the end of it, plunder and the light. A few steps farther and the gluttons emerged, one by one, from beneath the bridge. They distributed themselves about the enclosure. The maize was abundant; and the Turkeys' crops grew swollen.

When all was gathered, the band wished to retreat, but not one of the prisoners paid any attention to the central hole by which he had arrived. Gobbling uneasily, they passed again and again across the bridge whose arch was yawning beside them; they circled round against the palisade, treading a hundred times in their own footprints; they thrust their necks, with their crimson wattles, through the bars; and there, with their beaks in the open air, they fought and struggled until they were exhausted.

Remember, O inept one, what happened but a little while ago; think of the tunnel that led you hither! If that poor brain of yours contains an atom of ability, put two ideas together and remind yourself that the passage by which you entered is there and open for your escape! You will do nothing of the kind. The light, an irresistible attraction, holds you subjugated against the palisade; and the shadow of the yawning pit, which has but lately permitted you to enter and will quite as readily permit you to go out, leaves you indifferent. To recognize the use of this opening you would have to reflect a little, to recall the past; but this tiny retrospective calculation is beyond your powers. So the trapper, returning a few days later, will find a rich booty, the entire flock imprisoned!

Of poor intellectual repute, does the Turkey deserve his name for stupidity? He does not appear to be more limited than another. Audubon depicts him as endowed with certain useful ruses, in particular when he has to baffle the attacks of his nocturnal enemy, the Virginian Owl. As for his behaviour in the snare with the underground passage, any other bird, impassioned of the light, would do the same.

Under rather more difficult conditions, the Necrophorus repeats the ineptness of the Turkey. When he wishes to return to the daylight, after resting in a short burrow against the rim of the cover, the Beetle, seeing a little light filtering through the loose soil, reascends the entrance-well, incapable of telling himself that he has only to prolong the tunnel as far in the opposite direction to reach the outer world beyond the wall and gain his freedom. Here again is one in whom we shall seek in vain for any sign of reflection. Like the rest, in spite of his legendary renown, he has no guide but the unconscious promptings of instinct.

THE PSYCHE MOTH

« 32 »

As is true with many insects, it is the early part of the life of the Psyche Moth that is most interesting. In its larval form it is the familiar Bag-Worm. The adult moth is small and drab. In some American species, the Bag-Worm produces a house that resembles a pinecone. In others, bits of grass-stems are attached lengthwise to the bag. One species in the South constructs a kind of miniature log-cabin by using bits of twigs on the exterior. This study of the strange life of the Psyche Moth originally appeared in Fabre's THE LIFE OF THE CATERPILLAR.

In the springtime, old walls and dusty roads harbour a surprise for whoso has eyes to see. Tiny faggots, for no apparent reason, set themselves in motion and make their way along by sudden jerks. The inanimate comes to life, the immovable stirs. How does this come about? Look closer and the motive power will stand revealed.

Enclosed within the moving bundle is a fairly well-developed caterpillar, prettily striped in black and white. Seeking for food or perhaps for a spot where the transformation can be effected, he hurries along timidly, attired in a queer rig-out of twigs from which nothing emerges except the head and the front part of the body, which is furnished with six short legs. At the least alarm he goes right in and does not budge again. This is the whole secret of the little roaming bundle of sticks.

The faggot caterpillar belongs to the Psyche group, whose name conveys an allusion to the classic Psyche, symbolical of the soul. We must not allow this phrase to carry our thoughts to loftier heights than is fitting. The nomenclator, with his rather circumscribed view of the world, did not trouble about the soul when inventing his descriptive label. He simply wanted a pretty name; and certainly he could have hit on nothing better.

To protect himself from the weather, our chilly, bare-skinned Psyche builds himself a portable shelter, a travelling cottage which the owner never leaves until he becomes a Moth. It is something better than a hut on wheels with a thatched roof to it: it is a hermit's frock, made of an unusual sort of frieze. In the valley of the Danube the peasant wears a goatskin cloak fastened with a belt of rushes. The Psyche dons an even more rustic apparel. He makes himself a suit of clothes out of hop-poles. It is true that, beneath this rude conglomeration, which would be a regular hair-shirt to a skin as delicate as his, he puts a thick lining of silk. The Clythra Beetle garbs himself in pottery; this one dresses himself in a faggot.

In April, on the walls of my chief observatory, that famous pebbly acre with its wealth of insect life, I find the Psyche who is to furnish me with my most circumstantial and detailed records. He is at this period in the torpor of the approaching metamorphosis. As we can ask him nothing else for the moment, let us look into the construction and composition of his faggot.

It is a not irregular structure, spindle-shaped and about an inch and a half long. The pieces that compose it are fixed in front and free at the back, are arranged anyhow and would form a rather ineffective shelter against the sun and rain if the recluse had no other protection than his thatched roof.

The word thatch is suggested to my mind by a summary inspection of what I see, but it is not an exact expression in this case. On the contrary, graminaceous straws are rare, to the great advantage of the future family, which, as we shall learn presently, would find nothing to suit them in jointed planks. What predominates is remnants of very small stalks, light, soft and rich in pith, such as are possessed by various Chicoriaceæ. I recognize in particular the floral stems of the mouse-ear hawkweed and the Nimes pterotheca. Next come bits of grass-leaves, scaly twigs provided by the cypress-tree and all sorts of little sticks, coarse materials adopted for the lack of anything better. Lastly, if the favourite cylindrical pieces fall short, the mantle is sometimes finished off with an ample flounced tippet, that is to say, with fragments of dry leaves of any kind.

Incomplete as it is, this list shows us that the caterpillar apart from his preference for pithy morsels, has no very exclusive tastes. He employs indifferently anything that he comes upon, provided that it be light, very dry, softened by long exposure to the air and of suitable dimensions. All his finds, if they come anywhere near his estimates, are used just as they are, without any alterations or sawing to reduce them to the proper length. The Psyche does not trim the laths that go to form his roof; he gathers them as he finds them. His work is limited to imbricating them one after the other by fixing them at the fore-end.

In order to lend itself to the movements of the journeying caterpillar and in particular to facilitate the action of the head and legs when a new piece is to be placed in position, the front part of the sheath requires a special structure. Here a casing of beams is no longer allowable, for their length and stiffness would hamper the artisan and even make his work impossible; what is essential here is a flexible neck, able to bend in all directions. The assemblage of stakes does, in fact, end suddenly at some distance from the fore-part and is there replaced by a collar in which the silken woof is merely hardened with very tiny ligneous particles, tending to strengthen the material without impairing its flexibility. This collar, which gives free movement, is so important that all the Psyches make equal use of it, however much the rest of the work may differ. All carry, in front of the faggot of sticks, a yielding neck, soft to the touch, formed inside of a web of pure silk and velveted outside with a fine sawdust which the caterpillar obtains by crushing with his mandibles any sort of dry straw.

A similar velvet, but lustreless and faded, apparently through age, finishes the sheath at the back, in the form of a rather long, bare appendix, open at the end.

Let us now remove the outside of the straw envelope, shredding it piecemeal. The demolition gives us a varying number of joists: I have counted as many as eighty and more. The ruin that remains is a cylindrical sheath wherein we discover, from one end to the other, the structure which we perceived at the front and rear, the two parts which are naturally bare. The tissue everywhere is of very stout silk, which resists without breaking when pulled by the fingers, a smooth tissue, beautifully white inside, drab and wrinkled outside, where it bristles with encrusted woody particles.

There will be an opportunity later to discover by what means the caterpillar makes himself so complicated a garment, in which are laid one upon the other, in a definite order, first, the extremely fine satin which is in direct contact with the skin; next, the mixed stuff, a sort of frieze dusted with ligneous matter, which saves the silk and gives consistency to the work; lastly, the surtout of overlapping laths.

While retaining this general threefold arrangement, the scabbard offers notable variations of structural detail in the different species. Here, for instance, is a second Psyche, the most belated of the three which I have chanced to come upon. I meet him towards the end of June, hurrying across some dusty path near the houses. His cases surpass those of the previous species both in size and in regularity of arrangement. They form a thick coverlet, of many pieces, in which I recognize here fragments of hollow stalks, there bits of fine straw, with perhaps straps formed of blades of grass. In front there is never any mantilla of dead leaves, a troublesome piece of finery which, without being in regular use, is pretty frequent in the costume of the first-named species. At the back, no long, denuded vestibule. Save for the indispensable collar at the aperture, all the rest is cased in logs. There is not much variety about the thing, but, when all is said, there is a certain elegance in its stern faultlessness.

The smallest in size and simplest in dress is the third, who is very common at the end of winter on the walls, as well as in the furrows of the barks of gnarled old trees, be they olive-trees, holm-oaks, elms or almost any other. His case, a modest little bundle, is hardly more than two-fifths of an inch in length. A dozen rotten straws, gleaned at random and fixed close to one another in a parallel direction, represent, with the silk sheath, his whole outlay on dress. It would be difficult to clothe one's self more economically.

This pigmy, apparently so uninteresting, shall supply us with our first records of the curious life-story of the Psyches. I gather him in profusion in April and instal him in a wire bell-jar. What he eats I know not. My ignorance would be grievous under other conditions; but at present I need not trouble about provisions. Taken from their walls and trees, where they had suspended themselves for their transformation, most of my little Psyches are in the chrysalis state. A few of them are still active. They hasten to clamber to the top of the trellis-work; they fix themselves there perpendicularly by means of a little silk cushion; then everything is still.

June comes to an end; and the male Moths are hatched, leaving the chrysalid wrapper half caught in the case, which remains fixed where it is and will remain there indefinitely until dismantled by the weather. The emergence is effected through the hinder end of the bundle of sticks, the only way by which it can be effected. Having permanently closed the top opening, the real door of the house, by fastening it to the support which he has chosen, the caterpillar therefore has turned the other way round and undergone his transformation in a reversed position, which enables the adult insect to emerge through the outlet made at the back, the only one now free.

For that matter, this is the method followed by all the Psyches. The case has two apertures. The front one, which is more regular and more carefully constructed, is at the caterpillar's service so long as larval activity lasts. It is closed and firmly fastened to its support at the time of the nymphosis. The hinder one, which is faulty and even hidden by the sagging of the sides, is at the Moth's service. It does not really open until right at the end, when pushed by the chrysalis or the adult insect.

In their modest pearl-grey dress, with their insignificant wing-equipment, hardly exceeding that of a Common Fly, our little Moths are still not without elegance. They have handsome feathery plumes for antennæ; their wings are edged with delicate fringes. They whirl very fussily inside the bell-jar; they skim the ground, fluttering their wings; they crowd eagerly around certain sheaths which nothing on the outside distinguishes from the others. They alight upon them and sound them with their plumes.

This feverish agitation marks them as lovers in search of their brides. This one here, that one there, each of them finds his mate. But the coy one does not leave her home. Things happen very discreetly through the wicket left open at the free end of the case. The male stands on the threshold of this back-door for a little while; and then it is over: the wedding is finished. There is no need for us to linger over these nuptials in which the parties concerned do not know, do not see each other.

I hasten to place in a glass tube the few cases in which the mysterious events have happened. Some days later, the recluse comes out of the sheath and shows herself in all her wretchedness. Call that little fright a Moth! One cannot easily get used to the idea of such poverty. The caterpillar of the start was no humbler-looking. There are no wings, none at all; no silky fur either. At the tip of the abdomen, a round, tufty pad, a crown of dirty-white velvet; on each segment, in the middle of the back, a large rectangular dark patch: these are the sole attempts at ornament. The mother Psyche renounces all the beauty which her name of Moth promised.

From the centre of the hairy coronet a long ovipositor stands out, consisting of two parts, one stiff, forming the base of the implement, the other soft and flexible, sheathed in the first just as a telescope fits in its tube. The laying mother bends herself into a hook, grips the lower end of her case with her six feet and drives her probe into the back-window, a window which serves manifold purposes, allowing of the consummation of the clandestine marriage, the emergence of the fertilized bride, the installation of the eggs and, lastly, the exodus of the young family.

There, at the free end of her case, the mother remains for a long time, bowed and motionless. What can she be doing in this contemplative attitude? She is lodging her eggs in the house which she has just left, she is bequeathing the maternal cottage to her heirs. Some thirty hours pass and the ovipositor is at last withdrawn. The laying is finished.

A little wadding, supplied by the coronet on the hind-quarters, closes the door and allays the dangers of invasion. The fond mother makes a barricade for her brood of the sole ornament which, in her extreme indigence, she possesses. Better still, she makes a rampart of her body. Bracing herself convulsively on the threshold of her home, she dies there, dries up there, devoted to her family even after death. It needs an accident, a breath of air, to make her fall from her post.

Let us now open the case. It contains the chrysalid wrapper, intact except for the front breach through which the Psyche emerged. The male, because of his wings and his plumes, very cumbersome articles when he is about to make his way through the narrow pass, takes advantage of his chrysalis state to make a start for the door and come out half-way. Then, bursting his amber tunic, the delicate Moth finds an open space, where flight is possible, right in front of him. The mother, unprovided with wings and plumes, is not compelled to observe any such precautions. Her cylindrical form, bare and differing but little from that of the caterpillar, allows her to crawl, to slip into the narrow passage and to come forth without obstacle. Her cast chrysalid skin is, therefore, left right at the back of the case, well covered by the thatched roof.

And this is an act of prudence marked by exquisite tenderness. The eggs, in fact, are packed in the barrel, in the parchmentlike wallet formed by the slough. The mother has thrust her telescopic ovipositor to the bottom of that receptacle and has methodically gone on laying until it is full. Not satisfied with bequeathing her home and her velvet coronet to her offspring, as a last sacrifice she leaves them her skin.

With a view to observing at my ease the events which are soon to happen, I extract one of these chrysalid bags, stuffed with eggs, from its faggot and place it by itself, beside its case, in a glass tube. I have not long to wait. In the first week of July, I find myself all of a sudden in possession of a large family. The quickness of the hatching balked my watchfulness. The new-born caterpillars, about forty in number, have already had time to garb themselves.

They wear a Persian head-dress, a mage's tiara in dazzling white plush. Or, to abandon high-flown language, let us say a cotton night-cap without a tassel; only the cap does not stand up from the head: it covers the hind-quarters. Great animation reigns in the tube, which is a spacious residence for such vermin. They roam about gaily, with their caps sticking up almost perpendicular to the floor. With a tiara like that and things to eat, life must be sweet indeed.

But what do they eat? I try a little of everything that grows on the bare stone and the gnarled old trees. Nothing is welcomed. More eager to dress than to feed themselves, the Psyches scorn what I set before them. My ignorance as an insect-breeder will not matter, provided that I succeed in seeing with what materials and in what manner the first out-lines of the cap are woven.

I may fairly hope to achieve this ambition, as the chrysalid bag is far from having exhausted its contents. I find in it, teeming amid the rumpled wrapper of the eggs, an additional family as numerous as the swarm that is already out. The total laying must therefore amount to five or six dozen. I transfer to another receptacle the precocious band which is already dressed and keep only the naked laggards in the tube. They have bright red heads, with the rest of their bodies dirty white; and they measure hardly a twenty-fifth of an inch in length.

My patience is not long put to the test. Next day, little by little, singly or in groups, the belated grubs quit the chrysalid bag. They come out without breaking the frail wallet, through the front breach made by the liberation of the mother. Not one of them utilizes it as a dress-material, though it has the delicacy and amber colouring of an onion-skin; nor do any of them make use of a fine quilting which lines the in-side of the bag and forms an exquisitely soft bed for the eggs. This down, whose origin we shall have to investigate presently, ought, one would say, to make an excellent blanket for these chilly ones, impatient to cover themselves up. Not a single one uses it; there would not be enough to go round.

All go straight to the coarse faggot, which I left in contact with the wallet that was the chrysalis. Time presses. Before making your en-trance into the world and going agrazing, you must first be clad. All therefore, with equal fury, attack the old sheath and hastily dress them-selves in the mother's cast clothes. Some turn their attention to bits that happen to be open lengthwise and scrape the soft, white inner layer; others, greatly daring, penetrate into the tunnel of a hollow stalk and go and collect their cotton goods in the dark. At such times the materials are first-class; and the garment woven is of a dazzling white. Others bite deep into the piece which they select and make themselves a motley garment, in which dark-coloured particles mar the snowy whiteness of the rest.

The tool which they use for their gleaning consists of the mandibles, shaped like wide shears with five strong teeth apiece. The two planes fit into each other and form an implement capable of seizing and slicing any fibre, however small. Seen under the microscope, it is a wonderful specimen of mechanical precision and power. Were the Sheep similarly equipped in proportion to her size, she would browse upon the bottom of the trees instead of cropping the grass.

A very instructive workshop is that of the Psyche-vermin toiling to make themselves a cotton night-cap. There are numbers of things to remark in both the finish of the work and the ingenuity of the methods employed. To avoid repeating ourselves, we will say nothing about these yet, but wait for a little and return to the subject when setting forth the talents of a second Psyche, of larger stature and easier to observe. The two weavers observe exactly the same procedure.

Nevertheless let us take a glance at the bottom of the egg-cup, a general workyard in which I instal my dwarfs as the cases turn them out. There are some hundreds of them, with the sheaths from which they came and an assortment of clipped stalks, chosen from among the driest and richest in pith. What a whirl! What bewildering animation!

In order to see man, Micromégas cut himself a lens out of a diamond of his necklace; he held his breath lest the storm from his nostrils should blow the mite away. I in my turn will be the good giant, newly arriving from Sirius; I screw a magnifying-glass into my eye and am careful not to breathe for fear of overturning and sweeping out of existence my cotton-workers. If I need one of them, to focus him under a stronger glass, I lime him as it were, seizing him with the fine point of a needle which I have passed over my lips. Taken away from his work, the tiny caterpillar struggles at the end of the needle, shrivels up, makes himself, small as he is, still smaller; he strives to withdraw as far as possible into his clothing, which as yet is incomplete, the merest flannel vest or even a narrow scarf, covering nothing but the top of his shoulders. Let us leave him to complete his coat. I give a puff; and the creature is swallowed up in the crater of the egg-cup.

And this speck is alive. It is industrious; it is versed in the art of blanket-making. An orphan, born that moment, it knows how to cut itself out of its dead mother's old clothes the wherewithal to clothe itself in its turn. Soon it will become a carpenter, an assembler of timber, to make a defensive covering for its delicate fabric. What must instinct be, to be capable of awakening such industries in an atom!

It is at the end of June also that I obtain, in his adult shape, the Psyche whose scabbard is continued underneath by a long, naked vestibule. Most of the cases are fastened by a silk pad to the trelliswork of the cage and hang vertically, like stalactites. Some few of them have never left the ground. Half immersed in the sand, they stand erect, with their rear in the air and their fore-part buried and firmly anchored to the side of the pan by means of a silky paste.

This inverted position excludes any idea of weight as a guide in the caterpillar's preparations. An adept at turning round in his cabin, he is careful, before he sinks into the immobility of pupadom, to turn his head now upwards, now downwards, towards the opening, so that the adult insect, which is much less free than the larva in its movements, may reach the outside without obstacle.

Moreover, it is the pupa itself, the unbending chrysalis, incapable of turning and obliged to move all in one piece, which, stubbornly crawling, carries the male to the threshold of the case. It emerges half way at the end of the uncovered silky vestibule and there breaks, obstructing the opening with its slough as it does so. For a time the Moth stands still on the roof of the cottage, allowing his humours to evaporate, his wings to spread and gather strength; then at last the gallant takes flight, in search of her for whose sake he has made himself so spruce.

He wears a costume of deepest black, all except the edges of the wings, which, having no scales, remain diaphanous. His antennæ, likewise black, are wide and graceful plumes. Were they on a larger scale, they would throw the feathered beauty of the Marabou and Ostrich into the shade. The bravely beplumed one visits case after case in his tortuous flight, prying into the secrets of those alcoves. If things go as he wishes, he settles, with a quick flutter of his wings, on the extremity of the denuded vestibule. Comes the wedding, as discreet as that of the smaller Psyche. Here is yet another who does not see or at most catches a fleeting glimpse of her for whose sake he has donned Marabou-feathers and a black-velvet cloak.

The recluse on her side is equally impatient. The lovers are short-lived; they die in my cages within three or four days, so that, for long intervals, until the hatching of some late-comer, the female population is short of suitors. So, when the morning sun, already hot, strikes the cage, a very singular spectacle is repeated many times before my eyes. The entrance to the vestibule swells imperceptibly, opens and emits a mass of infinitely delicate down. A Spider's web, carded and made into wadding, would give nothing of such gossamer fineness. It is a vaporous cloud. Then, from out of this incomparable eiderdown, appear the head and fore-part of a very different sort of caterpillar from the original collector of straws.

It is the mistress of the house, the marriageable Moth, who, feeling her hour about to come and failing to receive the expected visit, herself makes the advances and goes, as far as she can, to meet her plumed swain. He does not come hastening up and for good reason: there is not a male left in the establishment. For two or three hours the poor forsaken one leans, without moving, from her window. Then, tired of waiting, very gently she goes indoors again, backwards, and returns to her cell.

Next day, the day after and later still, as long as her strength permits, she reappears on her balcony, always in the morning, in the soft rays of a warm sun and always on a sofa of that incomparable down, which disperses and turns to vapour if I merely fan it with my hand. Again no one comes. For the last time the disappointed Moth goes back to her boudoir, never to leave it again. She dies in it, dries up, a useless thing. I hold my bell-jars responsible for this crime against motherhood. In the open fields, without a doubt, sooner or later wooers would have appeared, coming from the four winds.

The said bell-jars have an even more pitiful catastrophe on their conscience. Sometimes, leaning too far from her window, miscalculating the balance between the front of the body, which is at liberty, and the back, which remains sheathed in its case, the Moth allows herself to drop to the ground. It is all up now with the fallen one and her lineage. Still, there is one good thing about it. Accidents such as this lay bare the mother Psyche, without our having to break into her house.

What a miserable creature she is, a great deal uglier than the original caterpillar! Here transfiguration spells disfigurement, progress means retrogression. What we have before our eyes is a wrinkled satchel, an earthy-yellow sausage; and this horror, worse than a maggot, is a Moth in the full bloom of life, a genuine adult Moth. She is the betrothed of the elegant black Bombyx, all plumed with Marabou-feathers, and represents to him the last word in beauty. As the proverb says, beauty lies in lovers' eyes: a profound truth which the Psyche confirms in striking fashion.

Let us describe the ugly little sausage. A very small head, a paltry globule, disappearing almost entirely in the folds of the first segment. What need is there of cranium and brains for a germ-bag! And so the tiny creature almost does without them, reduces them to the simplest expression. Nevertheless, there are two black ocular specks. Do these vestigial eyes see their way about? Not very clearly, we may be sure. The pleasures of light must be very small for this stay-at-home, who appears at her window only on rare occasions, when the male Moth is late in arriving.

The legs are well-shaped, but so short and weak that they are of no use at all for locomotion. The whole body is a pale yellow, semitransparent in front, opaque and stuffed with eggs behind. Underneath the first segments is a sort of neck-band, that is to say, a dark stain, the vestige of a crop showing through the skin. A pad of short down ends the oviferous part at the back. It is all that remains of a fleece, of a thin velvet which the insect rubs off as it moves backwards and forwards in its narrow lodging. This forms the flaky mass which whitens the trysting-window at the wedding-time and also lines the inside of the sheath with down. In short, the creature is little more than a bag swollen with eggs for the best part of its length. I know nothing lower in the scale of wretchedness.

The germ-bag moves, but not, of course, with those vestiges of legs which form too short and feeble supports; it gets about in a way that allows it to progress on its back, belly or side indifferently. A groove is hollowed out at the hinder end of the bag, a deep, dividing groove which cuts the insect into two. It runs to the front part, spreading like a wave, and gently and slowly reaches the head. This undulation constitutes a step. When it is done, the animal has advanced about a twenty-fifth part of an inch.

To go from one end to the other of a box two inches long and filled with fine sand, the living sausage takes nearly an hour. It is by crawling like this that it moves about in its case, when it comes to the threshold to meet its visitor and goes in again.

For three or four days, exposed to the roughness of the soil, the oviferous bag leads a wretched life, creeping about at random, or, more often, standing still. No Moth pays attention to the poor thing, who possesses no attractions outside her home; the lovers pass by with an indifferent air. This coolness is logical enough. Why should she become a mother, if her family is to be abandoned to the inclemencies of the public way? And so, after falling by accident from her case, which would have been the cradle of the youngsters, the wanderer withers in a few days and dies childless.

The fertilized ones—and these are the more numerous—the prudent ones who have saved themselves from a fall by being less lavish with their appearances at the window, reenter the sheath and do not show themselves again once the Moth's visit to the threshold is over. Let us wait a fortnight and then open the case lengthwise with our scissors. At the end, in the widest part, opposite the vestibule, is the slough of the chrysalis, a long, fragile, amber-coloured sack, open at the end that contains the head, the end facing the exit-passage. In this sack, which she fills like a mould, lies the mother, the egg-bladder, now giving no sign of life.

From this amber sheath, which presents all the usual characteristics of a chrysalis, the adult Psyche emerged, in the guise of a shapeless Moth, looking like a big maggot; at the present time, she has slipped back into her old jacket, moulding herself into it in such a way that it becomes difficult to separate the container from the contents. One would take the whole thing for a single body.

It seems very likely that this cast skin, which occupies the best place in the home, formed the Psyche's refuge when, weary of waiting on the threshold of her hall, she retired to the back room. She has there-fore gone in and out repeatedly. This constant going and coming, this continual rubbing against the sides of a narrow corridor, just wide enough for her to pass through, ended by stripping her of her down. She had a fleece to start with, a very light and scanty fleece, it is true, but still a vestige of the costume which Moths are wont to wear. This fluff she has lost. What has she done with it?

The Eider robs herself of her down to make a luxurious bed for her brood; the new-born Rabbits lie on a mattress which their mother cards for them with the softest part of her fur, shorn from the belly and neck, wherever the shears of her front teeth can reach it. This fond tenderness is shared by the Psyche, as you will see.

In front of the chrysalid bag is an abundant mass of extra-fine wad-ding, similar to that of which a few flocks used to fall outside on the oc-casions when the recluse went to her window. Is it silk? Is it spun muslin? No; but it is something of incomparable delicacy. The micro-scope recognizes it as the scaly dust, the impalpable down in which every Moth is clad. To give a snug shelter to the little caterpillars who will soon be swarming in the case, to provide them with a refuge in which they can play about and gather strength before entering the wide world, the Psyche has stripped herself of her fur like the mother Rabbit.

This denudation may be a mere mechanical result, an unintentional effect of repeated rubbing against the low-roofed walls; but there is nothing to tell us so. Maternity has its foresight, even among the humblest. I therefore picture the hairy Moth twisting about, going to and fro in the narrow passage in order to get rid of the fleece and prepare bedding for her offspring. It is even possible that she manages to use her lips, that vestige of a mouth, in order to pull out the down that refuses to come away of itself.

No matter what the method of shearing may be, a mound of scales and hairs fills up the case in front of the chrysalid bag. For the moment, it is a barricade preventing access to the house, which is open at the hinder end; soon, it will be a downy couch on which the little caterpillars will rest for a while after leaving the egg. Here, warmly ensconced in a rug of extreme softness, they call a halt as a preparation for the emergence and the work that follows it.

Not that silk is lacking: on the contrary, it abounds. The caterpillar lavished it during his time as a spinner and a picker-up of straws. The whole interior of the case is padded with thick white satin. But how greatly preferable to this too-compact and luxurious upholstery is the delightful eiderdown bedding of the new-born youngsters!

We know the preparations made for the coming family. Now, where are the eggs? At what spot are they laid? The smallest of my three Psyches, who is less misshapen than the others and freer in her movements, leaves her case altogether. She possesses a long ovipositor and inserts it, through the exit-hole, right into the chrysalid slough, which is left where it was in the form of a bag. This slough receives the laying. When the operation is finished and the bag of eggs is full, the mother dies outside, hanging on to the case.

The two other Psyches, who do not carry telescopic ovipositors and whose only method of changing their position is a dubious sort of crawling, have more singular customs to show us. One might quote with regard to them what used to be said of the Roman matrons, those model mothers:

"*Domi mansit, lanam fecit.*"

Yes, *lanam fecit*. The Psyche does not really work the wool on the distaff; but at least she bequeathes to her sons her own fleece converted into a heap of wadding. Yes, *domi mansit*. She never leaves her house, not even for her wedding, not even for the purpose of laying her eggs.

We have seen how, after receiving the visit of the male, the shapeless Moth, that uncouth sausage, retreats to the back of her case and withdraws into her chrysalid slough, which she fills exactly, just as though she had never left it. The eggs are in their place then and there; they occupy the regulation sack favoured by the various Psyches. Of what use would a laying be now? Strictly speaking, there is none, in fact; that is to say, the eggs do not leave the mother's womb. The living pouch which has engendered them keeps them within itself.

Soon this bag loses its moisture by evaporation; it dries up and at the same time remains sticking to the chrysalid wrapper, that firm support. Let us open the thing. What does the magnifying glass show us? A few trachean threads, lean bundles of muscles, nervous ramifications, in short, the relics of a form of vitality reduced to its simplest expression. Taken all around, very nearly nothing. The rest of the contents is a mass of eggs, an agglomeration of germs numbering close upon three hundred. In a word, the insect is one enormous ovary, assisted by just so much as enables it to perform its functions.

THE LEAF-CUTTERS

« 33 »

During the summer, leaves and flower petals sometimes exhibit round-ish holes as though someone had snipped out little disks with a pair of small scissors. The shears are the jaws of the Megachile bees, the Leaf-Cutters of this selection from Fabre's BRAMBLE-BEES AND OTHERS. *Numerous species of these insects are found in the United States. One American species uses as many as 1,000 leaf-disks in the formation of the series of cells that hold the eggs and the stored-up food for the larvae. The Cerambyx, referred to by Fabre in this chapter, is a beetle; the Anthophorae are wild bees that burrow in the ground; the Capricorn is a long-horned beetle whose grub tunnels through the wood of oaks.*

One winter evening, as we were sitting round the fire, whose cheerful blaze unloosed our tongues, I put the problem of the Leaf-cutter to my family:

"Among your kitchen-utensils," I said, "you have a pot in daily use; but it has lost its lid, which was knocked over and broken by the cat playing on the shelves. To-morrow is market-day and one of you will be going to Orange to buy the week's provisions. Would she undertake, without a measure of any kind, with the sole aid of memory, which we would allow her to refresh by a careful examination of the object before starting, to bring back exactly what the pot wants, a lid neither too large nor too small, in short the same size as the top?"

It was admitted with one accord that nobody would accept such a commission without taking a measure with her, or at least a bit of string giving the width. Our memory for sizes is not accurate enough. She would come back from the town with something that "might do"; and it would be the merest chance if this turned out to be the right size.

Well, the Leaf-cutter is even less well-off than ourselves. She has no mental picture of her pot, because she has never seen it; she is not able to pick and choose in the crockery-dealer's heap, which acts as something of a guide to our memory by comparison; she must, without hesitation, far away from her home, cut out a disk that fits the top of her jar. What is impossible to us is child's-play to her. Where we could not do without a measure of some kind, a bit of string, a pattern or a scrap of paper with figures upon it, the little Bee needs nothing at all. In housekeeping matters she is cleverer than we are.

One objection was raised. Was it not possible that the Bee, when at work on the shrub, should first cut a round piece of an approximate diameter, larger than that of the neck of the jar, and that afterwards, on returning home, she should gnaw away the superfluous part until the lid exactly fitted the pot? These alterations made with the model in front of her would explain everything.

That is perfectly true; but are there any alterations? To begin with, it seems to me hardly possible that the insect can go back to the cutting once the piece is detached from the leaf: it lacks the necessary support to gnaw the flimsy disk with any precision. A tailor would spoil his cloth if he had not the support of a table when cutting out the pieces for a coat. The Megachile's scissors, so difficult to wield on anything not firmly held, would do equally bad work.

Besides, I have better evidence than this for my refusal to believe in the existence of alterations when the Bee has the cell in front of her. The lid is composed of a pile of disks whose number sometimes reaches half a score. Now the bottom part of all these disks is the under surface of the leaf, which is paler and more strongly veined; the top part is the upper surface, which is smooth and greener. In other words, the insect places them in the position which they occupy when gathered. Let me explain. In order to cut out a piece, the Bee stands on the upper surface of the leaf. The piece detached is held in the feet and is therefore laid with its top surface against the insect's chest at the moment of departure. There is no possibility of its being turned over on the way. Consequently, the piece is laid as the Bee has just picked it, with the lower surface towards the inside of the cell and the upper surface towards the outside. If alterations were necessary to reduce the lid to the diameter of the pot, the disk would be bound to get turned over: the piece, manipulated, set upright, turned round, tried this way and that, would, when finally laid in position, have its top or bottom surface inside just as it happened to come. But this is exactly what does not take place.

Therefore, as the order of stacking never changes, the disks are cut, from the first clip of the scissors, with their proper dimensions. The insect excels us in practical geometry. I look upon the Leaf-cutter's pot and lid as an addition to the many other marvels of instinct that cannot be explained by mechanics; I submit it to the consideration of science; and I pass on.

The Silky Leaf-cutter (*Megachile sericans,* FONSCOL.; *M. Dufourii,* LEP.) makes her nests in the disused galleries of the Anthophoræ. I know her to occupy another dwelling which is more elegant and affords a more roomy installation: I mean the old dwelling of the fat Capricorn, the denizen of the oaks. The metamorphosis is effected in a spacious chamber lined with soft felt. When the long-horned Beetle reaches the adult stages, he releases himself and emerges from the tree by following a vestibule which the larva's powerful tools have prepared beforehand. If the deserted cabin, owing to its position, remains wholesome and there is no sign of any running from its walls, no brown stuff smelling of the tanyard, it is soon visited by the Silky Megachile, who finds in it the most sumptuous of the apartments inhabited by the Leaf-cutters. It combines every condition of comfort: perfect safety, an even temperature, freedom from damp, ample room; and so the mother who is fortunate enough to become the possessor of such a lodging uses it entirely, vestibule and drawing-room alike. Accommodation is found for all her family of eggs; at least, I have nowhere seen nests as populous as here.

One of them provides me with seventeen cells, the highest number appearing in my census of the Megachile genus. Most of them are lodged in the nymphal chamber of the Capricorn; and, as the spacious recess is too wide for a single row, the cells are arranged in three parallel series. The remainder, in a single string, occupy the vestibule, which is completed and filled up by the terminal barricade. In the materials employed, hawthorn- and paliurus-leaves predominate. The pieces, both in the cells and in the stopper, vary in size. It is true that the hawthorn-leaves, with their deep indentations, do not lend themselves to the cutting of neat oval pieces. The insect seems to have detached each morsel without troubling overmuch about the shape of the piece, so long as it was big enough. Nor has it been very particular about arranging the pieces according to the nature of the leaf: after a few bits of paliurus come bits of vine and hawthorn; and these again are followed by bits of bramble and paliurus. The Bee has collected her pieces anyhow, taking a bit here and there, just as her fancy dictated. Nevertheless, paliurus is the commonest, perhaps for economical reasons.

I notice, in fact, that the leaves of this shrub, instead of being used piecemeal, are employed whole, when they do not exceed the proper dimensions. Their oval form and their moderate size suit the insect's requirements; and there is therefore no necessity to cut them into pieces. The leaf-stalk is clipped with the scissors and, without more ado, the Megachile retires the richer by a first-rate bit of material.

Split up into their component parts, two cells give me altogether eighty-three pieces of leaves, whereof eighteen are smaller than the others and of a round shape. The last-named come from the lids. If they average forty-two each, the seventeen cells of the nest represent seven hundred and fourteen pieces. These are not all: the nest ends, in the Capricorn's vestibule, with a stout barricade in which I count three hundred and fifty pieces. The total therefore amounts to one thousand and sixty-four. All those journeys and all that work with the scissors to furnish the deserted chamber of the Cerambyx! If I did not know the Leaf-cutter's solitary and jealous disposition, I should attribute the huge structure to the collaboration of several mothers; but there is no question of communism in this case. One dauntless creature and one alone, one solitary, inveterate worker, has produced the whole of the prodigious mass. If work is the best way to enjoy life, this one certainly has not been bored during the few weeks of her existence.

I gladly award her the most honourable of eulogies, that due to the industrious; and I also compliment her on her talent for closing the honey-pots. The pieces stacked into lids are round and have nothing to suggest those of which the cells and the final barricade are made. Excepting the first, those nearest the honey, they are perhaps cut a trifle less neatly than the disks of the White-girdled Leaf-cutter; no matter: they stop the jar perfectly, especially when there are some ten of them one above the other. When cutting them, the Bee was as sure of her scissors as a dressmaker guided by a pattern laid on the stuff; and yet she was cutting without a model, without having in front of her the mouth to be closed. To enlarge on this interesting subject would mean to repeat one's self. All the Leaf-cutters have the same talent for making the lids of their pots.

THE THISTLE WEEVIL

« 34 »

More pages of Fabre's writings are devoted to the beetles than to any other group of insects. In this extract from THE LIFE OF THE WEEVIL *he tells of the Bear Larinus, the exploiter of the carline thistles of Provence. By hollowing out a little chamber in the head of the thistle, the immature beetle prepares for itself a shelter that protects it throughout the winter.*

I sally forth in the night, with a lantern, to spy out the land. Around me, a circle of faint light enables me to recognize the broad masses fairly well, but leaves the fine details unperceived. At a few paces' distance, the modest illumination disperses, dies away. Farther off still, everything is pitch-dark. The lantern shows me—and but very indistinctly—just one of the innumerable pieces that compose the mosaic of the ground.

To see some more of them, I move on. Each time there is the same narrow circle, of doubtful visibility. By what laws are these points, inspected one by one, correlated in the general picture? The candle-end cannot tell me; I should need the light of the sun.

Science too proceeds by lantern-flashes; it explores nature's inexhaustible mosaic piece by piece. Too often the wick lacks oil; the glass panes of the lantern may not be clean. No matter: his work is not in vain who first recognizes and shows to others one speck of the vast unknown.

However far our ray of light may penetrate, the illuminated circle is checked on every side by the barrier of the darkness. Hemmed in by the unfathomable depths of the unknown, let us be satisfied if it be vouchsafed to us to enlarge by a span the narrow domain of the known. Seekers, all of us, tormented by the desire for knowledge, let us move our lantern from point to point: with the particles explored we shall perhaps be able to piece together a fragment of the picture.

To-day the shifting of the lantern's rays leads us to the Bear Larinus (*L. ursus,* FABR.), the exploiter of the carline thistles. We must not let this inappropriate name of Bear give us an unfavourable notion of the insect. It is due to the whim of a nomenclator who, having exhausted his vocabulary, baffled by the never-ending stream of things already named, uses the first word that comes to hand.

Others, more happily inspired, perceiving a vague resemblance between the sacerdotal ornament, the stole, and the white bands that run down the Weevil's back, have proposed the name of Stoled Larinus (*L. stolatus,* GMEL.). This term would please me; it gives a very good picture of the insect. The Bear, making nonsense, has prevailed. So be it: *non nobis tantas componere lites.*

The domain of this Weevil is the corymbed carlina (*C. corymbosa,* LIN.), a slender thistle, not devoid of elegance, harsh-looking though it be. Its heads, with their tough, yellow-varnished spokes, expand into a fleshy mass, a genuine heart, like an artichoke's, which is defended by a hedge of savage folioles broadly welded at the base. It is at the centre of this palatable heart that the larva is established, always singly.

Each has its exclusive demesne, its inviolable ration. When an egg, a single egg, has been entrusted to the mass of florets, the mother moves on, to continue elsewhere; and, should some new-comer by mistake take possession of it, her grub, arriving too late and finding the place occupied, will die.

This isolation tells us how the larva feeds. The carlina's foster-child cannot live on a clear broth, as does the echinops'; for, if the drops trickling from a wound were sufficient, there would be victuals for several here. The blue thistle feeds three or four boarders without any loss of solid material beyond that resulting from a slight gash. Given such coy-toothed feeders, the heart of the carline thistle would support quite as many.

It is always, on the contrary, the portion of one alone. Thus we already guess that the grub of the Bear Larinus does not confine itself to lapping up discharges of sap and that it likewise feeds upon its artichoke-heart, the standing dish.

The adult also feeds upon it. On the cone covered with imbricated folioles it makes spacious excavations in which the sweet milk of the plant hardens into white beads. But these broken victuals, these cut cakes off which the Weevil has made her meal, are disdained when the egg-laying comes into question, in June and July. A choice is then made of untouched heads, not as yet developed, not yet expanded and still contracted into prickly globules. The interior will be tenderer than after they are full-blown.

The method is the same as that of the Spotted Larinus. With her rostral gimlet the mother bores a hole through the scales, on a level with the base of the florets; then, with the aid of her guiding probe, she installs her opalescent white egg at the bottom of the shaft. A week later, the grub makes its appearance.

Some time in August let us open the thistle-heads. Their contents are very diverse. There are larvæ here of all ages; nymphs covered with reddish ridges, above all on the last segments, twitching violently and spinning round when disturbed; lastly, perfect insects, not yet adorned with their stoles and other ornaments of the final costume. We have before our eyes the means of following the whole development of the Weevil at the same time.

The folioles of the blossom, those stout halberds, are welded together at their base and enclose within their rampart a fleshy mass, with a flat upper surface and cone-shaped underneath. This is the larder of the Bear Larinus.

From the bottom of its cell the new-born grub dives forthwith into this fleshy mass. It cuts into it deep. Unreservedly, respecting only the walls, it digs itself, in a couple of weeks, a recess shaped like a sugar-loaf and prolonged until it touches the stalk. The canopy of this recess is a dome of florets and hairs forced upwards and held in place by an adhesive. The artichoke-heart is completely emptied; nothing is respected save the scaly walls.

As its isolation led us to expect, the grub of the Bear Larinus therefore eats solid food. There is, however, nothing to prevent it from adding to this diet the milky exudations of the sap.

This fare, in which solid matter predominates, necessarily involves solid excreta, which are unknown in the inmate of the blue thistle. What does the hermit of the carline thistle do with them, cooped up in a narrow cell from which nothing can be shot outside? It employs them as the other does its viscous drops; it upholsters its cell with them.

I see it curved into a circle with its mouth applied to the opposite orifice, carefully collecting the granules as these are evacuated by the intestinal factory. It is precious stuff, this, very precious; and the grub will be careful not to lose a scrap of it, for it has naught else wherewith to plaster its dwelling.

The dropping seized is therefore placed in position at once, spread with the tips of the mandibles and compressed with the forehead and rump. A few waste chips and flakes, a few bits of down are torn from the uncemented ceiling overhead; and the plasterer incorporates them, atom by atom, with the still moist putty.

This gives, as the inmate increases in size, a coat of rough-cast which, smoothed with meticulous care, lines the whole of the cell. Together with the natural wall furnished by the prickly rind of the artichoke, it makes a powerful bastion, far superior, as a defensive system, to the thatched huts of the Spotted Larinus.

The plant, moreover, lends itself to protracted residence. It is slightly built but slow to decay. The winds do not prostrate it in the mire, supported as it is by brushwood and sturdy grasses, its habitual environment. When the handsome thistle with the blue spheres has long been mouldering on the edge of the roads, the carlina, with its rot-proof base, still stands erect, dead and brown but not dilapidated. Another excellent quality is this: the scales of its heads contract and make a roof which the rain has difficulty in penetrating.

In such a shelter there is no occasion to fear the dangers which make the Spotted Larinus quit her pitchers at the approach of winter: the dwelling is securely founded and the cell is dry. The Bear Larinus is well aware of these advantages; she is careful not to imitate the other in wintering under the cover of dead leaves and stone-heaps. She does not stir abroad, assured beforehand of the efficiency of her roof.

On the roughest days of the year, in January, if the weather permits me to go out, I open the heads of the carline thistles which I come across. I always find the Larinus there, in all the freshness of her striped costume. She is waiting, benumbed, until the warmth and animation of May return. Then only will she break the dome of her cabin and go to take part in the festival of spring.

THE CRICKET

« 35 »

*Active, omnivorous, musical, the field cricket is well-known wherever
it is found. Fabre's account of some phases of this insect's life, which
is given here, has been taken from* THE LIFE OF THE GRASSHOPPER. *The
Crested Lark, which he mentions, is a relative of the British Skylark.
The Decticus is a long-horned grasshopper.*

Who does not know the Cricket's abode! Who has not, as a child playing in the fields, stopped in front of the hermit's cabin! However light
your footfall, he has heard you coming and has abruptly withdrawn to
the very bottom of his hiding-place. When you arrive, the threshold of
the house is deserted.

Everybody knows the way to bring the skulker out. You insert a straw
and move it gently about the burrow. Surprised at what is happening
above, tickled and teased, the Cricket ascends from his secret apartment;
he stops in the passage, hesitates and enquires into things by waving his
delicate antennæ; he comes to the light and, once outside, he is easy to
catch, so greatly have events puzzled his poor head. Should he be missed
at the first attempt, he may become more suspicious and obstinately resist the titillation of the straw. In that case, we can flood him out with a
glass of water.

O those adorable times when we used to cage our Crickets and feed
them on a leaf of lettuce, those childish hunting-trips along the grassy
paths! They all come back to me to-day, as I explore the burrows in
search of subjects for my studies; they appear to me almost in their
pristine freshness when my companion, little Paul, already an expert in
the tactical use of the straw, springs up suddenly, after a long trial of
skill and patience with the recalcitrant, and, brandishing his closed hand
in the air, cries, excitedly:

281

"I've got him, I've got him!"

Quick, here's a bag; in you go, my little Cricket! You shall be petted and pampered; but mind you teach us something and, first of all, show us your house.

It is a slanting gallery, situated in the grass, on some sunny bank which soon dries after a shower. It is nine inches long at most, hardly as thick as one's finger and straight or bent according to the exigencies of the ground. As a rule, a tuft of grass, which is respected by the Cricket when he goes out to browse upon the surrounding turf, half-conceals the home, serving as a porch and throwing a discreet shade over the entrance. The gently-sloping threshold, scrupulously raked and swept, is carried for some distance. This is the belvedere on which, when everything is peaceful round about, the Cricket sits and scrapes his fiddle.

The inside of the house is devoid of luxury, with bare and yet not coarse walls. Ample leisure allows the inhabitant to do away with any unpleasant roughness. At the end of the passage is the bedroom, the terminal alcove, a little more carefully smoothed than the rest and slightly wider. All said, it is a very simple abode, exceedingly clean, free from damp and conforming with the requirements of a well-considered system of hygiene. On the other hand, it is an enormous undertaking, a regular Cyclopean tunnel, when we consider the modest means of excavation. Let us try to be present at the work. Let us also enquire at what period the enterprise begins. This obliges us to go back to the egg.

Any one wishing to see the Cricket lay her eggs can do so without making great preparations: all that he wants is a little patience, which, according to Buffon, is genius, but which I, more modestly, will describe as the observer's chief virtue. In April, or at latest in May, we establish isolated couples of the insect in flower-pots containing a layer of heaped-up earth. Their provisions consist of a lettuce-leaf renewed from time to time. A square of glass covers the retreat and prevents escape.

Some extremely interesting facts can be obtained with this simple installation, supplemented, if need be, with a wire-gauze cover, the best of all cages. We shall return to this matter. For the moment, let us watch the laying and make sure that the propitious hour does not evade our vigilance.

It is in the first week in June that my assiduous visits begin to show satisfactory results. I surprise the mother standing motionless, with her ovipositor planted perpendicularly in the soil. For a long time she remains stationed at the same point, heedless of her indiscreet caller. At last she withdraws her dibble, removes, more or less perfunctorily, the traces of the boring-hole, takes a moment's rest, walks away and starts again somewhere else, now here, now there, all over the area at her disposal. Her behaviour, though her movements are slower, is a repetition of what the Decticus has shown us. Her egg-lying appears to me to be ended within the twenty-four hours. For greater certainty, I wait a couple of days longer.

I then dig up the earth in the pot. The straw-coloured eggs are cylinders rounded at both ends and measuring about one-ninth of an inch in length. They are placed singly in the soil, arranged vertically and grouped in more or less numerous patches, which correspond with the successive layings. I find them all over the pot, at a depth of three-quarters of an inch. There are difficulties in examining a mass of earth through a magnifying-glass; but, allowing for these difficulties, I estimate the eggs laid by one mother at five or six hundred. So large a family is sure to undergo a drastic purging before long.

The Cricket's egg is a little marvel of mechanism. After hatching, it appears as an opaque white sheath, with a round and very regular aperture at the top; to the edge of this a cap adheres, forming a lid. Instead of bursting anyhow under the thrusts or cuts of the new-born larva, it opens of its own accord along a specially prepared line of least resistance.

It became important to observe the curious hatching. About a fortnight after the egg is laid, two large, round, rusty-black eye-dots darken the front end. A little way above these two dots, right at the apex of the cylinder, you see the outline of a thin circular swelling. This is the line of rupture which is preparing. Soon the translucency of the egg enables the observer to perceive the delicate segmentation of the tiny creature within. Now is the time to redouble our vigilance and multiply our visits, especially in the morning.

Fortune, which loves the persevering, rewards me for my assiduity. All round this swelling where, by a process of infinite delicacy, the line of least resistance has been prepared, the end of the egg, pushed back by the inmate's forehead, becomes detached, rises and falls to one side like the top of a miniature scent-bottle. The Cricket pops out like a Jack-in-the-box.

When he is gone, the shell remains distended, smooth, intact, pure white, with the cap or lid hanging from the opening. A bird's egg breaks clumsily under the blows of a wart that grows for the purpose at the end of the chick's beak; the Cricket's egg, endowed with a superior mechanism, opens like an ivory case. The thrust of the inmate's head is enough to work the hinge.

The hatching of the eggs is hastened by the glorious weather; and the observer's patience is not much tried, the rapidity rivalling that of the Dung-beetles. The summer solstice has not yet arrived when the ten couples interned under glass for the benefit of my studies are surrounded by their numerous progeny. The egg-stage, therefore, lasts just about ten days.

I said above that, when the lid of the ivory case is lifted, a young Cricket pops out. This is not quite accurate. What appears at the opening is the swaddled grub, as yet unrecognizable in a tight-fitting sheath. I expected to see this wrapper, this first set of baby-clothes, for the same reasons that made me anticipate it in the case of the Decticus:

"The Cricket," said I to myself, "is born underground. He also sports two very long antennæ and a pair of overgrown hind-legs, all of which are cumbrous appendages at the time of the emergence. He must therefore possess a tunic in which to make his exit."

My forecast, correct enough in principle, was only partly confirmed. The new-born Cricket does in fact possess a temporary structure; but, so far from employing it for the purpose of hoisting himself outside, he throws off his clothes as he passes out of the egg.

To what circumstances are we to attribute this departure from the usual practice? Perhaps to this: the Cricket's egg stays in the ground for only a few days before hatching; the egg of the Decticus remains there for eight months. The former, save for rare exceptions in a season of drought, lies under a thin layer of dry, loose, unresisting earth; the latter, on the contrary, finds itself in soil which has been caked together by the persistent rains of autumn and winter and which therefore presents serious difficulties. Moreover, the Cricket is shorter and stouter, less long-shanked than the Decticus. These would appear to be the reasons for the difference between the two insects in respect of their methods of emerging. The Decticus, born lower down, under a close-packed layer, needs a climbing-costume with which the Cricket is able to dispense, being less hampered and nearer to the surface and having only a powdery layer of earth to pass through.

Then what is the object of the tights which the Cricket flings aside as soon as he is out of the egg? I will answer this question with another: what is the object of the two white stumps, the two pale-coloured embryo wings carried by the Cricket under his wing-cases, which are turned into a great mechanism of sound? They are so insignificant, so feeble that the insect certainly makes no use of them, any more than the Dog utilizes the thumb that hangs limp and lifeless at the back of his paw.

Sometimes, for reasons of symmetry, the walls of a house are painted with imitation windows to balance the other windows, which are real. This is done out of respect for order, the supreme condition of the beautiful. In the same way, life has its symmetries, its repetitions of a general prototype. When abolishing an organ that has ceased to be employed, it leaves vestiges of it to maintain the primitive arrangement.

The Dog's rudimentary thumb predicates the five-fingered hand that characterizes the higher animals; the Cricket's wing-stumps are evidence that the insect would normally be capable of flight; the moult undergone on the threshold of the egg is reminiscent of the tight-fitting wrapper needed for the laborious exit of the Locustidæ born underground. They are so many symmetrical superfluities, so many remains of a law that has fallen into disuse but never been abrogated.

As soon as he is deprived of his delicate tunic, the young Cricket, pale all over, almost white, begins to battle with the soil overhead. He hits out with his mandibles; he sweeps aside and kicks behind him the powdery obstruction, which offers no resistance. Behold him on the surface, amidst the joys of the sunlight and the perils of conflict with the living, poor, feeble creature that he is, hardly larger than a Flea. In twenty-four hours he colours and turns into a magnificent blackamoor, whose ebon hue vies with that of the adult insect. All that remains of his original pallor is a white sash that girds his chest and reminds us of a baby's leading-string. Very nimble and alert, he sounds the surrounding space with his long, quivering antennæ, runs about and jumps with an impetuosity in which his future obesity will forbid him to indulge.

This is also the age when the stomach is still delicate. What sort of food does he need? I do not know. I offer him the adult's treat, tender lettuce-leaves. He scorns to touch them, or perhaps he takes mouthfuls so exceedingly small that they escape me.

In a few days, with my ten households, I find myself overwhelmed with family cares. What am I to do with my five or six thousand Crickets, a pretty flock, no doubt, but impossible to rear in my ignorance of the treatment required? I will set you at liberty, my little dears; I will entrust you to nature, the sovran nurse.

Thus it comes to pass. I release my legions in the enclosure, here, there and everywhere, in the best places. What a concert I shall have outside my door next year, if they all turn out well! But no, the symphony will probably be one of silence, for the savage pruning due to the mother's fertility is bound to come. All that I can hope for is that a few couples may survive extermination.

As in the case of the young Praying Mantis, the first that hasten to this manna and the most eager for the slaughter are the little Grey Lizard and the Ant. The latter, loathsome freebooter that she is, will, I fear, not leave me a single Cricket in the garden. She snaps up the poor little creatures, eviscerates them and gobbles them down at frantic speed.

Oh, the execrable wretch! And to think that we place the Ant in the front rank of insects! Books are written in her honour and the stream of eulogy never ceases; the naturalists hold her in the greatest esteem and add daily to her reputation, so true is it, among animals as among men, that of the various ways of making history, the surest way is to do harm to others.

Nobody asks after the Dung-beetle and the Necrophorus, invaluable scavengers both, whereas everybody knows the Gnat, that drinker of men's blood; the Wasp, that hot-tempered swashbuckler, with her poisoned dagger; and the Ant, that notorious evil-doer, who, in our southern villages, saps and imperils the rafters of a dwelling with the same zest with which she devours a fig. I need not trouble to say more: every one will discover in the records of mankind similar instances of usefulness ignored and frightfulness exalted.

The massacre instituted by the Ants and other exterminators is so great that my erstwhile populous colonies in the enclosure become too small to enable me to continue my observations; and I am driven to have recourse to information outside. In August, among the falling leaves, in those little oases where the grass has not been wholly scorched by the sun, I find the young Cricket already rather big, black all over like the adult, with not a vestige of the white girdle of his early days. He has no domicile. The shelter of a dead leaf, the cover of a flat stone are enough for him; they represent the tents of a nomad who cares not where he lays his head.

This vagabond life continues until the middle of autumn. It is then that the yellow-winged Sphex hunts down the wanderers, an easy prey, and stores her bag of Crickets underground. She decimates those who have survived the Ants' devastating raids. A settled dwelling, dug a few weeks before the usual time, would save them from the spoilers. The sorely-tried victims do not think of it. The bitter experience of the centuries has taught them nothing. Though already strong enough to dig a protecting burrow, they remain invincibly faithful to their ancient customs and would go on roaming though the Sphex stabbed the last of their race.

It is at the close of October, when the first cold weather threatens, that the burrow is taken in hand. The work is very simple, judging by the little that my observation of the caged insect has shown me. The digging is never done at a bare point in the pan, but always under the shelter of a withered lettuce-leaf, some remnant of the food provided. This takes the place of the grass screen that seems indispensable to the secrecy of the establishment.

The miner scrapes with his fore-legs and uses the pincers of his mandibles to extract the larger bits of gravel. I see him stamping with his powerful hind-legs, furnished with a double row of spikes; I see him raking the rubbish, sweeping it backwards and spreading it slantwise. There you have the method in its entirety.

The work proceeds pretty quickly at first. In the yielding soil of my cages, the digger disappears underground after a spell that lasts a couple of hours. He returns to the entrance at intervals, always backwards and always sweeping. Should he be overcome with fatigue, he takes a rest on the threshold of his half-finished home, with his head outside and his antennæ waving feebly. He goes in again and resumes work with pincers and rakes. Soon the periods of repose become longer and wear out my patience.

The most urgent part of the work is done. Once the hole is a couple of inches deep, it suffices for the needs of the moment. The rest will be a long-winded business, resumed in a leisurely fashion, a little one day and a little the next; the hole will be made deeper and wider as demanded by the inclemencies of the weather and the growth of the insect. Even in winter, if the temperature be mild and the sun playing over the entrance to the dwelling, it is not unusual to see the Cricket shooting out rubbish, a sign of repairs and fresh excavations. Amidst the joys of spring, the upkeep of the building still continues. It is constantly undergoing improvements and repairs until the owner's decease.

April comes to an end and the Cricket's song begins, at first in rare and shy solos, soon developing into a general symphony in which each clod of turf boasts its performer. I am more than inclined to place the Cricket at the head of the spring choristers. In our waste lands, when the thyme and the lavender are gaily flowering, he has as his partner the Crested Lark, who rises like a lyrical rocket, his throat swelling with notes, and from the sky, invisible in the clouds, sheds his sweet music upon the fallows. Down below the Crickets chant the responses. Their song is monotonous and artless, but so well-suited, in its very crudity, to the rustic gladness of renascent life! It is the hosanna of the awakening, the sacred alleluia understood by swelling seed and sprouting blade. Who deserves the palm in this duet? I should award it to the Cricket. He surpasses them all, thanks to his numbers and his unceasing note. Were the Lark to fall silent, the fields blue-grey with lavender, swinging its fragrant censers before the sun, would still receive from this humble chorister a solemn celebration.

THE CRAB SPIDER

« 36 »

The ballooning of baby spiders is a dramatic event in the lives of these creatures. Drifting through the air for miles, buoyed up by silken threads, they are widely dispersed by the breeze. Such an aerial journey is part of the life-story of most species including the sedentary crab spiders that spend their adult lives lurking among flowers in wait for prey. Numerous species of these spiders, all belonging to the family, Thomisidae, live in the United States. The Banded and Silky Epeira, of which Fabre speaks in this selection from THE LIFE OF THE SPIDER, *are orb-weavers. The Narbonne Lycosa is a wolf spider which pursues and captures its prey on the ground.*

The Spider with the Crab-like figure does not know how to manufacture nets for catching game. Without springs or snares, she lies in ambush, among the flowers, and awaits the arrival of the quarry, which she kills by administering a scientific stab in the neck. The Thomisus, in particular, the subject of this chapter, is passionately addicted to the pursuit of the Domestic Bee. I have described the contests between the victim and her executioner, at greater length, elsewhere.

The Bee appears, seeking no quarrel, intent upon plunder. She tests the flowers with her tongue; she selects a spot that will yield a good return. Soon she is wrapped up in her harvesting. While she is filling her baskets and distending her crop, the Thomisus, that bandit lurking under cover of the flowers, issues from her hiding-place, creeps round behind the bustling insect, steals up close and, with a sudden rush, nabs her in the nape of the neck. In vain, the Bee protests and darts her sting at random; the assailant does not let go.

Besides, the bite in the neck is paralyzing, because the cervical nerve-centres are affected. The poor thing's legs stiffen; and all is over in a second. The murderess now sucks the victim's blood at her ease and, when she has done, scornfully flings the drained corpse aside. She hides herself once more, ready to bleed a second gleaner should the occasion offer.

This slaughter of the Bee engaged in the hallowed delights of labour has always revolted me. Why should there be workers to feed idlers, why sweated to keep sweaters in luxury? Why should so many admirable lives be sacrificed to the greater prosperity of brigandage? These hateful discords amid the general harmony perplex the thinker, all the more as we shall see the cruel vampire become a model of devotion where her family is concerned.

The ogre loved his children; he ate the children of others. Under the tyranny of the stomach, we are all of us, beasts and men alike, ogres. The dignity of labour, the joy of life, maternal affection, the terrors of death: all these do not count, in others; the main point is that the morsel be tender and savoury.

According to the etymology of her name—θώμιγξ, a cord—the Thomisus should be like the ancient lictor, who bound the sufferer to the stake. The comparison is not inappropriate as regards many Spiders who tie their prey with a thread to subdue it and consume it at their ease; but it just happens that the Thomisus is at variance with her label. She does not fasten her Bee, who, dying suddenly of a bite in the neck, offers no resistance to her consumer. Carried away by his recollection of the regular tactics, our Spider's godfather overlooked the exception; he did not know of the perfidious mode of attack which renders the use of a bowstring superfluous.

Nor is the second name of *onustus*—loaded, burdened, freighted— any too happily chosen. The fact that the Bee-huntress carries a heavy paunch is no reason to refer to this as a distinctive characteristic. Nearly all Spiders have a voluminous belly, a silk-warehouse where, in some cases, the rigging of the net, in others, the swan's down of the nest is manufactured. The Thomisus, a first-class nest-builder, does like the rest: she hoards in her abdomen, but without undue display of obesity, the wherewithal to house her family snugly.

Can the expression *onustus* refer simply to her slow and sidelong walk? The explanation appeals to me, without satisfying me fully. Except in the case of a sudden alarm, every Spider maintains a sober gait and a wary pace. When all is said, the scientific term is composed of a misconception and a worthless epithet. How difficult it is to name animals rationally! Let us be indulgent to the nomenclator: the dictionary is becoming exhausted and the constant flood that requires cataloguing mounts incessantly, wearing out our combinations of syllables.

As the technical name tells the reader nothing, how shall he be informed? I see but one means, which is to invite him to the May festivals, in the waste-lands of the South. The murderess of the Bees is of a chilly constitution; in our parts, she hardly ever moves away from the olive-districts. Her favourite shrub is the white-leaved rock-rose (*Cistus albidus*), with the large, pink, crumpled, ephemeral blooms that last but a morning and are replaced, next day, by fresh flowers, which have blossomed in the cool dawn. This glorious efflorescence goes on for five or six weeks.

Here, the Bees plunder enthusiastically, fussing and bustling in the spacious whorl of the stamens, which beflour them with yellow. Their persecutrix knows of this affluence. She posts herself in her watch-house, under the rosy screen of a petal. Cast your eyes over the flower, more or less everywhere. If you see a Bee lying lifeless, with legs and tongue outstretched, draw nearer: the Thomisus will be there, nine times out of ten. The thug has struck her blow; she is draining the blood of the departed.

After all, this cutter of Bees' throats is a pretty, a very pretty creature, despite her unwieldy paunch fashioned like a squat pyramid and embossed on the base, on either side, with a pimple shaped like a camel's hump. The skin, more pleasing to the eye than any satin, is milk-white in some, in others lemon-yellow. There are fine ladies among them who adorn their legs with a number of pink bracelets and their back with carmine arabesques. A narrow pale-green ribbon sometimes edges the right and left of the breast. It is not so rich as the costume of the Banded Epeira, but much more elegant because of its soberness, its daintiness and the artful blending of its hues. Novice fingers, which shrink from touching any other Spider, allow themselves to be enticed by these attractions; they do not fear to handle the beauteous Thomisus, so gentle in appearance.

Well, what can this gem among Spiders do? In the first place she makes a nest worthy of its architect. With twigs and horse-hair and bits of wool, the Goldfinch, the Chaffinch and other masters of the builder's art construct an aerial bower in the fork of the branches. Herself a lover of high places, the Thomisus selects as the site of her nest one of the upper twigs of the rock-rose, her regular hunting-ground, a twig withered by the heat and possessing a few dead leaves, which curl into a little cottage. This is where she settles with a view to her eggs.

Ascending and descending with a gentle swing in more or less every direction, the living shuttle, swollen with silk, weaves a bag whose outer casing becomes one with the dry leaves around. The work, which is partly visible and partly hidden by its supports, is a pure dead-white, its shape, moulded in the angular interval between the bent leaves, is that of a cone and reminds us, on a smaller scale, of the nest of the Silky Epeira.

When the eggs are laid, the mouth of the receptacle is hermetically closed with a lid of the same white silk. Lastly, a few threads, stretched like a thin curtain, form a canopy above the nest and, with the curved tips of the leaves, frame a sort of alcove wherein the mother takes up her abode.

It is more than a place of rest after the fatigues of her confinement: it is a guard-room, an inspection-post where the mother remains sprawling until the youngsters' exodus. Greatly emaciated by the laying of her eggs and by her expenditure of silk, she lives only for the protection of her nest.

Should some vagrant pass near by, she hurries from her watch-tower, lifts a limb and puts the intruder to flight. If I tease her with a straw, she parries with big gestures, like those of a prize-fighter. She uses her fists against my weapon. When I propose to dislodge her in view of certain experiments, I find some difficulty in doing so. She clings to the silken floor, she frustrates my attacks, which I am bound to moderate lest I should injure her. She is no sooner attracted outside than she stubbornly returns to her post. She declines to leave her treasure.

Even so does the Narbonne Lycosa struggle when we try to take away her pill. Each displays the same pluck and the same devotion; and also the same denseness in distinguishing her property from that of others. The Lycosa accepts without hesitation any strange pill which she is given in exchange for her own; she confuses alien produce with the produce of her ovaries and her silk-factory. Those hallowed words, maternal love, were out of place here: it is an impetuous, an almost mechanical impulse, wherein real affection plays no part whatever. The beautiful Spider of the rock-roses is no more generously endowed. When moved from her nest to another of the same kind, she settles upon it and never stirs from it, even though the different arrangement of the leafy fence be such as to warn her that she is not really at home. Provided that she have satin under her feet, she does not notice her mistake; she watches over another's nest with the same vigilance which she might show in watching over her own.

The Lycosa surpasses her in maternal blindness. She fastens to her spinnerets and dangles, by way of a bag of eggs, a ball of cork polished with my file, a paper pellet, a little ball of thread. In order to discover if the Thomisus is capable of a similar error, I gathered some broken pieces of silk-worm's cocoon into a closed cone, turning the fragments so as to bring the smoother and more delicate inner surface outside. My attempt was unsuccessful. When removed from her home and placed on the artificial wallet, the mother Thomisus obstinately refused to settle there. Can she be more clear-sighted than the Lycosa? Perhaps so. Let us not be too extravagant with our praise, however; the imitation of the bag was a very clumsy one.

The work of laying is finished by the end of May, after which, lying flat on the ceiling of her nest, the mother never leaves her guard-room, either by night or day. Seeing her look so thin and wrinkled, I imagine that I can please her by bringing her a provision of Bees, as I was wont to do. I have misjudged her needs. The Bee, hitherto her favourite dish, tempts her no longer. In vain does the prey buzz close by, an easy capture within the cage: the watcher does not shift from her post, takes no notice of the windfall. She lives exclusively upon maternal devotion, a commendable but unsubstantial fare. And so I see her pining away from day to day, becoming more and more wrinkled. What is the withered thing waiting for, before expiring? She is waiting for her children to emerge; the dying creature is still of use to them.

When the Banded Epeira's little ones issue from their balloon, they have long been orphans. There is none to come to their assistance; and they have not the strength to free themselves unaided. The balloon has to split automatically and to scatter the youngsters and their flossy mattress all mixed up together. The Thomisus' wallet, sheathed in leaves over the greater part of its surface, never bursts; nor does the lid rise, so carefully is it sealed down. Nevertheless, after the delivery of the brood, we see, at the edge of the lid, a small, gaping hole, an exit-window. Who contrived this window, which was not there at first?

The fabric is too thick and tough to have yielded to the twitches of the feeble little prisoners. It was the mother, therefore, who, feeling her offspring shuffle impatiently under the silken ceiling, herself made a hole in the bag. She persists in living for five or six weeks, despite her shattered health, so as to give a last helping hand and open the door for her family. After performing this duty, she gently lets herself die, hugging her nest and turning into a shrivelled relic.

When July comes, the little ones emerge. In view of their acrobatic habits, I have placed a bundle of slender twigs at the top of the cage in which they were born. All of them pass through the wire gauze and form a group on the summit of the brushwood, where they swiftly weave a spacious lounge of criss-cross threads. Here they remain, pretty quietly, for a day or two; then foot-bridges begin to be flung from one object to the next. This is the opportune moment.

I put the bunch laden with beasties on a small table, in the shade, before the open window. Soon, the exodus commences, but slowly and unsteadily. There are hesitations, retrogressions, perpendicular falls at the end of a thread, ascents that bring the hanging Spider up again. In short, much ado for a poor result.

As matters continue to drag, it occurs to me, at eleven o'clock, to take the bundle of brushwood swarming with the little Spiders, all eager to be off, and place it on the window-sill, in the glare of the sun. After a few minutes of heat and light, the scene assumes a very different aspect. The emigrants run to the top of the twigs, bustle about actively. It becomes a bewildering rope-yard, where thousands of legs are drawing the hemp from the spinnerets. I do not see the ropes manufactured and sent floating at the mercy of the air; but I guess their presence.

Three or four Spiders start at a time, each going her own way in directions independent of her neighbours'. All are moving upwards, all are climbing some support, as can be perceived by the nimble motion of their legs. Moreover, the road is visible behind the climber, it is of double thickness, thanks to an added thread. Then, at a certain height, individual movement ceases. The tiny animal soars in space and shines, lit up by the sun. Softly it sways, then suddenly takes flight.

What has happened? There is a slight breeze outside. The floating cable has snapped and the creature has gone off, borne on its parachute. I see it drifting away, showing, like a spot of light, against the dark foliage of the near cypresses, some forty feet distant. It rises higher, it crosses over the cypress-screen, it disappears. Others follow, some higher, some lower, hither and thither.

But the throng has finished its preparations; the hour has come to disperse in swarms. We now see, from the crest of the brushwood, a continuous spray of starters, who shoot up like microscopic projectiles and mount in a spreading cluster. In the end, it is like the bouquet at the finish of a pyrotechnic display, the sheaf of rockets fired simultaneously. The comparison is correct down to the dazzling light itself. Flaming in the sun like so many gleaming points, the little Spiders are the sparks of that living firework. What a glorious send-off! What an entrance into the world! Clutching its aeronautic thread, the minute creature mounts in an apotheosis.

Sooner or later, nearer or farther, the fall comes. To live, we have to descend, often very low, alas! The Crested Lark crumbles the mule-droppings in the road and thus picks up his food, the oaten grain which he would never find by soaring in the sky, his throat swollen with song. We have to descend; the stomach's inexorable claims demand it. The Spiderling, therefore, touches land. Gravity, tempered by the parachute, is kind to her.

The rest of her story escapes me. What infinitely tiny Midges does she capture before possessing the strength to stab her Bee? What are the methods, what the wiles of atom contending with atom? I know not. We shall find her again in spring, grown quite large and crouching among the flowers whence the Bee takes toll.

A BUILDER OF WEBS

« 37 »

Henry D. Thoreau described himself in WALDEN *as a self-appointed inspector of snowstorms and rainstorms. In this selection from* THE LIFE OF THE SPIDER, *J. Henri Fabre declares that for many years he was a self-appointed inspector of spiders' webs. The marvelous mechanics of the web fascinated him. Here he tells the story of its production by the Epeira, orb-weavers having numerous relatives among the web-making spiders of America.*

The fowling-snare is one of man's ingenious villainies. With lines, pegs and poles, two large, earth-coloured nets are stretched upon the ground, one to the right, the other to the left of a bare surface. A long cord, pulled, at the right moment, by the fowler, who hides in a brushwood hut, works them and brings them together suddenly, like a pair of shutters.

Divided between the two nets are the cages of the decoy-birds—Linnets and Chaffinches, Greenfinches and Yellowhammers, Buntings and Ortolans—sharp-eared creatures which, on perceiving the distant passage of a flock of their own kind, forthwith utter a short calling note. One of them, the *Sambé,* an irresistible tempter, hops about and flaps his wings in apparent freedom. A bit of twine fastens him to his convict's stake. When, worn with fatigue and driven desperate by his vain attempts to get away, the sufferer lies down flat and refuses to do his duty, the fowler is able to stimulate him without stirring from his hut. A long string sets in motion a little lever working on a pivot. Raised from the ground by this diabolical contrivance, the bird flies, falls down and flies up again at each jerk of the cord.

The fowler waits, in the mild sunlight of the autumn morning. Suddenly, great excitement in the cages. The Chaffinches chirp their rallying-cry:

"Pinck! Pinck!"

There is something happening in the sky. The *Sambé,* quick! They are coming, the simpletons; they swoop down upon the treacherous floor. With a rapid movement, the man in ambush pulls his string. The nets close and the whole flock is caught.

Man has wild beast's blood in his veins. The fowler hastens to the slaughter. With his thumb, he stifles the beating of the captives' hearts, staves in their skulls. The little birds, so many piteous heads of game, will go to market, strung in dozens on a wire passed through their nostrils.

For scoundrelly ingenuity, the Epeira's net can bear comparison with the fowler's; it even surpasses it when, on patient study, the main features of its supreme perfection stand revealed. What refinement of art for a mess of Flies! Nowhere, in the whole animal kingdom, has the need to eat inspired a more cunning industry. If the reader will meditate upon the description that follows, he will certainly share my admiration.

First of all, we must witness the making of the net; we must see it constructed and see it again and again, for the plan of such a complex work can only be grasped in fragments. To-day, observation will give us one detail; to-morrow, it will give us a second, suggesting fresh points of view; as our visits multiply, a new fact is each time added to the sum total of the acquired data, confirming those which come before or directing our thoughts along unsuspected paths.

The snow-ball rolling over the carpet of white grows enormous, however scanty each fresh layer be. Even so with truth in observational science: it is built up of trifles patiently gathered together. And, while the collecting of these trifles means that the student of Spider industry must not be chary of his time, at least it involves no distant and speculative research. The smallest garden contains Epeiræ, all accomplished weavers.

I am able, at the proper hours, all through the fine season, to question them, to watch them at work, now this one, anon that, according to the chances of the day. What I did not see very plainly yesterday I can see the next day, under better conditions, and on any of the following days, until the phenomenon under observation is revealed in all clearness.

Let us go every evening, step by step, from one border of tall rosemaries to the next. Should things move too slowly, we will sit down at the foot of the shrubs, opposite the rope-yard, where the light falls favourably, and watch with unwearying attention. Each trip will be

good for a fact that fills some gap in the ideas already gathered. To ap-
point one's self, in this way, an inspector of Spiders' webs, for many
years in succession and for long seasons, means joining a not over-
crowded profession, I admit. Heaven knows, it does not enable one to
put money by! No matter: the meditative mind returns from that school
fully satisfied.

My subjects, in the first instance, are young and boast but a slight
corporation, very far removed from what it will be in the late autumn.
The belly, the wallet containing the rope-works, hardly exceeds a pep-
percorn in bulk. This slenderness on the part of the spinstresses must
not prejudice us against their work: there is no parity between their
skill and their years. The adult Spiders, with their disgraceful paunches,
can do no better.

Moreover, the beginners have one very precious advantage for the observer: they work by day, work even in the sun, whereas the old ones
weave only at night, at unseasonable hours. The first show us the secrets
of their looms without much difficulty; the others conceal them from
us. Work starts in July, a couple of hours before sunset.

The spinstresses of my enclosure then leave their daytime hiding-
places, select their posts and begin to spin, one here, another there.
There are many of them; we can choose where we please. Let us stop
in front of this one, whom we surprise in the act of laying the foundations of the structure. Without any appreciable order, she runs about
the rosemary-hedge, from the tip of one branch to another, within the
limits of some eighteen inches. Gradually, she puts a thread in position,
drawing it from her wire-mill with the combs attached to her hind-
legs. This preparatory work presents no appearance of a concerted plan.
The Spider comes and goes impetuously, as though at random; she goes
up, comes down, goes up again, dives down again and each time strengthens the points of contact with intricate moorings distributed here and
there. The result is a scanty and disordered scaffolding.

Is disordered the word? Perhaps not. The Epeira's eye, more experienced in matters of this sort than mine, has recognized the general
lie of the land; and the rope-fabric has been erected accordingly: it is
very inaccurate in my opinion, but very suitable for the Spider's designs.
What is it that she really wants? A solid frame to contain the network
of the web. The shapeless structure which she has just built fulfils the
desired conditions: it marks out a flat, free and perpendicular area. This
is all that is necessary.

The whole work, for that matter, is now soon completed; it is done all over again, each evening, from top to bottom, for the incidents of the chase destroy it in a night. The net is as yet too delicate to resist the desperate struggles of the captured prey. On the other hand, the adults' net, which is formed of stouter threads, is adapted to last some time; and the Epeira gives it a more carefully-constructed frame-work, as we shall see elsewhere.

A special thread, the foundation of the real net, is stretched across the area so capriciously circumscribed. It is distinguished from the others by its isolation, its position at a distance from any twig that might interfere with its swaying length. It never fails to have, in the middle, a thick white point, formed of a little silk cushion. This is the beacon that marks the centre of the future edifice, the post that will guide the Epeira and bring order into the wilderness of twists and turns.

The time has come to weave the hunting-snare. The Spider starts from the centre, which bears the white sign-post, and, running along the transversal thread, hurriedly reaches the circumference, that is to say, the irregular frame enclosing the free space. Still with the same sudden movement, she rushes from the circumference to the centre; she starts again backwards and forwards, makes for the right, the left, the top, the bottom; she hoists herself up, dives down, climbs up again, runs down and always returns to the central landmark by roads that slant in the most unexpected manner. Each time a radius or spoke is laid, here, there, or elsewhere, in what looks like mad disorder.

The operation is so erratically conducted that it takes the most unremitting attention to follow it at all. The Spider reaches the margin of the area by one of the spokes already placed. She goes along this margin for some distance from the point at which she landed, fixes her thread to the frame and returns to the centre by the same road which she has just taken.

The thread obtained on the way in a broken line, partly on the radius and partly on the frame, is too long for the exact distance between the circumference and the central point. On returning to this point, the Spider adjusts her thread, stretches it to the correct length, fixes it and collects what remains on the central sign-post. In the case of each radius laid, the surplus is treated in the same fashion, so that the sign-post continues to increase in size. It was first a speck; it is now a little pellet, or even a small cushion of a certain breadth.

We shall see presently what becomes of this cushion whereon the Spider, that niggardly housewife, lays her saved-up bits of thread; for the moment, we will note that the Epeira works it up with her legs after placing each spoke, teazles it with her claws, mats it into felt with noteworthy diligence. In so doing, she gives the spokes a solid common support, something like the hub of our carriage-wheels.

The eventual regularity of the work suggests that the radii are spun in the same order in which they figure in the web, each following immediately upon its next neighbour. Matters pass in another matter, which at first looks like disorder, but which is really a judicious contrivance. After setting a few spokes in one direction, the Epeira runs across to the other side to draw some in the opposite direction. These sudden changes of course are highly logical; they show us how proficient the Spider is in the mechanics of rope-construction. Were they to succeed one another regularly, the spokes of one group, having nothing as yet to counteract them, would distort the work by their straining, would even destroy it for lack of a stabler support. Before continuing, it is necessary to lay a converse group which will maintain the whole by its resistance. Any combination of forces acting in one direction must be forthwith neutralized by another in the opposite direction. This is what our statics teach us and what the Spider puts into practice; she is a past mistress of the secrets of rope-building, without serving an apprenticeship.

One would think that this interrupted and apparently disordered labour must result in a confused piece of work. Wrong: the rays are equidistant and form a beautifully-regular orb. Their number is a characteristic mark of the different species. The Angular Epeira places 21 in her web, the Banded Epeira 32, the Silky Epeira 42. These numbers are not absolutely fixed; but the variation is very slight.

Now which of us would undertake, off-hand, without much preliminary experiment and without measuring-instruments, to divide a circle into a given quantity of sectors of equal width? The Epeiræ, though weighted with a wallet and tottering on threads shaken by the wind, effect the delicate division without stopping to think. They achieve it by a method which seems mad according to our notions of geometry. Out of disorder they evolve order.

We must not, however, give them more than their due. The angles are only approximately equal; they satisfy the demands of the eye, but cannot stand the test of strict measurement. Mathematical precision would be superfluous here. No matter, we are amazed at the result obtained. How does the Epeira come to succeed with her difficult problem, so strangely managed? I am still asking myself the question.

The laying of the radii is finished. The Spider takes her place in the centre, on the little cushion formed of the inaugural signpost and the bits of thread left over. Stationed on this support, she slowly turns round and round. She is engaged on a delicate piece of work. With an extremely thin thread, she describes from spoke to spoke, starting from the centre, a spiral line with very close coils. The central space thus worked attains, in the adults' webs, the dimensions of the palm of one's hand; in the younger Spiders' webs, it is much smaller, but it is never absent. For reasons which I will explain in the course of this study, I shall call it, in future, the "resting-floor."

The thread now becomes thicker. The first could hardly be seen; the second is plainly visible. The Spider shifts her position with great slanting strides, turns a few times, moving farther and farther from the centre, fixes her line each time to the spoke which she crosses and at last comes to a stop at the lower edge of the frame. She has described a spiral with coils of rapidly-increasing width. The average distance between the coils, even in the structures of the young Epeiræ, is one centimetre.

Let us not be misled by the word "spiral," which conveys the notion of a curved line. All curves are banished from the Spiders' work; nothing is used but the straight line and its combinations. All that is aimed at is a polygonal line drawn in a curve as geometry understands it. To this polygonal line, a work destined to disappear as the real toils are woven, I will give the name of the "auxiliary spiral." Its object is to supply crossbars, supporting rungs, especially in the outer zone, where the radii are too distant from one another to afford a suitable groundwork. Its object is also to guide the Epeira in the extremely delicate business which she is now about to undertake.

But, before that, one last task becomes essential. The area occupied by the spokes is very irregular, being marked out by the supports of the branch, which are infinitely variable. There are angular niches which, if skirted too closely, would disturb the symmetry of the web about to be constructed. The Epeira needs an exact space wherein gradually to lay her spiral thread. Moreover, she must not leave any gaps through which her prey might find an outlet.

An expert in these matters, the Spider soon knows the corners that have to be filled up. With an alternating movement, first in this direction, then in that, she lays, upon the support of the radii, a thread that forms two acute angles at the lateral boundaries of the faulty part and describes a zigzag line not wholly unlike the ornament known as the fret.

The sharp corners have now been filled with frets on every side; the time has come to work at the essential part, the snaring-web for which all the rest is but a support. Clinging on the one hand to the radii, on the other to the chords of the auxiliary spiral, the Epeira covers the same ground as when laying the spiral, but in the opposite direction: formerly, she moved away from the centre; now she moves towards it and with closer and more numerous circles. She starts from the base of the auxiliary spiral, near the frame.

What follows is difficult to observe, for the movements are very quick and spasmodic, consisting of a series of sudden little rushes, sways and bends that bewilder the eye. It needs continuous attention and repeated examination to distinguish the progress of the work however slightly.

The two hind-legs, the weaving implements, keep going constantly. Let us name them according to their position on the work-floor. I call the leg that faces the centre of the coil, when the animal moves, the "inner leg"; the one outside the coil the "outer leg."

The latter draws the thread from the spinneret and passes it to the inner leg, which, with a graceful movement, lays it on the radius crossed. At the same time, the first leg measures the distance; it grips the last coil placed in position and brings within a suitable range that point of the radius whereto the thread is to be fixed. As soon as the radius is touched, the thread sticks to it by its own glue. There are no slow operations, no knots: the fixing is done of itself.

Meanwhile, turning by narrow degrees, the spinstress approaches the auxiliary chords that have just served as her support. When, in the end, these chords become too close, they will have to go; they would impair the symmetry of the work. The Spider, therefore, clutches and holds on to the rungs of a higher row; she picks up, one by one, as she goes along, those which are of no more use to her and gathers them into a fine-spun ball at the contact-point of the next spoke. Hence arises a series of silky atoms marking the course of the disappearing spiral.

The light has to fall favourably for us to perceive these specks, the only remains of the ruined auxiliary thread. One would take them for grains of dust, if the faultless regularity of their distribution did not remind us of the vanished spiral. They continue, still visible, until the final collapse of the net.

And the Spider, without a stop of any kind, turns and turns and turns, drawing nearer to the centre and repeating the operation of fixing her thread at each spoke which she crosses. A good half-hour, an hour even among the full-grown Spiders, is spent on spiral circles, to the number of about fifty for the web of the Silky Epeira and thirty for those of the Banded and the Angular Epeira.

At last, at some distance from the centre, on the borders of what I have called the resting-floor, the Spider abruptly terminates her spiral when the space would still allow of a certain number of turns. Next, the Epeira, no matter which, young or old, hurriedly flings herself upon the little central cushion, pulls it out and rolls it into a ball which I expected to see thrown away. But no: her thrifty nature does not permit this waste. She eats the cushion, at first an inaugural landmark, then a heap of bits of thread; she once more melts in the digestive crucible what is no doubt intended to be restored to the silken treasury. It is a tough mouthful, difficult for the stomach to elaborate; still, it is precious and must not be lost. The work finishes with the swallowing. Then and there, the Spider installs herself, head downwards, at her hunting-post in the centre of the web.

THE OIL-BEETLE'S JOURNEY

« 38 »

The Sunken Road near Carpentras, that site of so many of Fabre's early observations, is again celebrated in this account of the oil-beetle and its fantastic piggy-back ride through the air to its future home. The material originally appeared in Chapter Four of THE GLOW-WORM AND OTHER BEETLES. *There are a number of species of Meloe beetles in the United States. Their common name of oil-beetles arises from their habit of giving off a disagreeable oily fluid when disturbed. As many as 10,000 eggs are laid by the females of some species.*

A vertical bank on the road from Carpentras to Bédoin is this time the scene of my observations. This bank, baked by the sun, is exploited by numerous swarms of Anthophoræ, who, more industrious than their congeners, are in the habit of building, at the entrance to their corridors, with serpentine fillets of earth, a vestibule, a defensive bastion in the form of an arched cylinder. In a word, they are swarms of *A. parietina*. A sparse carpet of turf extends from the edge of the road to the foot of the bank. The more comfortably to follow the work of the Bees, in the hope of wresting some secret from them, I had been lying for a few moments upon this turf, in the very heart of the inoffensive swarm, when my clothes were invaded by legions of little yellow lice, running with desperate eagerness through the hairy thickets of the nap of the cloth. In these tiny creatures, with which I was powdered here and there as with yellow dust, I soon recognized an old acquaintance, the young Oil-beetles, whom I now saw for the first time elsewhere than in the Bees' fur or the interior of their cells. I could not lose so excellent an opportunity of learning how these larvæ manage to establish themselves upon the bodies of their foster-parents.

In the grass where, after lying down for a moment, I had caught these lice were a few plants in blossom, of which the most abundant were three composites: *Hedypnois polymorpha, Senecio gallicus* and *Anthemis arvensis*. Now it was on a composite, a dandelion, that Newport seemed to remember seeing some young Oil-beetles; and my attention therefore was first of all directed to the plants which I have named. To my great satisfaction, nearly all the flowers of these three plants, especially those of the camomile (*Anthemis*) were occupied by young Oil-beetles in greater or lesser numbers. On one head of camomile I counted forty of these tiny insects, cowering motionless in the centre of the florets. On the other hand, I could not discover any on the flowers of the poppy or of a wild rocket (*Diplotaxis muralis*) which grew promiscuously among the plants aforesaid. It seems to me, therefore, that it is only on the composite flowers that the Meloe-larvæ await the Bees' arrival.

In addition to this population encamped upon the heads of the composites and remaining motionless, as though it had achieved its object for the moment, I soon discovered yet another, far more numerous, whose anxious activity betrayed a fruitless search. On the ground, in the grass, numberless little larvæ were running in a great flutter, recalling in some respects the tumultuous disorder of an overturned Ant-hill; others were hurriedly climbing to the tip of a blade of grass and descending with the same haste; others again were plunging into the downy fluff of the withered everlastings, remaining there a moment and quickly reappearing to continue their search. Lastly, with a little attention, I was able to convince myself that within an area of a dozen square yards there was perhaps not a single blade of grass which was not explored by several of these larvæ.

I was evidently witnessing the recent emergence of the young Oil-beetles from their maternal lairs. Part of them had already settled on the groundsel- and camomile-flowers to await the arrival of the Bees; but the majority were still wandering in search of this provisional refuge. It was by this wandering population that I had been invaded when I lay down at the foot of the bank. It was impossible that all these larvæ, the tale of whose alarming thousands I would not venture to define, should form one family and recognize a common mother; despite what Newport has told us of the Oil-beetles' astonishing fecundity, I could not believe this, so great was their multitude.

Though the green carpet was continued for a considerable distance along the side of the road, I could not detect a single Meloe-larva elsewhere than in the few square yards lying in front of the bank inhabited by the Mason-bee. These larvæ therefore could not have come far; to find themselves near the Anthophoræ they had had no long pilgrimage to make, for there was not a sign of the inevitable stragglers and laggards that follow in the wake of a travelling caravan. The burrows in which the eggs were hatched were therefore in that turf opposite the Bees' abode. Thus the Oil-beetles, far from laying their eggs at random, as their wandering life might lead one to suppose, and leaving their young to the task of approaching their future home, are able to recognize the spots haunted by the Anthophoræ and lay their eggs in the near neighbourhood of those spots.

With such a multitude of parasites occupying the composite flowers in close proximity to the Anthophora's nests, it is impossible that the majority of the swarm should not become infested sooner or later. At the time of my observations, a comparatively tiny proportion of the starving legion was waiting on the flowers; the others were still wandering on the ground, where the Anthophoræ very rarely alight; and yet I detected the presence of several Meloe-larvæ in the thoracic down of nearly all the Anthophoræ which I caught and examined.

I have also found them on the bodies of the Melecta- and Cœlioxys-bees, who are parasitic on the Anthophoræ. Suspending their audacious patrolling before the galleries under construction, these spoilers of the victualled cells alight for an instant on a camomile-flower and lo, the thief is robbed! A tiny, imperceptible louse has slipped into the thick of the downy fur and, at the moment when the parasite, after destroying the Anthophora's egg, is laying her own upon the stolen honey, will creep upon this egg, destroy it in its turn and remain sole mistress of the provisions. The mess of honey amassed by the Anthophora will thus pass through the hands of three owners and remain finally the property of the weakest of the three.

And who shall say whether the Meloe, in its turn, will not be dispossessed by a fresh thief; or even whether it will not, in the state of a drowsy, fat and flabby larva, fall a prey to some marauder who will munch its live entrails? As we meditate upon this deadly, implacable struggle which nature imposes, for their preservation, on these different creatures, which are by turns possessors and dispossessed, devourers and devoured, a painful impression mingles with the wonder aroused by the means employed by each parasite to attain its end; and, forgetting for a moment the tiny world in which these things happen, we are seized with terror at this concatenation of larceny, cunning and brigandage which forms part, alas, of the designs of *alma parens rerum!*

The young Meloe-larvæ established in the down of the Anthophoræ or in that of the Melecta- and the Cœlioxys-bees, their parasites, had adopted an infallible means of sooner or later reaching the desired cell. Was it, so far as they were concerned, a choice dictated by the fore-sight of instinct, or just simply the result of a lucky chance? The question was soon decided. Various Flies—Drone-flies and Bluebottles (*Eristalis tenax* and *Calliphora vomitoria*)—would settle from time to time on the groundsel- or camomile-flowers occupied by the young Meloes and stop for a moment to suck the sweet secretions. On all these Flies, with very few exceptions, I found Meloe-larvæ, motionless in the silky down of the thorax. I may also mention, as infested by these larvæ, an Ammophila (*A. hirsuta*), who victuals her burrows with a caterpillar in early spring, while her kinswomen build their nests in autumn. This Wasp merely grazes, so to speak, the surface of a flower; I catch her; there are Meloes moving about her body. It is clear that neither the Drone-flies nor the Bluebottles, whose larvæ live in putrefying matter, nor yet the Ammophilæ who victual theirs with caterpillars, could ever have carried the larvæ which invaded them into cells filled with honey. These larvæ therefore had gone astray; and instinct, as does not often happen, was here at fault.

Let us now turn our attention to the young Meloes waiting expectant upon the camomile-flowers. There they are, ten, fifteen or more, lodged half-way down the florets of a single blossom or in their interstices; it therefore needs a certain degree of scrutiny to perceive them, their hiding-place being the more effectual in that the amber colour of their bodies merges in the yellow hue of the florets. So long as nothing unusual happens upon the flower, so long as no sudden shock announces the arrival of a strange visitor, the Meloes remain absolutely motionless and give no sign of life. To see them dipping vertically, head downwards, into the florets, one might suppose that they were seeking some sweet liquid, their food; but in that case they ought to pass more frequently from one floret to another, which they do not, except when, after a false alarm, they regain their hiding-places and choose the spot which seems to them the most favourable. This immobility means that the florets of the camomile serve them only as a place of ambush, even as later the Anthophora's body will serve them solely as a vehicle to convey them to the Bee's cell. They take no nourishment, either on the flowers or on the Bees; and, as with the Sitares, their first meal will consist of the Anthophora's egg, which the hooks of their mandibles are intended to rip open.

Their immobility is, as we have said, complete; but nothing is easier than to arouse their suspended activity. Shake a camomile-blossom lightly with a bit of straw: instantly the Meloes leave their hiding-places, come up and scatter in all directions on the white petals of the circumference, running over them from one end to the other with all the speed which the smallness of their size permits. On reaching the extreme end of the petals, they fasten to it either with their caudal appendages, or perhaps with a sticky substance similar to that furnished by the anal button of the Sitares; and, with their bodies hanging outside and their six legs free, they bend about in every direction and stretch as far out as they can, as though striving to touch an object out of their reach. If nothing offers for them to seize upon, after a few vain attempts they regain the centre of the flower and soon resume their immobility.

But, if we place near them any object whatever, they do not fail to catch on to it with surprising agility. A blade of grass, a bit of straw, the handle of my tweezers which I hold out to them: they accept anything in their eagerness to quit the provisional shelter of the flower. It is true that, after finding themselves on these inanimate objects, they soon recognize that they have gone astray, as we see by their bustling movements to and fro and their tendency to go back to the flower if there still be time. Those which have thus giddily flung themselves upon a bit of straw and are allowed to return to their flower do not readily fall a second time into the same trap. There is therefore, in these animated specks, a memory, an experience of things.

After these experiments I tried others with hairy materials imitating more or less closely the down of the Bees, with little pieces of cloth or velvet cut from my clothes, with plugs of cotton wool, with pellets of flock gathered from the everlastings. Upon all these objects, offered with the tweezers, the Meloes flung themselves without any difficulty; but, instead of keeping quiet, as they do on the bodies of the Bees, they soon convinced me, by their restless behaviour, that they found themselves as much out of their element on these furry materials as on the smooth surface of a bit of straw. I ought to have expected this: had I not just seen them wandering without pause upon the everlastings enveloped with cottony flock? If reaching the shelter of a downy surface were enough to make them believe themselves safe in harbour, nearly all would perish, without further attempts, in the down of the plants.

Let us now offer them live insects and, first of all, Anthophoræ. If the Bee, after we have rid her of the parasites which she may be carrying, be taken by the wings and held for a moment in contact with the flower, we invariably find her, after this rapid contact, overrun by Meloes clinging to her hairs. The larvæ nimbly take up their position on the thorax, usually on the shoulders or sides, and once there they remain motionless: the second stage of their strange journey is compassed.

After the Anthophoræ, I tried the first live insects that I was able to procure at once: Drone-flies, Bluebottles, Hive-bees, small Butterflies. All were alike overrun by the Meloes, without hesitation. What is more, there was no attempt made to return to the flowers. As I could not find any Beetles at the moment, I was unable to experiment with them. Newport, experimenting, it is true, under conditions very different from mine, since his observations related to young Meloes held captive in a glass jar, while mine were made in the normal circumstances, Newport, I was saying, saw Meloes fasten to the body of a Malachius and stay there without moving, which inclines me to believe that with Beetles I should have obtained the same results as, for instance, with a Drone-fly. And I did, in fact, at a later date, find some Meloe-larvæ on the body of a big Beetle, the Golden Rose-chafer (*Cetonia aurata*), an assiduous visitor of the flowers.

After exhausting the insect class, I put within their reach my last resource, a large black Spider. Without hesitation they passed from the flower to the arachnid, made for places near the joints of the legs and settled there without moving. Everything therefore seems to suit their plans for leaving the provisional abode where they are waiting; without distinction of species, genus, or class, they fasten to the first living creature that chance brings within their reach. We now understand how it is that these young larvæ have been observed upon a host of different insects and especially upon the early Flies and Bees pillaging the flowers; we can also understand the need for that prodigious number of eggs laid by a single Oil-beetle, since the vast majority of the larvæ which come out of them will infallibly go astray and will not succeed in reaching the cells of the Anthophoræ. Instinct is at fault here; and fecundity makes up for it.

But instinct recovers its infallibility in another case. The Meloes, as we have seen, pass without difficulty from the flower to the objects within their reach, whatever these may be, smooth or hairy, living or inanimate. This done, they behave very differently, according as they have chanced to invade the body of an insect or some other object. In the first case, on a downy Fly or Butterfly, on a smooth-skinned Spider or Beetle, the larvæ remain motionless after reaching the point which suits them. Their instinctive desire is therefore satisfied. In the second case, in the midst of the nap of cloth or velvet, or the filaments of cotton, or the flock of the everlasting, or, lastly, on the smooth surface of a leaf or a straw, they betray the knowledge of their mistake by their continual coming and going, by their efforts to return to the flower imprudently abandoned.

How then do they recognize the nature of the object to which they have just moved? How is it that this object, whatever the quality of its surface, will sometimes suit them and sometimes not? Do they judge their new lodging by sight? But then no mistake would be possible; the sense of sight would tell them at the outset whether the object within reach was suitable or not; and emigration would or would not take place according to its decision. And then how can we suppose that, buried in the dense thicket of a pellet of cotton-wool or in the fleece of an Anthophora, the imperceptible larva can recognize, by sight, the enormous mass which it is perambulating?

Is it by touch, by some sensation due to the inner vibrations of living flesh? Not so, for the Meloes remain motionless on insect corpses that have dried up completely, on dead Anthophoræ taken from cells at least a year old. I have seen them keep absolutely quiet on fragments of an Anthophora on a thorax long since nibbled and emptied by the Mites. By what sense then can they distinguish the thorax of an Anthophora from a velvety pellet, when sight and touch are out of the question? The sense of smell remains. But in that case what exquisite subtlety must we not take for granted? Moreover, what similarity of smell can we admit between all the insects which, dead or alive, whole or in pieces, fresh or dried, suit the Meloes, while anything else does not suit them? A wretched louse, a living speck, leaves us mightily perplexed as to the sensibility which directs it. Here is yet one more riddle added to all the others.

After the observations which I have described, it remained for me to search the earthen surface inhabited by the Anthophoræ: I should then have followed the Meloe-larva in its transformations. It was certainly *cicatricosus* whose larvæ I had been studying; it was certainly this insect which ravaged the cells of the Mason-bee, for I found it dead in the old galleries which it had been unable to leave. This opportunity, which did not occur again, promised me an ample harvest. I had to give it all up. My Thursday was drawing to a close; I had to return to Avignon, to resume my lessons on the electrophorus and the Toricellian tube. O happy Thursdays! What glorious opportunities I lost because you were too short!

We will go back a year to continue this history. I collected, under far less favourable conditions, it is true, enough notes to map out the biography of the tiny creature which we have just seen migrating from the camomile-flowers to the Anthophora's back. From what I have said of the Sitaris-larvæ, it is plain that the Meloe-larvæ perched, like the former, on the back of a Bee, have but one aim: to get themselves conveyed by this Bee to the victualled cells. Their object is not to live for a time on the body that carries them.

Were it necessary to prove this, it would be enough to say that we never see these larvæ attempt to pierce the skin of the Bee, or else to nibble at a hair or two, nor do we see them increase in size so long as they are on the Bee's body. To the Meloes, as to the Sitares, the Anthophora serves merely as a vehicle which conveys them to their goal, the victualled cell.

It remains for us to learn how the Meloe leaves the down of the Bee which has carried it, in order to enter the cell. With larvæ collected from the bodies of different Bees, before I was fully acquainted with the tactics of the Sitares, I undertook, as Newport had done before me, certain investigations intended to throw light on this leading point in the Oil-beetle's history. My attempts, based upon those which I had made with the Sitares, resulted in the same failure. The tiny creatures, when brought into contact with Anthophora-larvæ or -nymphs, paid no attention whatever to their prey; others, placed near cells which were open and full of honey, did not enter them, or at most ventured to the edge of the orifice; others, lastly, put inside the cell, on the dry wall or on the surface of the honey, came out again immediately or else got stuck and died. The touch of the honey is as fatal to them as to the young Sitares.

Searches made at various periods in the nests of the Hairy-footed Anthophora had taught me some years earlier that *Meloe cicatricosus,* like the Sitares, is a parasite of that Bee; indeed I had at different times discovered adult Meloes, dead and shrivelled, in the Bee's cells. On the other hand, I knew from Léon Dufour that the little yellow animal, the Louse found in the Bee's down, had been recognized, thanks to Newport's investigations, as the larva of the Oil-beetle. With these data, rendered still more striking by what I was learning daily on the subject of the Sitares, I went to Carpentras, on the 21st of May, to inspect the nests of the Anthophoræ, then building, as I have described. Though I was almost certain of succeeding, sooner or later, with the Sitares, who were excessively abundant, I had very little hope of the Meloes, which on the contrary are very scarce in the same nests. Circumstances, however, favoured me more than I dared hope and, after six hours' labour, in which the pick played a great part, I became the possessor, by the sweat of my brow, of a considerable number of cells occupied by Sitares and two other cells appropriated by Meloes.

While my enthusiasm had not had time to cool at the sight, momentarily repeated, of a young Sitaris perched upon an Anthophora's egg floating in the centre of the little pool of honey, it might well have burst all restraints on beholding the contents of one of these cells. On the black, liquid honey a wrinkled pellicle is floating; and on this pellicle, motionless, is a yellow louse. The pallicle is the empty envelope of the Anthophora's egg; the louse is a Meloe-larva.

The story of this larva becomes self-evident. The young Meloe leaves the down of the Bee at the moment when the egg is laid; and, since contact with the honey would be fatal to the grub, it must, in order to save itself, adopt the tactics followed by the Sitaris, that is to say, it must allow itself to drop on the surface of the honey with the egg which is in the act of being laid. There, its first task is to devour the egg which serves it for a raft, as is attested by the empty envelope on which it still remains; and it is after this meal, the only one that it takes so long as it retains its present form, that it must commence its long series of transformations and feed upon the honey amassed by the Anthophora.

THE GLOW-WORM

« 39 »

The centuries-old mystery of the cold light produced by the Firefly and the Glow-Worm is still partially a riddle. Since Fabre's day, however, scientists have discovered that two substances, luciferin and luciferase, are essential. Luciferin oxidizes in the presence of the enzyme-like luciferase when light is produced. If a Firefly or Glow-Worm is placed in an airtight container and the oxygen gradually diminished, its light grows dimmer and dimmer; if oxygen is added, the light increases. Fabre's account of his studies of the Glow-Worm and of its unique method of overcoming the snail originally formed the first chapter of THE GLOW-WORM AND OTHER BEETLES.

Few insects in our climes vie in popular fame with the Glow-worm, that curious little animal which, to celebrate the little joys of life, kindles a beacon at its tail-end. Who does not know it, at least by name? Who has not seen it roam amid the grass, like a spark fallen from the moon at its full? The Greeks of old called it λάμπουρις, meaning, the bright-tailed. Science employs the same term: it calls the lantern-bearer, *Lampyris noctiluca,* LIN. In this case, the common name is inferior to the scientific phrase, which, when translated, becomes both expressive and accurate.

In fact, we might easily cavil at the word "worm." The Lampyris is not a worm at all, not even in general appearance. He has six short legs, which he well knows how to use; he is a gad-about, a trot-about. In the adult state, the male is correctly garbed in wing-cases, like the true Beetle that he is. The female is an ill-favoured thing who knows naught of the delights of flying: all her life long, she retains the larval shape, which, for the rest, is similar to that of the male, who himself is imperfect so long as he has not achieved the maturity that comes with

313

pairing-time. Even in this initial stage, the word "worm" is out of place. We French have the expression "Naked as a worm," to point to the lack of any defensive covering. Now the Lampyris is clothed, that is to say, he wears an epidermis of some consistency; moreover, he is rather richly coloured: his body is dark brown all over, set off with pale pink on the thorax, especially on the lower surface. Finally, each segment is decked at the hinder edge with two spots of a fairly bright red. A costume like this was never worn by a worm.

Let us leave this ill-chosen denomination and ask ourselves what the Lampyris feeds upon. That master of the art of gastronomy, Brillat-Savarin, said:

"Show me what you eat and I will tell you what you are."

A similar question should be addressed, by way of a preliminary, to every insect whose habits we propose to study, for, from the least to the greatest in the zoological progression, the stomach sways the world; the data supplied by food are the chief of all the documents of life. Well, in spite of his innocent appearance, the Lampyris is an eater of flesh, a hunter of game; and he follows his calling with rare villainy. His regular prey is the Snail.

This detail has long been known to entomologists. What is not so well-known, what is not known at all yet, to judge by what I have read, is the curious method of attack, of which I have seen no other instance anywhere.

Before he begins to feast, the Glow-worm administers an anæsthetic: he chloroforms his victim, rivalling in the process the wonders of our modern surgery, which renders the patient insensible before operating on him. The usual game is a small Snail hardly the size of a cherry, such as, for instance, *Helix variabilis,* DRAP., who, in the hot weather, collects in clusters on the stiff stubble and on other long, dry stalks, by the roadside, and there remains motionless, in profound meditation, throughout the scorching summer days. It is in some such resting-place as this that I have often been privileged to light upon the Lampyris banqueting on the prey which he had just paralyzed on its shaky support by his surgical artifices.

But he is familiar with other preserves. He frequents the edges of the irrigating-ditches, with their cool soil, their varied vegetation, a favourite haunt of the mollusc. Here, he treats the game on the ground; and, under these conditions, it is easy for me to rear him at home and to follow the operator's performance down to the smallest detail.

I will try to make the reader a witness of the strange sight. I place a little grass in a wide glass jar. In this I install a few Glow-worms and a provision of Snails of a suitable size, neither too large nor too small, chiefly *Helix variabilis*. We must be patient and wait. Above all, we must keep an assiduous watch, for the desired events come unexpectedly and do not last long.

Here we are at last. The Glow-worm for a moment investigates the prey, which, according to its habit, is wholly withdrawn in the shell, except the edge of the mantle, which projects slightly. Then the hunter's weapon is drawn, a very simple weapon, but one that cannot be plainly perceived without the aid of a lens. It consists of two mandibles bent back powerfully into a hook, very sharp and as thin as a hair. The microscope reveals the presence of a slender groove running throughout the length. And that is all.

The insect repeatedly taps the Snail's mantle with its instrument. It all happens with such gentleness as to suggest kisses rather than bites. As children, teasing one another, we used to talk of "tweaksies" to express a slight squeeze of the finger-tips, something more like a tickling than a serious pinch. Let us use that word. In conversing with animals, language loses nothing by remaining juvenile. It is the right way for the simple to understand one another.

The Lampyris doles out his tweaks. He distributes them methodically, without hurrying, and takes a brief rest after each of them, as though he wished to ascertain the effect produced. Their number is not great: half-a-dozen, at most, to subdue the prey and deprive it of all power of movement. That other pinches are administered later, at the time of eating, seems very likely, but I cannot say anything for certain, because the sequel escapes me. The first few, however—there are never many— are enough to impart inertia and loss of all feeling to the mollusc, thanks to the prompt, I might almost say, lightning methods of the Lampyris, who, beyond a doubt, instils some poison or other by means of his grooved hooks.

Here is the proof of the sudden efficacity of those twitches, so mild in appearance: I take the Snail from the Lampyris, who has operated on the edge of the mantle some four or five times. I prick him with a fine needle in the fore-part, which the animal, shrunk into its shell, still leaves exposed. There is no quiver of the wounded tissues, no reaction against the brutality of the needle. A corpse itself could not give fewer signs of life.

Here is something even more conclusive: chance occasionally gives me Snails attacked by the Lampyris while they are creeping along, the foot slowly crawling, the tentacles swollen to their full extent. A few disordered movements betray a brief excitement on the part of the mollusc and then everything ceases: the foot no longer slugs; the front-part loses its graceful swan-neck curve; the tentacles become limp and give way under their weight, dangling feebly like a broken stick. This condition persists.

Is the Snail really dead? Not at all, for I am free to resuscitate the seeming corpse. After two or three days of that singular condition which is no longer life and yet not death, I isolate the patient and, although this is not really necessary to success, I give him a douche which will represent the shower so dear to the able-bodied mollusc. In about a couple of days, my prisoner, but lately injured by the Glow-worm's treachery, is restored to his normal state. He revives, in a manner; he recovers movement and sensibility. He is affected by the stimulus of a needle; he shifts his place, crawls, puts out his tentacles, as though nothing unusual had occurred. The general torpor, a sort of deep drunkenness, has vanished outright. The dead returns to life. What name shall we give to that form of existence which, for a time, abolishes the power of movement and the sense of pain? I can see but one that is approximately suitable: anæsthesia. The exploits of a host of Wasps whose flesh-eating grubs are provided with meat that is motionless though not dead have taught us the skilful art of the paralyzing insect, which numbs the locomotory nerve-centres with its venom. We have now a humble little animal that first produces complete anæsthesia in its patient. Human science did not in reality invent this art, which is one of the wonders of our latter-day surgery. Much earlier, far back in the centuries, the Lampyris and, apparently, others knew it as well. The animal's knowledge had a long start of ours; the method alone has changed. Our operators proceed by making us inhale the fumes of ether or chloroform; the insect proceeds by injecting a special virus that comes from the mandibular fangs in infinitesimal doses. Might we not one day be able to benefit by this hint? What glorious discoveries the future would have in store for us, if we understood the beastie's secrets better!

What does the Lampyris want with anæsthetical talent against a harmless and moreover eminently peaceful adversary, who would never begin the quarrel of his own accord? I think I see. We find in Algeria a Beetle known as *Drilus maroccanus,* who, though non-luminous, approaches our Glow-worm in his organization and especially in his habits. He too feeds on land molluscs. His prey is a Cyclostome with a graceful spiral shell, tight-closed with a stony lid which is attached to the animal by a powerful muscle. The lid is a movable door which is quickly shut by the inmate's mere withdrawal into his house and as easily opened when the hermit goes forth. With this system of closing, the abode becomes inviolable; and the Drilus knows it.

Fixed to the surface of the shell by an adhesive apparatus whereof the Lampyris will presently show us the equivalent, he remains on the look-out, waiting, if necessary, for whole days at a time. At last, the need of air and food oblige the besieged non-combatant to show himself; at least, the door is set slightly ajar. That is enough. The Drilus is on the spot and strikes his blow. The door can no longer be closed and the assailant is henceforth master of the fortress. Our first impression is that the muscle moving the lid has been cut with a quick-acting pair of shears. This idea must be dismissed. The Drilus is not well enough equipped with jaws to gnaw through a fleshy mass so promptly. The operation has to succeed at once, at the first touch: if not, the animal attacked would retreat, still in full vigour, and the siege must be recommenced, as arduous as ever, exposing the insect to fasts indefinitely prolonged. Although I have never come across the Drilus, who is a stranger to my district, I conjecture a method of attack very similar to that of the Glow-worm. Like our own Snail-eater, the Algerian insect does not cut its victim into small pieces: it renders it inert, chloroforms it by means of a few tweaks which are easily distributed, if the lid but half-opens for a second. That will do. The besieger thereupon enters and, in perfect quiet, consumes a prey incapable of the least muscular effort. That is how I see things by the unaided light of logic.

Let us now return to the Glow-worm. When the Snail is on the ground, creeping, or even shrunk into his shell, the attack never presents any difficulty. The shell possesses no lid and leaves the hermit's fore-part to a great extent exposed. Here, on the edges of the mantle contracted by the fear of danger, the mollusc is vulnerable and incapable of defence. But it also frequently happens that the Snail occupies a raised position, clinging to the tip of a grass-stalk or perhaps to the smooth surface of a stone. This support serves him as a temporary lid; it wards off the aggression of any churl who might try to molest the inhabitant of the cabin, always on the express condition that no slit show itself anywhere on the protecting circumference. If, on the other hand, in the frequent case when the shell does not fit its support quite closely, some point, however tiny, be left uncovered, this is enough for the subtle tools of the Lampyris, who just nibbles at the mollusc and at once plunges him into that profound immobility which favours the tranquil proceedings of the consumer.

These proceedings are marked by extreme prudence. The assailant has to handle his victim gingerly, without provoking contractions which would make the Snail let go his support and, at the very least, precipitate him from the tall stalk whereon he is blissfully slumbering. Now any game falling to the ground would seem to be so much sheer loss, for the Glow-worm has no great zeal for hunting-expeditions: he profits by the discoveries which good luck sends him, without undertaking assiduous searches. It is essential, therefore, that the equilibrium of a prize perched on the top of a stalk and only just held in position by a touch of glue should be disturbed as little as possible during the onslaught; it is necessary that the assailant should go to work with infinite circumspection and without producing pain, lest any muscular reaction should provoke a fall and endanger the prize. As we see, sudden and profound anæsthesia is an excellent means of enabling the Lampyris to attain his object, which is to consume his prey in perfect quiet.

What is his manner of consuming it? Does he really eat, that is to say, does he divide his food piecemeal, does he carve it into minute particles, which are afterwards ground by a chewing-apparatus? I think not. I never see a trace of solid nourishment on my captives' mouths. The Glow-worm does not eat in the strict sense of the word: he drinks his fill; he feeds on a thin gruel into which he transforms his prey by a method recalling that of the maggot. Like the flesh-eating grub of the Fly, he too is able to digest before consuming; he liquefies his prey before feeding on it.

This is how things happen: a Snail has been rendered insensible by the Glow-worm. The operator is nearly always alone, even when the prize is a large one, like the Common Snail, *Helix aspersa*. Soon a number of guests hasten up—two, three or more—and, without any quarrel with real proprietor, all alike fall to. Let us leave them to themselves for a couple of days and then turn the shell, with the opening downwards. The contents flow out as easily as would soup from an overturned saucepan. When the sated diners retire from this gruel, only insignificant leavings remain.

The matter is obvious: by repeated tiny bites, similar to the tweaks which we saw distributed at the outset, the flesh of the mollusc is converted into a gruel on which the various banqueters nourish themselves without distinction, each working at the broth by means of some special pepsine and each taking his own mouthfuls of it. In consequence of this method, which first converts the food into a liquid, the Glow-worm's mouth must be very feebly armed apart from the two fangs which sting the patient and inject the anæsthetic poison and, at the same time, no doubt, the serum capable of turning the solid flesh into fluid. These two tiny implements, which can just be examined through the lens, must, it seems, have some other object. They are hollow and in this resemble those of the Ant-lion, which sucks and drains its capture without having to divide it; but there is this great difference, that the Ant-lion leaves copious remnants, which are afterwards flung outside the funnel-shaped trap dug in the sand, whereas the Glow-worm, that expert liquefier, leaves nothing, or next to nothing. With similar tools, the one simply sucks the blood of its prey and the other turns every morsel of his to account, thanks to a preliminary liquefaction.

And this is done with exquisite precision, though the equilibrium is sometimes anything but steady. My rearing-glasses supply me with magnificent examples. Crawling up the sides, the Snails imprisoned in my apparatus sometimes reach the top, which is closed with a glass pane, and fix themselves to it by means of a speck of glair. This is a mere temporary halt, in which the mollusc is miserly with its adhesive product, and the merest shake is enough to loosen the shell and send it to the bottom of the jar.

Now it is not unusual for the Glow-worm to hoist himself to the top, with the help of a certain climbing-organ that makes up for his weak legs. He selects his quarry, makes a minute inspection of it to find an entrance-slit, nibbles it a little, renders it insensible and, without delay, proceeds to prepare the gruel which he will consume for days on end.

When he leaves the table, the shell is found to be absolutely empty; and yet this shell, which was fixed to the glass by a very faint stickiness, has not come loose, has not even shifted its position in the smallest degree: without any protest from the hermit gradually converted into broth, it has been drained on the very spot at which the first attack was delivered. These small details tell us how promptly the anæsthetic bite takes effect; they teach us how dexterously the Glow-worm treats his Snail without causing him to fall from a very slippery vertical support and without even shaking him on his slight line of adhesion.

Under these conditions of equilibrium, the operator's short, clumsy legs are obviously not enough; a special accessory apparatus is needed to defy the danger of slipping and to seize the unseizable. And this apparatus the Lampyris possesses. At the hinder end of the animal we see a white spot which the lens separates into some dozen short, fleshy appendages, sometimes gathered into a cluster, sometimes spread into a rosette. There is your organ of adhesion and locomotion. If he would fix himself somewhere, even on a very smooth surface, such as a grass-stalk, the Glow-worm opens his rosette and spreads it wide on the support, to which it adheres by its own stickiness. The same organ, rising and falling, opening and closing, does much to assist the act of progression. In short, the Glow-worm is a new sort of self-propelled cripple, who decks his hindquarters with a dainty white rose, a kind of hand with twelve fingers, not jointed, but moving in every direction: tubular fingers which do not seize, but stick.

The same organ serves another purpose: that of a toilet-sponge and brush. At a moment of rest, after a meal, the Glow-worm passes and repasses the said brush over his head, back, sides and hinder-parts, a performance made possible by the flexibility of his spine. This is done point by point, from one end of the body to the other, with a scrupulous persistency that proves the great interest which he takes in the operation. What is his object in thus sponging himself, in dusting and polishing himself so carefully? It is a question, apparently, of removing a few atoms of dust or else some traces of viscidity that remain from the evil contact with the snail. A wash and brush-up is not superfluous when one leaves the tub in which the mollusc has been treated.

If the Glow-worm possessed no other talent than that of chloroforming his prey by means of a few tweaks resembling kisses, he would be unknown to the vulgar herd; but he also knows how to light himself like a beacon; he shines, which is an excellent manner of achieving fame. Let us consider more particularly the female, who, while retaining her larval shape, becomes marriageable and glows at her best during the hottest part of summer. The lighting-apparatus occupies the last three segments of the abdomen. On each of the first two, it takes the form, on the ventral surface, of a wide belt covering almost the whole of the arch; on the third, the luminous part is much less and consists simply of two small crescent-shaped markings, or rather two spots which shine through to the back and are visible both above and below the animal. Belts and spots emit a glorious white light, delicately tinged with blue. The general lighting of the Glow-worm thus comprises two groups: first, the wide belts of the two segments preceding the last; secondly, the two spots of the final segments. The two belts, the exclusive attribute of the marriageable female, are the part richest in light: to glorify her wedding, the future mother dons her brightest gauds; she lights her two resplendent scarves. But, before that, from the time of the hatching, she had only the modest rush-light of the stern. This efflorescence of light is the equivalent of the final metamorphosis, which is usually represented by the gift of wings and flight. Its brilliance heralds the pairing-time. Wings and flight there will be none: the female retains her humble larval form, but she kindles her blazing beacon.

The male, on his side, is fully transformed, changes his shape, acquires wings and wing-cases; nevertheless, like the female, he possesses, from the time when he is hatched, the pale lamp of the end segment. This luminous aspect of the stern is characteristic of the entire Glow-worm tribe, independently of sex and season. It appears upon the budding grub and continues throughout life unchanged. And we must not forget to add that it is visible on the dorsal as well as on the ventral surface, whereas the two large belts peculiar to the female shine only under the abdomen.

My hand is not so steady nor my sight so good as once they were, but, as far as they allow me, I consult anatomy for the structure of the luminous organs. I take a scrap of the epidermis and manage to separate pretty neatly half of one of the shining belts. I place my preparation under the microscope. On the skin, a sort of white-wash lies spread, formed of a very fine, granular substance. This is certainly the light-producing matter. To examine this white layer more closely is beyond the power of my weary eyes. Just beside it is a curious air-tube, whose short and remarkably wide stem branches suddenly into a sort of bushy tuft of very delicate ramifications. These creep over the luminous sheet, or even dip into it. That is all.

The luminescence, therefore, is controlled by the respiratory organs and the work produced is an oxidization. The white sheet supplies the oxidizable matter and the thick air-tube spreading into a tufty bush distributes the flow of air over it. There remains the question of the substance whereof this sheet is formed. The first suggestion was phosphorus, in the chemist's sense of the word. The Glow-worm has been calcined and treated with the violent reagents that bring the simple substances to light; but no one, so far as I know, has obtained a satisfactory answer along these lines. Phosphorus seems to play no part here, in spite of the name of phosphorescence which is sometimes bestowed upon the Glow-worm's gleam. The answer lies elsewhere, no one knows where.

We are better informed as regards another question. Has the Glow-worm a free control of the light which he emits? Can he turn it on or down or put it out as he pleases? Has he an opaque screen which is drawn over the flame at will, or is that flame always left exposed? There is no need for any such mechanism: the insect has something better for its revolving light.

The thick tube supplying the light-producing sheet increases the flow of air and the light is intensified; the same air-tube, swayed by the animal's will, slackens or even suspends the passage of air and the light grows fainter or even goes out. It is, in short, the mechanism of a lamp which is regulated by the access of air to the wick.

Excitement can set the attendant air-duct in motion. We must here distinguish between two cases: that of the gorgeous scarves, the exclusive ornament of the female ripe for matrimony, and that of the modest fairy-lamp on the last segment, which both sexes kindle at any age. In the second case, the extinction caused by a flurry is sudden and complete, or nearly so. In my nocturnal hunts for young Glow-worms, measuring about 5 millimetres long, I can plainly see the glimmer on the blades of grass; but, should the least false step disturb a neighbouring twig, the light goes out at once and the coveted insect becomes invisible. Upon the full-grown females, lit up with their nuptial scarves, even a violent start has but a slight effect and often none at all.

I fire a gun beside a wire-gauze cage in which I am rearing my menagerie of females in the open air. The explosion produces no result. The illumination continues, as bright and placid as before. I take a spray and rain down a slight shower of cold water upon the flock. Not one of my animals puts out its light; at the very most, there is a brief pause in the radiance; and then only in some cases. I send a puff of smoke from my pipe into the cage. This time, the pause is more marked. There are even some extinctions, but these do not last long. Calm soon returns and the light is renewed as brightly as ever. I take some of the captives in my fingers, turn and return them, tease them a little. The illumination continues and is not much diminished, if I do not press too hard with my thumb. At this period, with the pairing close at hand, the insect is in all the fervour of its passionate splendour; and nothing short of very serious reasons would make it put out its signals altogether.

All things considered, there is not a doubt but that the Glow-worm himself manages his lighting-apparatus, extinguishing and rekindling it at will; but there is one point at which the voluntary agency of the insect is without effect. I detach a strip of the epidermis showing one of the luminescent sheets and place it in a glass tube, which I close with a plug of damp wadding, to avoid too rapid an evaporation. Well, this scrap of carcass shines away merrily, although not quite as brilliantly as on the living body.

Life's aid is now superfluous. The oxidizable substance, the luminescent sheet, is in direct communication with the surrounding atmosphere; the flow of oxygen through an air-tube is not necessary; and the luminous emission continues to take place, in the same way as when it is produced by the contact of the air with the real phosphorus of the chemists. Let us add that, in aerated water, the luminousness continues as brilliant as in the free air, but that it is extinguished in water deprived of its air by boiling. No better proof could be found of what I have already propounded, namely, that the Glowworm's light is the effect of a slow oxidization.

The light is white, calm and soft to the eyes and suggests a spark dropped by the full moon. Despite its splendour, it is a very feeble illuminant. If we move a Glow-worm along a line of print, in perfect darkness, we can easily make out the letters, one by one, and even words, when these are not too long; but nothing more is visible beyond a narrow zone. A lantern of this kind soon tires the reader's patience.

Suppose a group of Glow-worms placed almost touching one another. Each of them sheds its glimmer, which ought, one would think, to light up its neighbours by reflexion and give us a clear view of each individual specimen. But not at all: the luminous party is a chaos in which our eyes unable to distinguish any definite form at a medium distance. The collective lights confuse the link-bearers into one vague whole.

Photography gives us a striking proof of this. I have a score of females, all at the height of their splendour, in a wire-gauze cage in the open air. A tuft of thyme forms a grove in the centre of their establishment. When night comes, my captives clamber to this pinnacle and strive to show off their luminous charms to the best advantage at every point of the horizon, thus forming along the twigs marvellous clusters from which I expected magnificent effects on the photographer's plates and paper. My hopes are disappointed. All that I obtain is white, shapeless patches, denser here and less dense there according to the numbers forming the group. There is no picture of the Glow-worms themselves; not a trace either of the tuft of thyme. For want of satisfactory light, the glorious firework is represented by a blurred splash of white on a black ground.

The beacons of the female Glow-worms are evidently nuptial signals, invitations to the pairing; but observe that they are lighted on the lower surface of the abdomen and face the ground, whereas the summoned males, whose flights are sudden and uncertain, travel overhead, in the air, sometimes a great way up. In its normal position, therefore, the glittering lure is concealed from the eyes of those concerned; it is covered by the thick bulk of the bride. The lantern ought really to gleam on the back and not under the belly; otherwise the light is hidden under a bushel.

The anomaly is corrected in a very ingenious fashion, for every female has her little wiles of coquetry. At nightfall, every evening, my caged captives make for the tuft of thyme with which I have thoughtfully furnished the prison and climb to the top of the upper branches, those most in sight. Here, instead of keeping quiet, as they did at the foot of the bush just now, they indulge in violent exercises, twist the tip of their very flexible abdomen, turn it to one side, turn it to the other, jerk it in every direction. In this way, the search-light cannot fail to gleam, at one moment or another, before the eyes of every male who goes a-wooing in the neighbourhood, whether on the ground or in the air.

It is very like the working of the revolving mirror used in catching Larks. If stationary, the little contrivance would leave the bird indifferent; turning and breaking up its light in rapid flashes, it excites it.

While the female Glow-worm has her tricks for summoning her swains, the male, on his side, is provided with an optical apparatus suited to catch from afar the least reflection of the calling-signal. His corselet expands into a shield and overlaps his head considerably in the form of a peaked cap or eye-shade, the object of which appears to be to limit the field of vision and concentrate the view upon the luminous speck to be discerned. Under this arch are the two eyes, which are relatively enormous, exceedingly convex, shaped like a skull-cap and contiguous to the extent of leaving only a narrow groove for the insertion of the antennæ. This double eye, occupying almost the whole face of the insect and contained in the cavern formed by the spreading peak of the corselet, is a regular Cyclop's eyes.

At the moment of the pairing, the illumination becomes much fainter, is almost extinguished; all that remains alight is the humble fairy-lamp of the last segment. This discreet night-light is enough for the wedding, while, all around, the host of nocturnal insects, lingering over their respective affairs, murmur the universal marriage-hymn. The laying follows very soon. The round, white eggs are laid, or rather strewn at random, without the least care on the mother's part, either on the more or less cool earth or on a blade of grass. These brilliant ones know nothing at all of family-affection.

Here is a very singular thing: the Glow-worm's eggs are luminous even when still contained in the mother's womb. If I happen by accident to crush a female big with germs that have reached maturity, a shiny streak runs along my fingers, as though I had broken some vessel filled with a phosphorescent fluid. The lens shows me that I am wrong. The luminosity comes from the cluster of eggs forced out of the ovary. Besides, as laying-time approaches, the phosphorescence of the eggs is already made manifest without this clumsy midwifery. A soft opalescent light shines through the skin of the belly.

The hatching follows soon after the laying. The young of either sex have two little rush-lights on the last segment. At the approach of the severe weather, they go down into the ground, but not very far. In my rearing-jars, which are supplied with fine and very loose earth, they descend to a depth of three or four inches at most. I dig up a few in mid-winter. I always find them carrying their faint stern-light. About the month of April, they come up again to the surface, there to continue and complete their evolution.

From start to finish, the Glow-worm's life is one great orgy of light. The eggs are luminous; the grubs likewise. The full-grown females are magnificent light-houses, the adult males retain the glimmer which the grubs already possessed. We can understand the object of the feminine beacon; but of what use is all the rest of the pyrotechnic display? To my great regret, I cannot tell. It is and will be, for many a day to come, perhaps for all time, the secret of animal physics, which is deeper than the physics of the books.

THE EDGE OF THE UNKNOWN

« 40 »

"The more I observe and experiment," Fabre once wrote, "the more clearly I see rising out of the black mists of possibility an enormous note of interrogation." To a friend, he said on another occasion: "Because I have shifted a few grains of sand upon the shore, am I in a position to understand the abysmal depths of the ocean? Life has unfathomable secrets. Human knowledge will be erased from the world's archives before we possess the last word that a gnat has to say to us." His philosophy, his outlook on the insects—maintained through a lifetime of labor—his essential humility of mind, are reflected in this last selection. It is taken from the fifth chapter of THE MASON-WASPS.

Is it really worth while to spend our time, the time which escapes us so swiftly, this stuff of life, as Montaigne calls it, in gleaning facts of indifferent moment and of highly contestable utility? Is it not childish to enquire so minutely into an insect's actions? Too many interests of a graver kind hold us in their grasp to leave leisure for these amusements. That is how the harsh experience of age impels us to speak: that is how I should conclude, as I bring my investigations to a close, if I did not perceive, amid the chaos of my observations, a few gleams of light touching the loftiest problems which we are privileged to discuss.

What is life? Will it ever be possible for us to trace it to its sources? Shall we ever be permitted to excite, in a drop of albumen, the uncertain quiverings which are the preludes of organization? What is human intelligence? What is instinct? Are these two mental aptitudes irreducible, or can they both be traced back to a common factor? Are the species connected with one another, are they related by evolution? Or are they, as it were, so many unchanging medals, each struck

326

from a separate die upon which the tooth of time has no effect, except to destroy it sooner or later? These questions are and always will be the despair of every cultivated mind, even though the insanity of our efforts to solve them urges us to cast them into the limbo of the unknowable. The theorists, proudly daring, have an answer nowadays for every question; but as a thousand theoretical views are not worth a single fact, thinkers untrammelled by preconceived ideas are far from becoming convinced. Problems such as these, whether their scientific solution be possible or not, require an enormous mass of well-established data, to which entomology, despite its humble province, can contribute a quota of some value. And that is why I am an observer, why, above all, I am an experimenter.

It is something to observe; but it is not enough: we must experiment, that is to say, we must ourselves intervene and create artificial conditions which oblige the animal to reveal to us what it would not tell if left to the normal course of events. Its actions, marvellously contrived to attain the end pursued, are capable of deceiving us as to their real meaning and of making us accept, in their linked sequence, that which our own logic dictates to us. It is not the animal that we are now consulting upon the nature of its aptitudes, upon the primary motives of its activity, but our own opinions, which always yield a reply in favor of our cherished notions. As I have already repeatedly shown, observation in itself is often a snare: we interpret its data according to the exigencies of our theories. To bring out the truth, we must needs resort to experiment, which alone is able to some extent to fathom the obscure problem of animal intelligence. It has sometimes been denied that zoology is an experimental science. The accusation would be well-founded if zoology confined itself to describing and classifying; but this is the least important part of its function: it has higher aims than that; and, when it consults the animal upon some problem of life, its method of questioning lies in experiment. In my own modest sphere, I should be depriving myself of the most potent method of study if I were to neglect experiment. Observation sets the problem; experiment solves it, always presuming that it can be solved; or at least, if powerless to yield the full light of truth, it sheds a certain gleam over the edges of the impenetrable cloud.

(END)

INDEX